Cellular Metabolism in Health and Disease

Cellular Metabolism in Health and Disease

Editor: Christian Stokes

FOSTER ACADEMICS
www.fosteracademics.com

www.fosteracademics.com

FOSTER ACADEMICS

Cataloging-in-publication Data

Cellular metabolism in health and disease / edited by Christian Stokes.
p. cm.
Includes bibliographical references and index.
ISBN 978-1-64646-592-7
1. Cell metabolism. 2. Cell metabolism--Health aspects. 3. Metabolism--Disorders.
4. Cellular control mechanisms. 5. Cell respiration. I. Stokes, Christian.
QH634.5 .C45 2023
574.876 1--dc23

Foster Academics,
118-35 Queens Blvd., Suite 400,
Forest Hills, NY 11375, USA

ISBN 978-1-64646-592-7 (Hardback)

Contents

Preface VII

Chapter 1 **Hypothyroidism** 1
Mauricio Alvarez Andrade and Oscar Rosero Olarte

Chapter 2 **Cerebral Energy Metabolism: Measuring and Understanding its Rate** 13
Avital Schurr

Chapter 3 **Sports and McArdle Disease (Glycogen Storage Disease Type V): Danger or Therapy?** 31
Georg Bollig

Chapter 4 **Metabolic Syndrome: Impact of Dietary Therapy** 42
Suzanne Fouad Soliman

Chapter 5 **Dipeptidyl Peptidase-4 Inhibitor-Associated Bullous Pemphigoid** 57
Ágnes Kinyó

Chapter 6 **Lipid Disorders in Uremia** 68
Valdete Topçiu-Shufta and Valdete Haxhibeqiri

Chapter 7 **Brown Adipose Tissue Energy Metabolism** 88
Yuan Lu

Chapter 8 **Diabetes Mellitus: A Group of Genetic-Based Metabolic Diseases** 109
Lilian Sanhueza, Pilar Durruty, Cecilia Vargas, Paulina Vignolo and Karina Elgueta

Chapter 9 **Metabolic Syndrome** 128
Armindo Miguel de Jesus Sousa de Araújo Ribeiro

Chapter 10 **Pathogenesis of Insulin Resistance** 145
Gaffar S. Zaman

Chapter 11 **Emerging Knowledge from Noninvasive Imaging Studies: Is Ammonia Control Enough?** 164
Andrea L. Gropman

Chapter 12 **Cardiovascular Risk/Disease in Type 2 Diabetes Mellitus** 184
Gabriela Roman and Anca Pantea Stoian

Chapter 13 **Hypothyroidism Therapy** 202
Wissal Abassi, Nejmeddine Ouerghi and Anissa Bouassida

Chapter 14 **Epigenetic: New Insight in Gestational Diabetes Mellitus** **215**
Maria Grazia Dalfrà, Silvia Burlina and Annunziara Lapolla

Chapter 15 **Morphology Aspects of Hypothyroidism** **226**
Fernando Candanedo-Gonzalez, Javier Rios-Valencia, Dafne Noemi Pacheco-Garcilazo, Wilfredo Valenzuela-Gonzalez and Armando Gamboa-Dominguez

Permissions

List of Contributors

Index

Preface

Cellular metabolism refers to the set of chemical reactions that take place within the cells of living organisms for maintaining life. Complex sequences of well controlled biochemical reactions are a part of cellular metabolism. These processes enable organisms to develop and reproduce, preserve their structural integrity and adapt to changes in their environment. Cell metabolism reflects the state of the cell's health. The mitochondrion is the primary source of energy for the cell, where bioenergetic processes take place by absorbing fuel sources like fatty acids and glucose. Afterwards, it gets converted into energy through a series of enzymatic reactions. Cellular bioenergetics deregulation results in various pathologies, like neurodegenerative diseases, cancer and diabetes. Cancer is characterized by altered cellular metabolism which aids in the development, growth and maintenance of tumors, as well as malignant transformation. This book contains some path-breaking studies on the role of cellular metabolism in health and disease. Those in search of information to further their knowledge will be greatly assisted by it.

This book is a result of research of several months to collate the most relevant data in the field.

When I was approached with the idea of this book and the proposal to edit it, I was overwhelmed. It gave me an opportunity to reach out to all those who share a common interest with me in this field. I had 3 main parameters for editing this text:

1. Accuracy - The data and information provided in this book should be up-to-date and valuable to the readers.

2. Structure - The data must be presented in a structured format for easy understanding and better grasping of the readers.

3. Universal Approach - This book not only targets students but also experts and innovators in the field, thus my aim was to present topics which are of use to all.

Thus, it took me a couple of months to finish the editing of this book.

I would like to make a special mention of my publisher who considered me worthy of this opportunity and also supported me throughout the editing process. I would also like to thank the editing team at the back-end who extended their help whenever required.

Editor

1

Hypothyroidism

Mauricio Alvarez Andrade and Oscar Rosero Olarte

Abstract

Hypothyroidism is a condition that results from thyroid hormone deficiency that can range from an asymptomatic condition to a life-threatening disease. The prevalence of hypothyroidism varies according to the population, from up to 3 to 4% in some populations and in the case of subclinical hypothyroidism up to 5–10%. Clinical symptoms of hypothyroidism are diverse, broad, and non-specific and can be related to many systems, reflecting the systemic effects of thyroid hormones. The severity of the symptoms is usually related to the severity of the thyroid hormone deficit. The most common form of hypothyroidism, primary hypothyroidism, is diagnosed when there is elevation of TSH and decrease in the level of free T4 and Subclinical hypothyroidism is diagnosed when there is an elevation of TSH with normal levels of free T4. The most frequent cause of primary hypothyroid-ism in populations without iodine deficiency is Hashimoto's thyroiditis or chronic lymphocytic thyroiditis. Iodine deficiency is the main cause of hypothyroidism in populations with deficiency of iodine intake. The treatment of choice for hypothyroidism is thyroxine (T4), which has shown efficacy in multiple studies to restore the euthyroid state and improve the symptoms of hypothyroidism. In subclinical hypothyroidism, the treatment depends on the age, functionality, and comorbidities of the patients. The total replacement dose of levothyroxine in adults is approxi-mately 1.6 mcg/kg; however in elderly patients with heart disease or coronary heart disease, the starting dose should be from 0.3 to 0.4 mcg/kg/day with progressive increase of 10% of the dose monthly.

Keywords: hypothyroidism, autoimmune, Hashimoto's disease, thyroxine, thyroid diseases

1. Introduction

Hypothyroidism is a common disease in many populations, and prevalence varies depending on the sex and the age of the population studied, the iodine status of the population, and the cut points used to define overt and subclinical hypothyroid-ism [1–11].

Hypothyroidism is 10 times more prevalent in women than in men [2]. It is more prevalent in the elderly people, ranging from 2 to 5% of the population. In iodine-sufficient populations, hypothyroidism ranges from 1 to 2%, while in iodine deficient areas, the prevalence can be as high as 3–4% [2–14]. Prevalence in the USA in the Colorado trial can range from 0.3% and 3.7%, while in Europe can be 0.2 and 5.3% [12].

The prevalence of subclinical hypothyroidism is even higher; in one of the largest studies in the USA, it reaches 9% of the population [12].

2. Physiology

Thyroid word comes from the Greek shield because of its shape. The weight is from 10 to 20 g. Changes of thyroid shape and volume are dependent on age and sex. In the adult population, the usual length is 40–60 mm and the diameter 13–18 mm. The volume is 10–15 ml for females and 12–18 ml for males [15, 16].

The thyroid gland is composed of functional secretory units known as follicles where thyroid hormones are synthetized and storage in the follicular cells and the lumen of the follicles where colloid is stored. The colloid contains high amounts of thyroglobulin (Tg), a 660 kDa glycoprotein, where thyroid hormones are stored [17].

Thyroid hormones are composed by an inner ring, the tyrosine molecule, and the outer ring with a phenyl ring. The active forms of thyroid hormones are thyroxine or T4, with four iodine atoms, and triiodothyronine or T3, with three iodine atoms. Under physiologic states, 90% of thyroid gland output is T4 and 10% is T3. The half-life of thyroid hormones is of few hours for T3 and 7 days for T4 [17].

The follicular cells contain a sodium iodide active symporter responsible for iodine transport against a concentration gradient to synthesize thyroid hormones; it can increase iodine concentration in the follicular cell by more than 20 times above the serum concentration. Iodine is then transported by pendrin in the apical membrane to the colloid in a passive manner [17].

Iodine must be organified to be attached to the tyrosine residues of thyroglobu-lin. The thyroid peroxidase (TPO) is the enzyme of the selenoprotein group, with hydrogen peroxide organifying iodine to thyroglobulin. Once in the colloid, iodine is organifid to thyroglobuline to produce, monoiodothyrosine and diiodothyrosine wich finally are coupled to produce T4 and T3, this process is catalyzed by the TPO [17].

TPO activity is regulated by iodine concentration and can be blocked by an excess of iodine concentration, which is known as the Wolff-Chaikoff effect and can lead to a temporary hypothyroidism with an escape mechanism. On the other side, an iodine-depleted thyroid gland that is exposed to iodine can increase thyroid synthesis, which is known as the Jod-Basedow effect, and lead to hyperthyroidism [17].

TPO is the enzyme related to autoimmune hypothyroidism as most of the patients have positive anti-TPO antibodies [16, 17].

2.1 Regulation of thyroid hormones production

Thyroid hormone production is regulated by the hypothalamus pituitary axis. At the hypothalamus, thyrotropin-releasing hormone (TRH) is produced. TRH stimulates the pituitary to produce thyroid-stimulating hormone (TSH) [17].

What trigger thyroid hormone synthesis and release is the stimulation of the TSH in the basolateral receptor of the follicular cell, which is a receptor of the seven transmembrane domain G protein-coupled receptors proteins. The effects of the TSH receptor in the follicular cells are derived to the increased concentration in intracellular cyclic adenosine monophosphate (cAMP) that results in increased iodine uptake and increased protein synthesis to enhance the production of thyroid hormones as well as produce a trophic effect at the thyroid gland [17].

3. Definitions

Hypothyroidism is a condition that results from thyroid hormone deficiency which can range from an asymptomatic condition to a life-threatening condition of the patient life [17–19].

Hypothyroidism can be primary due to a decrease in the production of thyroid hormones in the thyroid gland, secondary due to deficit in the production of TSH in the pituitary or tertiary due to a deficit in the production of TRH in the hypothalamus.

3.1 Subclinical hypothyroidism

Subclinical hypothyroidism is a biochemical diagnosis in which there is an eleva-tion of TSH with a normal level of free thyroid hormones in plasma.

4. Manifestations of hypothyroidism

4.1 Adults

Clinical symptoms of hypothyroidism are diverse, broad, and neither sensitive nor specific to make the diagnosis of hypothyroidism and can be related to many systems, reflecting the systemic effects of thyroid hormones. The severity of the symptoms is usually related to the severity of the thyroid hormone deficit.

Systemic symptoms include lethargy, cold intolerance, goiter, and weight gain. At the cardiovascular system hypothyroidism can produce bradycardia, cardiac failure, angina, and pericardial effusion. Gastrointestinal symptoms are constipation and ileus, neuromúscular manifestations include myalgia, hoarse voice, slow relaxing reflexes, depression, emotional lability, psychosis an carpal tunnel syndrome. Hematologic changes include macrocytic anemia, pernicious anemia, and iron deficiency anemia. Skin manifestations include, myxoedema, hair loss, and coarse skin. In the reproductive system menorrhagia and infertility and finally, hyperlipidaemia as the main metabolic manifestation [17–21].

4.2 Myxedema

Myxedema coma is a severe state of hypothyroidism and an endocrine emergency. Manifestations are depressed mental state and hypothermia with a hypometabolic state with bradycardia. Decreased myocardial contractility and pericardial effusion lead to hypotension. Other features are anemia, hyponatremia, and renal dysfunction [22].

4.3 Congenital hypothyroidism

Thyroid hormones are necessary to have a normal neurodevelopment and growth. Symptoms and signs of congenital hypothyroidism are goiter, poor feeding, macroglossia, prolonged jaundice, developmental delay hypothermia, bradycardia, edema, large fontanelles, umbilical hernia, and poor growth [23].

5. Etiology

5.1 Diagnosis

The diagnosis of hypothyroidism is made biochemically. The elevation of TSH levels associated with low levels of free T4 confirms the diagnosis of primary hypothyroidism; in this case it is not necessary to measure levels of free T3 or T3 because in the state of hypothyroidism the peripheral conversion of T4 to T3 is increased so that T4 will be more diminished than T3 [18, 19, 24].

In secondary hypothyroidism, decreased T4 is found, and TSH is low or normal (not elevated), which means that the pituitary is not responding adequately to a deficit of thyroid hormones.

Subclinical hypothyroidism is diagnosed in patients with elevated TSH despite having normal levels of free T4. These patients may or may not be symptomatic [18, 19, 24]Although the differential diagnosis in hypothyroidism involves multiple pathologies, with symptoms and signs related to hypothyroidism such as anemia and hyponatremia, among others, the differential diagnosis must also be among the various pathologies that produce hypothyroidism that will be discussed in the etiology section [25, 26].

However, it is important to take into account the sick euthyroid syndrome that refers to alterations in thyroid function tests that can be found in patients with critical illness and can vary depending on the severity and duration of the disease.

In laboratory alterations, there is a decrease in T3 and a smaller proportion of T4, due to an increase in the activity of reverse T3. Subsequently, there is a progressive decrease in TSH followed by a progressive elevation and finally normalization of all thyroid function tests once the injury is resolved [25, 26].

6. Differential diagnosis

6.1 Primary hypothyroidism

6.1.1 Chronic autoimmune thyroiditis or Hashimoto's thyroiditis

It is the most prevalent cause of hypothyroidism in iodine-sufficient countries.

Chronic autoimmune thyroiditis can be goitrous or atrophic. Goitrous hypothyroidism is called Hashimoto thyroiditis [27].

More than 90% of patients have elevated anti-thyroglobulin or anti-peroxidase (microsomal antigen) or anti-sodium iodine transporter. Antibodies against thyroid gland produce chronic autoimmune thyroiditis with lymphocytic infiltration and fibrosis, leading to goiter or atrophy of the thyroid gland [27–29].

Women are five times more affected than men. After the age of 45, the rates of hypothyroidism increase [27].

Based on one of the most representative studies of the population of the USA that included 17,353 people, it found a prevalence of 4.6% of hypothyroidism, 0.3% frank hypothyroidism and 4.3% subclinical hypothyroidism [30].

The course of chronic autoimmune thyroiditis is a gradual loss of thyroid function. The spectrum ranges from subclinical hypothyroidism with positive antibodies to frank hypothyroidism, a process that affects approximately 5% patients per year.

The majority of patients present hypothyroidism for life; however it may be transient [28].

The risk factors for Hashimoto's thyroiditis are multiple; the female gender and the older age are two risk factors. There is a genetic factor associated with multiple polymorphisms in human leukocyte antigen (HLA) genes, T cell receptors, and immunomodulatory molecules. Patients with Down or Turner syndrome have a higher prevalence. Chronic autonomic thyroiditis can be part of the autoimmune type 2 polyglandular syndrome, and affected patients are more likely to develop other autoimmune diseases such as diabetes and adrenal insufficiency [27, 31, 32].

6.1.2 Iodine deficiency

Iodine deficiency, defined as the daily intake of less than 100 mcg of iodine, is the most common cause of hypothyroidism worldwide and the most prevalent cause of hypothyroidism in populations with iodine deficiency. As previously described, excess iodine can produce hypothyroidism due to Wolff-Chaikoff effect; however this effect is usually transient [33, 34].

6.1.3 Iatrogenic disease

Post-thyroidectomy state, radioactive iodine therapy for thyroid cancer, or hyperthyroidism and neck radiation at doses greater than 25 Gy are the main causes of iatrogenic hypothyroidism, and as the use of these therapies increases, it becomes a more prevalent etiology [35, 36].

6.1.4 Drugs

Thionamides, lithium, tetracyclines, thalidomide, ethionamide, and iodine-containing drugs like amiodarone can cause hypothyroidism. Thyrosine kinase inhibitors can cause thyroiditis; sorafenib can produce hypothyroidism by increased type 3 deiodination. Immune therapy including pembrolizumab, nivolumab, ipilimumab, alemtuzumab, interleukin 2, and interferon alfa can produce hypothy-roidism [37].

6.2 Thyroiditis

6.2.1 Subacute or granulomatous thyroiditis

Subacute or granulomatous thyroiditis is an acute inflammation of the thyroid gland of viral etiology that presents with hyperthyroidism, followed by hypothyroidism and subsequent recovery of thyroid function. Transient hypothyroidism usually lasts from a few weeks to a maximum of 3–6 months and may be permanent in 5% of patients [38, 39].

The presentation consists of acute pain in the thyroid region that increases when swallowing or moving the head and radiates to the jaw. Symptoms may vary depending on whether the patient is in the hyperthyroid, euthyroid, or hypothyroid category [38, 39].

The findings in thyroid gammagraphy are compatible with thyroiditis due to a decrease in iodine uptake diffusely [38, 39].

6.2.2 Silent thyroiditis or postpartum thyroiditis

Silent thyroiditis or postpartum thyroiditis corresponds to a thyroiditis of autoimmune etiology, which occurs in the first year postpartum. It presents with hyperthyroidism in up to 30% of patients, followed by hypothyroidism in up to 50% of patients; however it can only be present as hypothyroidism or hyperthyroidism [40–42].

It is more frequent in patients with type 1 diabetes mellitus and patients with positive anti-peroxidase antibodies. The prevalence can reach up to 17%. It has been associated with deterioration or onset of postpartum depression. Up to 30% of patients remain hypothyroid [40–42].

6.2.3 Infiltrative diseases

Riedel's thyroiditis is a fibrosclerosing thyroiditis of unknown etiology, with a probable primary anti-immune or fibrotic origin similar to retroperitoneal fibrosis, fibrosing mediastinitis, sclerosing cholangitis, and lacrimal fibrosis, among others [43, 44].

Riedel's thyroiditis is characterized by slow, non-painful growth, sensation of pressure in the neck, dysphagia, dysphonia, and hypoparathyroidism. From 30 to 60% of patients present clinical or subclinical hypothyroidism. It is one of the IgG4-related disease varieties, together with Fibrosing hashimoto thyroiditis, Igg4-related Hashimoto's disease, and Graves' disease associated with IgG4 [43–45].

Other infiltrative diseases like hemochromatosis, scleroderma, leukemia, cystinosis, *M. tuberculosis* infection, and *P. carinii* are less frequent causes of hypothyroidism.

6.3 Secondary hypothyroidism

Central hypothyroidism is a much less frequent form of hypothyroidism, with a prevalence of 1:16,000 to 1:100,000 in the general population. It can be congenital or acquired [46].

The causes of acquired central hypothyroidism are usually related to the causes of hypopituitarism like a pituitary sellar region mass, usually a pituitary adenoma, which produces secondary hypothyroidism by thyrotropic cell compression, or tertiary hypothyroidism with decreased production of TRH by the hypothalamus. Other lesions of the sellar region, such as meningiomas, cysts, abscesses, metastasis, craniopharyngiomas, and dysgerminomas, can produce central hypothyroidism [46, 47].

In addition to space-occupying injuries, radiation with doses greater than 40 Gray performed for brain, orbital, or nasal lesions can produce central hypothyroidism [46, 47].

Other less frequent causes of central hypothyroidism are hypophyseal infiltrative pathologies such as haemochromatosis, sarcoidosis or tuberculosis, cranial trauma, Sheehan syndrome, and the use of drugs as checkpoint inhibitors [46, 47].

7. Treatment

7.1 Clinical hypothyroidism

Since the nineteenth century, levothyroxine has been used for the treatment of hypothyroidism. Usually hypothyroidism requires treatment with lifelong hormone replacement with levothyroxine. In a few cases, transient hormonal substitution due to transient secondary hypothyroidism is required, for example, to subacute thyroiditis or drug-induced hypothyroidism [48–50].

The treatment of choice is thyroxine (T4), which has shown efficacy in multiple studies to restore the euthyroid state and improve the symptoms of hypothy-roidism [51–54].

The goal of treatment of primary hypothyroidism is to take the patient to the normal range of TSH. However, the normal range of TSH varies depending on the age and population studied. Most people are in the range of 0.5–4.5 mU/L; however as the age of people increases, the normal range of TSH increases, leading to values of up to 7.0 mU/L in those over 90 years. In contrast, most young and healthy patients are in the range of 0.5–2.5 mU/L [50].

For this reason, in the treatment of hypothyroidism, the dose of levothyroxine and the goal of TSH depend on the age of the patient and the comorbidities [49, 50].

Levothyroxine (T4) is a prohormone and requires deionization to T3 which is the active form of thyroid hormone. Levothyroxine is absorbed in the small intestine. The meal affects the time of maximum concentration, which in normal conditions is 2 h. The bioavailability is from 60 to 80%. The metabolism is catabolized by the thyroid deionidase enzyme that removes the iodine from carbon 5 of the outer ring to transform T4 into T3. Approximately half of T4 is deionized to rT3 (inac-tive form) and half to T3 (active form). Both T3 and reverse T3 are metabolized to diiodothyronine (T2) and monoiodothyronamine (T1) and T2 and T1 reverses [48].

Multiple medications interact with the function or pharmacokinetics of levothyroxine, amiodarone, androgens, calcium carbonate and citrate, carbamazepine, cholestyramine, ferrous sulfate, glucocorticoids, orlistat, phenytoin, proton pump inhibitors, salicylates, sucralfate, and tamoxifen, which are just some of the medications that alter bioavailability, metabolism, protein binding, or hormone levels [48].

The total replacement dose of levothyroxine in adults is approximately 1.6 mcg/kg, given that the body's requirements for thyroid hormones are proportional to weight. In healthy young patients, the starting dose could be 1.6 mcg/kg/day; however in elderly patients with heart disease or coronary heart disease, the starting dose should be from 0.3 to 0.4 mcg/kg/day with a progressive increase of 10% in the dose every 4–6 weeks [50]. Levothyroxine must be taken with empty stomach 30–60 min before the next meal, usually, breakfast.

The thyroid function is monitored with TSH at 4–6 weeks after starting treatment. If the TSH goal is not achieved, the dose of levothyroxine should be adjusted by increasing or decreasing 10% of the dose ideally, especially in older adults [50].

In the case of secondary hypothyroidism, TSH levels are low or inappropriately normal for a low free T4. Therefore, the follow-up is not done with TSH levels but with free T4 levels, to achieve a normal level for the reference range.

7.2 Subclinical hypothyroidism

In the case of subclinical hypothyroidism, the treatment also depends on the age, functionality, and comorbidities of the patients [50].

For patients younger than 75 years, with TSH greater than 10 mU/L, treatment is recommended. However, in those patients with TSH between 4.5 and 10 mU/L, treatment depends on the presence of symptoms and especially the presence of goiter or anti-TPO antibodies, which predict progression to clinical hypothyroidism [54–56].

In patients older than 75 years, treatment depends on the patient's frailty and should be limited to functional patients with TSH greater than 10 mU/L or patients with TSH of 6–10 mU/L in the presence of antithyroid antibodies, symptoms, and concomitant diseases in that they can be impaired by hypothyroidism such as heart failure. Fragile patients, more than 75 years old, may be advisable to be observed without treatment [50].

8. Conclusion

Hypothyroidism is a highly prevalent chronic disease, widely studied by medical science, with a wide spectrum of severity, ranging from subclinical hypothyroidism to the hypothyroid myxedematous state. In some cases of subclinical hypothyroidism, treatment may not be necessary; however in other cases such as in myxedematous states, the treatment may be lifesaving. There are multiple trials that evaluate the treatment of hypothyroidism in different populations, and there is still controversy regarding the treatment of subclinical hypothyroidism in some populations. It is very important for the primary care physicians to have a broad knowledge of hypothyroidism since they will face hypothyroid patients in the day-to-day clinical practice.

Author details

Mauricio Alvarez Andrade[1*] and Oscar Rosero Olarte[2]

1 Endocrinólogo, Organización Sanitas Internacional, Keralty, Bogotá, Colombia

2 Endocrinólogo, Instituto de Osteoporosis de los Llanos, Osteollanos, Regional Hospital, Villavicencio, Colombia

*Address all correspondence to: mauricioalvarez613@gmail.com

References

[1] Vanderpump MP. The epidemiology of thyroid disease. British Medical Bulletin. 2011;**99**:39-51

[2] Bjoro T et al. Prevalence of thyroid disease, thyroid dysfunction and thyroid peroxidase antibodies in a large, unselected population. The Health Study of Nord-Trondelag (HUNT). European Journal of Endocrinology. 2000;**143**:639-647

[3] Konno N et al. Screening for thyroid diseases in an iodine sufficient area with sensitive thyrotrophin assays, and serum thyroid autoantibody and urinary iodide determinations. Clinical Endocrinology. 1993;**38**:273-281

[4] Walsh JP. Managing thyroid disease in general practice. The Medical Journal of Australia. 2016;**205**:179-184

[5] Gopinath B et al. Five-year incidence and progression of thyroid dysfunction in an older population. Internal Medicine Journal. 2010;**40**:642-649

[6] Laurberg P, Pedersen KM, Vestergaard H, Sigurdsson G. High incidence of multinodular toxic goitre in the elderly population in a low iodine intake area versus high incidence of Graves' disease in the young in a high iodine intake area: Comparative surveys of thyrotoxicosis epidemiology in East-Jutland Denmark and Iceland. Journal of Internal Medicine. 1991;**229**:415-420

[7] Flynn RW, MacDonald TM, Morris AD, Jung RT, Leese GP. The thyroid epidemiology, audit, and research study: Thyroid dysfunction in the general population. The Journal of Clinical Endocrinology and Metabolism. 2004;**89**:3879-3884

[8] Valdes S et al. Population-based national prevalence of thyroid dysfunction in Spain and associated factors: Diabetes study. Thyroid. 2017;**27**:156-166

[9] Sgarbi JA, Matsumura LK, Kasamatsu TS, Ferreira SR, Maciel RM. Subclinical thyroid dysfunctions are independent risk factors for mortality in a 7.5-year follow-up: The Japanese-Brazilian thyroid study. European Journal of Endocrinology. 2010;**162**:569-577

[10] Amouzegar A et al. Natural course of euthyroidism and clues for early diagnosis of thyroid dysfunction: Tehran Thyroid Study. Thyroid. 2017;**27**:616-625

[11] Amouzegar A et al. The prevalence, incidence and natural course of positive antithyroperoxidase antibodies in a population-based study: Tehran Thyroid Study. PLoS One. 2017;**12**:e0169283

[12] Canaris GJ, Manowitz NR, Mayor G, Ridgway EC. The Colorado thyroid disease prevalence study. Archives of Internal Medicine. 2000;**160**:526-534

[13] Asvold BO, Vatten LJ, Bjoro T. Changes in the prevalence of hypothyroidism: The HUNT Study in Norway. European Journal of Endocrinology. 2013;**169**:613-620

[14] McGrogan A, Seaman HE, Wright JW, de Vries CS. The incidence of autoimmune thyroid disease: A systematic review of the literature. Clinical Endocrinology. 2008;**69**:687-696

[15] Mihai R. Physiology of the pituitary, thyroid, parathyroid and adrenal glands. Surgery (Oxford). 2014;**32**(10):504-512

[16] Chaudhary V, Bano S. Thyroid ultrasound. Indian Journal of Endocrinology and Metabolism. 2013;**17**(2):219-227. DOI: 10.4103/2230-8210.109667

[17] Stathatos N. Thyroid physiology. The Medical Clinics of North America. 2012;**96**(2):165-173. DOI: 10.1016/j.

mcna.2012.01.007. Epub: February 1, 2012

[18] Gilbert J. Hypothyroidism. Medicine. 2017;**45**(8):506-509

[19] Chaker L, Bianco AC, Jonklaas J, Peeters RP. Hypothyroidism. The Lancet. 2017;**390**(10101):1550-1562

[20] McLean RM, Podell DN. Bone and joint manifestations of hypothyroidism. Seminars in Arthritis and Rheumatism. 1995;**24**(4):282-290

[21] Arora H, Abdullakutty J, Mathew R, Joseph J. Cardiovascular manifestations of hypothyroidism. Indian Heart Journal. 2018;**70**(Supplement 2):100

[22] Verma V, Menon AS, Bahadur A, Sikarwar A. Clinical manifestation and images of a patient having cardiovascular involvement in myxoedema coma. Medical Journal, Armed Forces India. 2019:0377-1237. In Press. Corrected Proof Available Online

[23] Wassner AJ. Congenital hypothyroidism. Clinics in Perinatology. 2018;**45**:1-18

[24] Almandoz JP, Gharib H. Hypothyroidism: Etiology, diagnosis, and management. The Medical Clinics of North America. 2012;**96**(2):203-221

[25] Mebis L, Van den Berghe G. Thyroid axis function and dysfunction in critical illness. Best Practice & Research. Clinical Endocrinology & Metabolism. 2011;**25**(5):745-757

[26] Soh S-B, Aw T-C. Laboratory testing in thyroid conditions—Pitfalls and clinical utility. Annals of Laboratory Medicine. 2019;**39**(1):3-14

[27] Dayan CM, Daniels GH. Chronic autoimmune thyroiditis. The New England Journal of Medicine. 1996;**335**:99-107

[28] Vanderpump T, French A, Bates D, Clark F, Grimley Evans J, Hasan DM, et al. The incidence of thyroid disorders in the community: A twenty-year follow-up of the Whickham Survey. Clinical Endocrinology. 1995;**43**(1):55-68

[29] Rallison ML, Dobyns BM, Meikle AW, Bishop M, Lyon JL, Stevens W. Natural history of thyroid abnormalities: Prevalence, incidence, and regression of thyroid diseases in adolescents and young adults. The American Journal of Medicine. 1991;**91**(4):363-370

[30] Hollowell SN, Flanders W, Hannon WH, Gunter EW, Spencer CA, Braverman LE. Serum TSH, T(4), and thyroid antibodies in the United States population (1988 to 1994): National Health and Nutrition Examination Survey (NHANES III). The Journal of Clinical Endocrinology and Metabolism. 2002;**87**(2):489-499

[31] Tomer Y, Davies TF. Searching for the autoimmune thyroid disease susceptibility genes: From gene mapping to gene function. Endocrine Reviews. 2003;**24**(5):694-717

[32] Fallahi P, Ferrari SM, Ruffilli I, et al. The association of other autoimmune diseases in patients with autoimmune thyroiditis: Review of the literature and report of a large series of patients. Autoimmunity Reviews. 2016;**15**:1125

[33] Andersson M, Takkouche B, Egli I, et al. Current global iodine status and progress over the last decade towards the elimination of iodine deficiency. Bulletin of the World Health Organization. 2005;**83**:518

[34] Bourdoux P, Delange F, Gerard M, et al. Evidence that cassava ingestion increases thiocyanate formation: A possible etiologic factor in endemic goiter. The Journal of Clinical

Endocrinology and Metabolism. 1978;**46**:613

[35] Sridama V, McCormick M, Kaplan EL, et al. Long-term follow-up study of compensated low-dose 131I therapy for Graves' disease. The New England Journal of Medicine. 1984;**311**:426

[36] Franklyn JA, Daykin J, Drolc Z, et al. Long-term follow-up of treatment of thyrotoxicosis by three different methods. Clinical Endocrinology. 1991;**34**:71

[37] Rizzo LFL, Mana DL, Serra HA. Drug-induced hypothyroidism. Medicina (Buenos Aires). 2017;**77**(5):394-404

[38] Singer PA. Thyroiditis. Acute, subacute, and chronic. The Medical Clinics of North America. 1991;**75**(1):61-77

[39] Slatosky J, Shipton B, Wahba H. Thyroiditis: Differential diagnosis and management. American Family Physician. 2000;**61**(4):1047-1052; 1054

[40] Nicholson WK, Robinson KA, Smallridge RC, et al. Prevalence of postpartum thyroid dysfunction: A quantitative review. Thyroid. 2006;**16**:573

[41] De Groot L, Abalovich M, Alexander EK, et al. Management of thyroid dysfunction during pregnancy and postpartum: An Endocrine Society clinical practice guideline. The Journal of Clinical Endocrinology and Metabolism. 2012;**97**:2543

[42] Stagnaro-Green A. Approach to the patient with postpartum thyroiditis. The Journal of Clinical Endocrinology and Metabolism. 2012;**97**:334

[43] Falhammar H, Juhlin CC, Barner C, Catrina S-B, Karefylakis C, Calissendorff J. Riedel's thyroiditis: Clinical presentation, treatment and outcomes. Endocrine. 2018;**60**(1):185-192

[44] Dahlgren M, Khosroshahi A, Nielsen GP, Deshpande V, Stone JH. Riedel's thyroiditis and multifocal fibrosclerosis are part of the IgG4-related systemic disease spectrum. Arthritis Care & Research (Hoboken). 2010;**62**(9):1312-1318

[45] Kottahachchi D, Toplissa DJ. Immunoglobulin G4-related thyroid diseases. European Thyroid Journal. 2016;**5**(4):231-239

[46] Persani L. Clinical review: Central hypothyroidism: Pathogenic, diagnostic, and therapeutic challenges. The Journal of Clinical Endocrinology and Metabolism. 2012;**97**(9):3068-3078

[47] Persani L, Cangiano B, Bonomi M. The diagnosis and management of central hypothyroidism in 2018. Endocrine Connections. 2019;**8**(2):R44-R54

[48] Colucci P, Yue CS, Ducharme M, Benvenga S. A review of the pharmacokinetics of levothyroxine for the treatment of hypothyroidism. European Endocrinology. 2013;**9**(1):40-47

[49] Clarke N, Kabadi UM. Optimizing treatment of hypothyroidism. Treatments in Endocrinology. 2004;**3**(4):217-221

[50] Calsolaro V, Niccolai F, Pasqualetti G, Tognini S, Magno S, Riccioni T, et al. Hypothyroidism in the elderly: Who should be treated and how? Journal of the Endocrine Society. 2019;**3**(1):146-158

[51] Hennessey JV, Espaillat R. Current evidence for the treatment of hypothyroidism with levothyroxine/levotriiodothyronine combination therapy versus levothyroxine

monotherapy. International Journal of Clinical Practice. 2018;**72**(2):e13062

[52] Marina M, Ceda GP, Aloe R, Gnocchi C, Ceresini G. Circulating concentrations of free thyroxine after an oral intake of liquid LT4 taken either during fasting conditions or at breakfast. Acta Bio-Medica. 2017;**87**(3):247-252

[53] Grozinsky-Glasberg S, Fraser A, Nahshoni E, Weizman A, Leibovici L. Thyroxine-triiodothyronine combination therapy versus thyroxine monotherapy for clinical hypothyroidism: Meta-analysis of randomized controlled trials. The Journal of Clinical Endocrinology and Metabolism. 2006;**91**(7):2592-2599

[54] Pollock MA, Sturrock A, Marshall K, et al. Thyroxine treatment in patients with symptoms of hypothyroidism but thyroid function tests within the reference range: Randomised double blind placebo controlled crossover trial. BMJ. 2001;**323**:891

[55] McDermott MT, Ridgway EC. Subclinical hypothyroidism is mild thyroid failure and should be treated. The Journal of Clinical Endocrinology and Metabolism. 2001;**86**:4585

[56] Cooper DS. Clinical practice. Subclinical hypothyroidism. The New England Journal of Medicine. 2001;**345**:260

2

Cerebral Energy Metabolism: Measuring and Understanding its Rate

Avital Schurr

Abstract

The study of brain energy metabolism has taken second place to that of muscle ever since the dawn of this field of research. Consequently, each new discovery made using muscle tissue that advanced our understanding of the biochemistry of energy metabolic processes was attempted to be duplicated in brain tissue. It was only when the brain's high energy needs were recognized that researchers realized its vulnerability to any mishap in its energy supplies and that this vulnerability may play a role in various brain disorders. Understanding of the mechanisms by which the brain deals with energy shortage is of utmost importance in shedding light on the fundamentals of brain disorders and their potential treatment. To achieve such understanding, accurate measurement of brain energy metabolic rates is necessary. This chapter summarizes the history of the current knowledge of the biochemical processes responsible for the production of adenosine triphosphate (ATP) in the brain. It briefly reviews the various techniques used to measure cerebral metabolic rates of oxygen (CMR_{O_2}) and glucose ($CMR_{glucose}$), and elaborates on the potential of measuring the cerebral metabolic rate of lactate ($CMR_{lactate}$) to improve our under-standing of brain energy metabolism.

Keywords: BOLD fMRI, cerebral metabolic rate, glucose, glycolysis, lactate, oxygen, paradigm shift, polarography

1. A short review of brain energy metabolism research

1.1 The first eight decades (1900–1980)

Most human cells produce adenosine triphosphate (ATP) via two, mostly inter-connected biochemical pathways, glycolysis and mitochondrial oxidative phosphor-ylation. Erythrocytes (red blood cells, RBCs) produce their ATP via the glycolytic pathway alone, since they lack mitochondria. Under anaerobic conditions, the less efficient glycolytic pathway is the main source of ATP supply, since without oxygen the oxidative phosphorylation pathway cannot be maintained. Throughout the first half of the twentieth century the majority of the researchers in the field of energy metabolism made muscle their tissue of choice for the study of energy metabolism. While muscle was believed to require a great deal of energy to perform its work, the brain was assumed to be a low consumer of energy, as indicated by the follow-ing quote: "*the brain is not a seat of active combustion, and considering the very small increase in CO_2 in the torcular blood it seems to us very improbable that the temperature* *of the brain should be perceptibly greater than that of the blood*" [1]. It was Tashiro [2] who was the first to demonstrate that nerve produces CO_2 and ammonia during its metabolism. By 1924, Warburg et al. [3] demonstrated that brain tissue is able to convert large amount of glucose to lactic acid, and in 1927 it was shown that nerve produces a measurable amount of heat, an amount that increases upon electric stimulation [4]. Moreover, this increase in heat production was shown to correlate with the amount of oxygen consumed. According to Holmes [5] the above finding was the necessary proof that nerve impulse is a chemical process. All the major discoveries that have led to the elucidation of the biochemical pathways of energy metabolism were made using the tissue of choice in the field i.e., muscle. Both the tricarboxylic acid cycle and the glycolytic pathways were introduced in 1937 and 1940, respectively (the reader is directed to the many detailed reviews on the topic that are available). While there is a general agreement among biochemists, physiologists and neuroscientists as to the accuracy of the mitochondrial tricarboxylic acid (TCA) cycle and the oxidative phosphorylation pathway, disagreements exist on the accuracy of the glycolytic pathway. Hence, glycolysis is the main focus of this chapter, since the original drawing of the pathway stands in conflict with various research findings of the past three decades (for more detailed reviews of this topic the reader is directed to [6, 7]) . Although some research on brain energy metabolism was performed during 1920s and 1930s, it was limited to very few laboratories. Among them, that of Eric G. Holmes and Barbara E. Holmes pioneered important research in the field. The duo, who later joined by C.A. Ashford, published a series of papers [8–14] to demonstrate brain tissue production of lactate from glucose (similar to muscle metabolism), the involvement of phosphates and glycogen in this metabolism and the ability of brain tissue to oxidize lactate. Unfortunately, the latter never received neither the praise nor the scrutiny it deserved. This point is expanded upon in the next section, although it must be emphasized here that if the importance of that discovery would have been recognized at the time, our understanding of brain energy metabolism would be significantly accelerated and advanced. Nevertheless, Holmes and Ashford interpreted their finding of lactate oxidation simply as a process by which the brain rids itself of a waste product, since lactate was believed until the mid 1980s and even beyond, to be just that, a useless end-product of carbohydrate metabolism. Consequently, the research by Holmes and Ashford quickly became obscure. It would have stayed this way if not for a liter-ature search I carried out in 2005 working on an upcoming paper that dealt with the possible role of lactate as an oxidative brain energy substrate [15]. Spending several weeks in the basement of my university medical school library (at the time most of the old literature has not yet been digitized) was an experience akin to treasure hunt, an experience I still cherish today. Discovering Holmes and Ashford's papers in 75-year old, dust-covered, heavy volumes of the Biochemical Journal was almost as exciting as conducting our own research [16]. Three years prior to the publication of the latter on brain energy metabolism, Brooks published his controversial work on muscle energy metabolism [17]. His proposal that skeletal muscle both produces and consumes lactate met with major objections because such consumption would require lactate to be a mitochondrial substrate, which requires the existence of lactate dehydrogenase in mitochondria (mLDH), an enzyme his detractors strongly insisted does not exist. Schurr and colleagues demonstrated *in vitro* that brain tissue is capable of maintaining normal neuronal function when lactate is the sole oxidizable energy substrate [16]. Skeptics of this finding argued that the phenomenon does not occur *in vivo* and even if it does, lactate cannot replace glucose as the obligatory energy substrate in brain [18]. And thus began a long-lasting debate on the validity of these findings and the potential importance of lactate as an oxidative substrate of energy metabolism in brain an elsewhere. **Figure 1** is an illustration

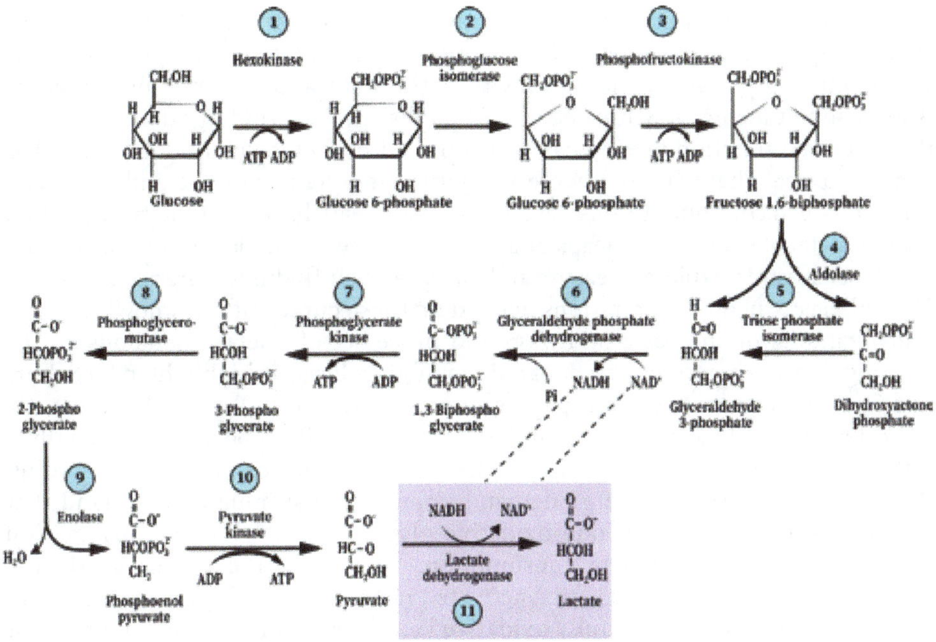

Figure 1.
The classical description of the 10 enzymatic steps of aerobic glycolysis, the pathway that converts glucose to pyruvate with a net production of two molecules of ATP and two molecules of pyruvate, its end-product and the substrate of the mitochondrial tricarboxylic acid (TCA) cycle. Under anaerobic conditions, an eleventh enzymatic step occurs where pyruvate in converted to lactate, which becomes the glycolytic end-product under these conditions.

of the glycolytic pathway as it has been conceived and taught everywhere since 1940. This concept has been based on an initial assumption, made not by those who proposed the original reaction sequence of the glycolytic pathway, but by Krebs and Johnston, who proposed the reaction sequence of the mitochondrial TCA cycle [19]. The latter postulated that pyruvate is the substrate of that cycle and assumed that glycolysis could be its origin. Such an assumption, when made by a leading scientist of the status Krebs had achieved, was persuasive enough to compel Gustav Embden, Otto Meyerhof, and Jakub Karol Parnas to decide that pyruvate is the end-product of aerobic glycolysis and the substrate of the TCA cycle. Hence, for almost eight decades the glycolytic pathway has been described as one that has two possible outcomes, an aerobic one, which ends with pyruvate and an anaerobic one that ends with lactate (**Figure 1**). Every biochemistry textbook published from the 1940s on, every biochemistry, physiology or neuroscience course being taught at any level and every online search, all display this very description of the glycolytic pathway. That, despite ample research data that clearly refute this dogmatic paradigm.

1.2 The last four decades (1980–2018)

As indicated above, questions as to the correctness of the old description of glycolysis began to appear in the late 1980s. The lactic acidosis hypothesis of delayed neuronal damage [20] had a strong following at the time. This hypothesis postulated that lactate accumulation in the ischemic brain is the cause of delayed neuronal damage, damage observed after a recovery from the original ischemic insult had occurred. The popularity of this hypothesis was so strong that any insinuation that lactate could be anything but a menacing factor in cerebral ischemia aroused great skepticism. Consequently, when Fox et al. [21] published a study, which indicated that neural activation does not require an increase in energy supply and is supported by a mere non-oxidative glucose utilization ("anaerobic" glycolysis), they met with both doubt and a degree of cynicism. Almost simultaneously, Schurr et al. [16] published their findings demonstrating that neuronal function *in vitro* can be supported by lactate as the sole oxidizable energy source These findings were met with even greater doubt and cynicism, some of which continues to this day. The past 30 years have seen the field of brain energy metabolism grow by leaps and bounds as new technologies and techniques enable scientists to explore, measure and interpret their findings more accurately. Nevertheless, such interpretations depend on the accuracy of our knowledge and understanding of the basic pathways and processes of energy metabolism. The ongoing debate about the correct paradigm of glycolysis, as highlighted by Schurr and Gozal [22] and many of the papers within this Research Topic volume, clearly indicates that such accuracy and understanding are still to be achieved. To illustrate this point consider on one hand, the conclusion of Fox et al. [21] regarding the very low increase in energy demand upon neural stimulation, demand that can be easily answered by non-oxidative glycolysis i.e., glucose consumption unac-companied by oxygen consumption, while on the other hand, the conclusion by Hyder et al. [23] that activated neural tissue exhibits an increase in energy production, which is fully oxidative i.e., the ratio of oxygen to glucose for this increase is 6:1. Could these completely opposing conclusions be explained by differences in the methodologies used in the two studies (the former made use of ^{18}F-2-fluoro-2-deoxy-D-glucose to measure glucose uptake and $^{15}O_2$ to measure oxygen consumption, while the latter made use of blood oxygen level dependent (BOLD) functional magnetic resonance imaging (fMRI))? Or maybe the measurements by both methods are correct, but their interpretation has relied on assumptions that emanated from older, dogmatic concept?

To answer this question, one must consider the mounting evidence supporting a paradigm shift in our comprehension of the glycolytic pathway [24]. The shift entails redrawing the glycolytic pathway as one consisting of 11 steps, beginning with glucose as its substrate and ending with lactate as its end-product, independent of the presence or absence of oxygen. From its inception, distinguishing between aerobic and anaerobic glycolysis was based not on specific evidence that the two pathways exist and produce two different products, pyruvate and lactate, respectively. That separation was an attempt by the pathway's elucidators to somehow fit it into, what they concluded, is an outcome that produces pyruvate as its main end-product since they accepted Kreb's suggestion that this monocarboxylate is the substrate of the TCA cycle. It must have been relatively easy to accept that suggestion considering lactate's negative reputation [6]. Hence, the glycolytic pathway should be considered one, uninterrupted chain of biochemical reactions that begins with glucose and ends with lactate (**Figure 1**). Accordingly, its last reaction (number 11), the reduction of pyruvate to lactate by the cytosolic lactate dehydrogenase (cLDH), plays a crucial role in keeping this pathway's cyclical nature operational as it provides a continuous supply of NAD^+. If pyruvate was the glycolytic end-product, NAD^+ would have to be imported from other sources and locations, a proposition that has offered a somewhat shaky resolution (see [25] and references within). This, of course, is not the only factor that justifies a paradigm shift. There are numerous studies published over the past two decades demonstrating the presence of lactate dehydrogenase in mitochondria (mLDH), an enzyme that converts lactate to pyruvate [26–32]. Brooks et al. [33] also demonstrated the presence of monocarboxylate transporter 1 (MCT1) in mitochondria, the transporter that is responsible for the transport of lactate along its gradient from the cytosol to the mitochondrion. Havel et al. showed that in blood and in other tissues the ratio lactate/pyruvate is >10, a value that is not consistent with the assumption that pyruvate is the glycolytic end-product [34]. Moreover, the proposal that aerobic glycolysis ends with pyruvate does not meet the known standard free-energy ($\Delta G^{0'}$) change of the reaction pyruvate → lactate, which is −6.0 kcal/mol, a value indicating that this reaction should proceed independently of the presence or absence of oxygen. In other words, glycolysis, whether aerobic or anaerobic, should always end up with lactate. **Figure 2A** demonstrates the free energy change profile of aerobic glycolysis that ends with the reaction phosphoenolpyruvate→ pyruvate, although the potential free-energy change of the conversion pyruvate → lactate (**Figure 2B**, anaerobic glycolysis) determines that glycolysis should end with lactate regardless of the oxygenation condition (**Figure 2C**). Last but not least, the reaction equilibrium of cLDH is tilted heavily in the direction of lactate production, which makes it unlikely for lactate to be converted back to pyruvate by that cytosolic enzyme. In contrast, the reaction equilibrium of mLDH tilts in the direction of lactate oxidation to pyruvate [35, 36].

The above points support the proposed paradigm shift in the glycolytic pathway [24], where lactate, not pyruvate, is its end-product and the oxidative mitochondrial substrate for the TCA cycle. Accordingly, is measuring the cerebral metabolic rates of oxygen (CMR_{O_2}) and glucose ($CMR_{glucose}$) sufficient in providing an accurate picture of brain energy metabolism during rest or activation, in health or disease? If lactate is an oxidative energy substrate, should not $CMR_{lactate}$ also be measured in order to have a more complete account of cerebral energy metabolism? How would the measurement of $CMR_{lactate}$ contribute to our understanding of the brain's ability to handle its energy demands under those conditions?

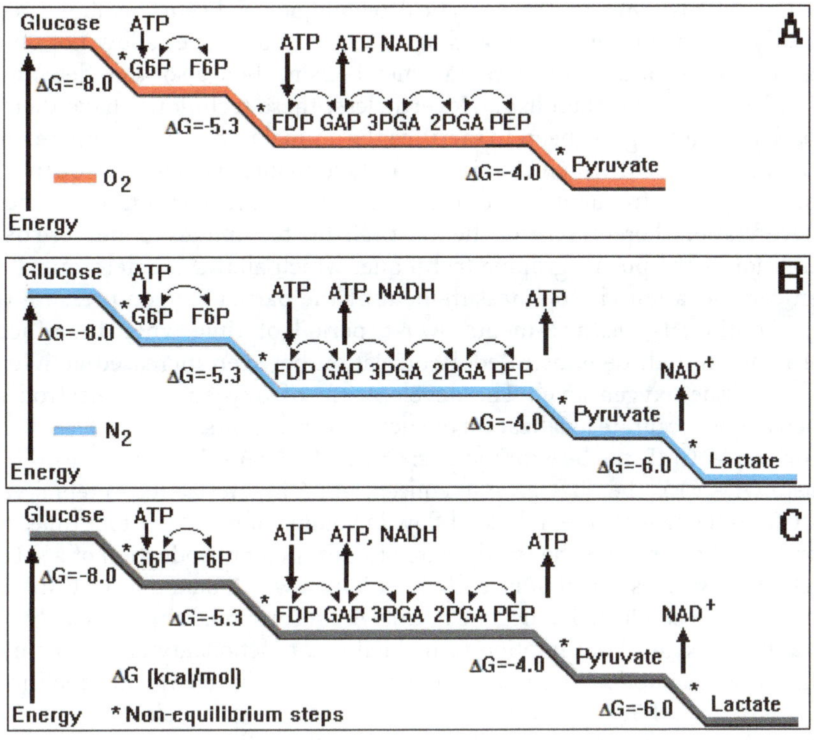

Figure 2.
A schematic illustration of the potential free-energy change profile of aerobic (A) and anaerobic glycolysis (B). The potential free-energy change of the reaction pyruvate → lactate dictates that it should proceed regardless of the oxygenation conditions (C).

2. Measurement of cerebral energy metabolic rates

At the basis of each technology designed to measure the rate of brain energy metabolism is the idea that measuring the consumption rate of the main two substrates of glycolysis and mitochondrial respiration, glucose and oxygen (O_2), should provide a complete picture of the brain's energy use. Theoretically, under normal conditions, each glucose molecule that enters the glycolytic pathway requires six molecules of oxygen to be fully oxidized via the mitochondrial TCA cycle and the electron transport chain. Thus, simultaneous measurements of glu-cose and oxygen consumption during rest or activation supposedly produces accu-rate estimate of the energy needs for the brain region under observation. However, the ratio CMR_{O_2} / $CMR_{glucose}$ values calculated are often significantly smaller than the expected 6/1. Such discrepancies have attributed to other glucose-consuming reactions not accompanies with oxygen consumption. Consequently, it has been a common understanding that a value of $CMR_{O_2}/CMR_{glucose} < 6$ indicates that a partial non-oxidative glucose consumption. The smaller the value of $CMR_{O_2}/CMR_{glucose}$, the greater is the non-oxidative consumption of glucose. This understanding makes sense when one assumes that a fully coupled glycolytic-mitochondrial respiratory apparatus should produce a $CMR_{O_2}/CMR_{glucose}$ value of 6 and an uncoupled apparatus (non-oxidative) should produce a $CMR_{O_2}/CMR_{glucose}$ value of ~0.

As indicated above, myriad techniques and technologies have been developed during the past six decades to measure both CMR_{O_2} and $CMR_{glucose}$. To measure cerebral energy metabolism *in vivo* one can analyze chemical changes in the blood entering and exiting the brain and/or in the cerebrospinal fluid. Of course, brain tissue samples can also be taken for analysis before and after physiological activity, although this approach would lend itself only to experimental animals. The introduction of radioisotopes to the analytical techniques of brain metabolic activity has greatly improved their speed and accuracy. Radioisotopes allow not only the tracing of end-products of cerebral metabolism, but also the detection of intermediates of that metabolism. Nevertheless, these techniques have their own drawbacks, includ-ing the need to sacrifice the animal under study only to receive a single measure-ment which provides mainly a qualitative value. A quantitative measurement is frequently confounded by compartmentation and its misinterpretation thereof. One of the most reliable techniques to measure oxygen consumption is the polaro-graphic technique, which allows the determination of oxygen concentration via the measurement of the partial oxygen pressure (PO_2) locally. Continuous measure-ments over a period of time when brain activity (EEG) is monitored, demonstrated a correlation between increased activity and decreased tissue oxygen level. The development of oxygen microelectrodes has afforded a more accurate localization of such measurements.

In principle, CMR can be expressed as: $CMR = CBF\ (A - V)$.

where $(A - V)$ is the difference in concentration in arterial and cerebral venous blood, CBF is the rate of cerebral blood flow in volume of blood per unit time, a CMR (cerebral metabolic rate) is the steady state of utilization or production of a substance by the brain [18]. This equation is the foundation on which quantitative CMR studies *in vivo* have been conducted. Since the normal brain consumed approximately 20% of the total body oxygen consumption to maintain its functionality and structure, it is clear that any interruption in this high demand for oxidative energy metabolism could have far reaching survivability consequences. Clearly, a non-oxidative energy metabolism (glycolysis) is incapable of answering the high energy demands of the brain. That a stimulated brain has still higher energy demands than the resting one would be an inevitable conclusion. Hence, when studying the energy demands of a specific activated brain region, such activation is expected to produce an increase in both CMR_{O_2} and $CMR_{glucose}$. Consequently, when Fox et al. published their study under the title "Nonoxidative glucose consumption during focal physiologic neural activity" [21] they stirred a small tempest among scientists in the community that studies cerebral blood flow and metabolism. These investigators employed ^{18}F-labeled 2-fluoro-2-deoxy-D-glucose to measure $CMR_{glucose}$, a method originally developed over a decade earlier [37], and ^{15}O-labeled molecular O_2 to measure CMR_{O_2}. They demonstrated that transient increases in neural activity elevated glucose tissue uptake in excess of that consumed by oxidative metabolism. They concluded their findings to indicate that stimulated brain activity requires significantly less energy than previously thought. Also, since they measured a corresponding increase in CBF along with the increase in glucose consumption, the investigators argued that this increase is for purposes other than oxidative metabolism. These conclusions stemmed from the prevailing postulate that over 90% of resting brain's glucose consumption is oxidative and less than 5% of that consumption ends in glycolytic lactate production. Since the oxidative consumption of one molecule of glucose produces approximately 36 molecules of ATP, while the glycolytic consumption of one molecule of glucose produces only 2 molecules of ATP, one can easily appreciate how oxidative consumption of glucose is responsible for 90% of the resting brain ATP production. Hence, the finding by [21] that brain stimulation increased glucose consumption without a corresponding increase in oxygen consumption unsettled the established understanding according to which increased brain activity must be accompanied by a corresponding increase in energy supply. This seminal paper was originated from the laboratory of Marcus Raichle, a laboratory that has become a leading center for functional brain imaging [38]. Imaging technologies such as X-ray computed tomography (CT), positron emission tomography (PET), near-infrared spectroscopy (NIRS) and magnetic resonance imaging (MRI) are the main techniques available for the measurement of brain energy metabolism during rest and activity. The most popular technology for this purpose today is the blood oxygen level dependent (BOLD) functional magnetic resonance imaging (fMRI), which was developed by [39]. BOLD fMRI measures changes in blood oxygenation in relation to brain activity, although that relationship is somewhat ambiguous, since it is not accompanied by a direct neural activity measurement such as that allowed by electrophysiology. When the latter is combined with direct oxygen concentration measurements, using an oxygen microelectrode (polarography), a higher resolution than BOLD fMRI can be achieved [40]. Besides tissue oxygen measurements using microelectrodes, tissue glucose and lactate concentrations can also be assessed using specific microelectrodes (sensors). Of course, this approach does not lend itself for regular use in humans, however, for the purpose of *in vivo* studies in experimental animals, the approach proved itself to be very useful and an eye opener.

In this respect, Hu and Wilson [41] studied the coupling of a temporary local energy pool to neuronal activity in the rat brain (**Figure 3**). They were the first to combine the use of three separate rapid response sensors (microelectrodes) to measure tissue oxygen, glucose and lactate concentrations. The investigators placed them in the dentate gyrus of the rat hippocampus, observing how they fluctuate in response to 10 consecutive electrical stimulations of the perforant pathway (each stimulus lasted 5 s and applied every 2 min). Their results were analyzed by Schurr and Gozal [36] (**Figure 3**). A literature search shows that Hu and Wilson's interpre-tation of their findings has its supporters [15, 42–47] and detractors [48–50]. The former group argued that these findings are strengthening the concept that lactate is the energy substrate that is utilized oxidatively upon neuronal activation. The latter group disagreed with this conclusion. Schurr [24] further analyzed the results of Hu and Wilson [41] beyond an earlier analysis [36]. The more recent analysis was prompted for two reasons. First, two decades have passed since the publication

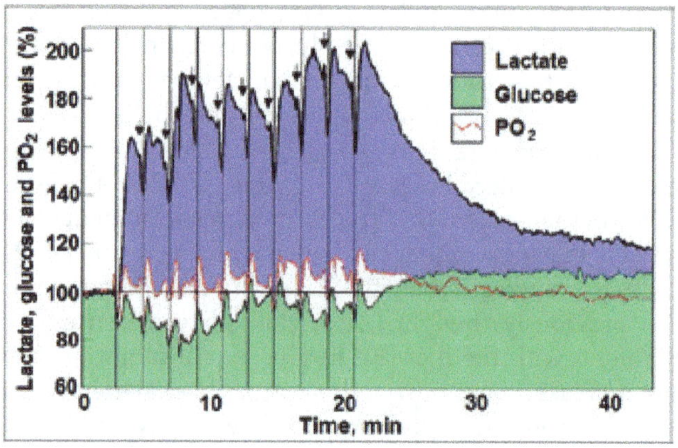

Figure3.
Profiles of time course and dynamic relationships of local extracellular lactate, glucose, and PO_2 levels in the rat hippocampal dentate gyrus during a series of 5 s electrical stimulations (arrows) of the perforant pathway at 2 min rest intervals (reproduced with permission from Hu and Wilson, copyright 1997, Blackwell, Oxford). The changes in the mean concentration of glucose were always in opposite direction to the changes in mean lactate concentration. The vertical lines were drawn to indicate the simultaneous dip in all three analytes in response to each of the electrical stimulations. For additional details see [41] from where the figure and the legend have been reproduced with permission and [36].

of the paper by Hu and Wilson [41], a period in which numerous studies added much support to the idea that lactate is a mitochondrial oxidative energy substrate. Second, many other studies on cerebral energy metabolism continue to conclude that neural activity is supported by "anaerobic" glycolysis and not by oxidative utili-zation of glucose, while ignoring the possibility that such activity may be supported by oxidative utilization of lactate.

3. Lactate cerebral metabolic rate and the importance of its measurement

When the results of the study by Hu and Wilson [41] were analyzed before [36], the analysis showed that upon a series of 10 stimulation of the rat hippocampal perforant pathway a steady glucose consumption was accompanied by a gradual increase in lactate consumption. Considering the conclusion of Fox et al. [21] that aerobic glycolytic ATP production is sufficient to answer the energy needs of activated neural tissue, one could assume that it should be sufficient to provide the energy needs of the stimulated hippocampal dentate gyrus. In addition, this analysis points out that if the conclusion of Fox et al. [21] is correct, the energy needs of the activated dentate gyrus declined with each stimulation or stayed the same at a very low level of ATP production (0.8–0.3 mM). However, if lactate oxidative consumption is postulated to be responsible for the ATP production that sustains the energy needs of the stimulated tissue, the increased lactate consumption with each consecutive stimulation signals a concomitant increased ATP production. The calculation shows that the response to the first stimulation produced 3 mM ATP, while the response to the last stimulation produced almost 11 mM (**Figure 4**).

The more recent analysis [24] also indicates that the increased levels of tissue lactate following each stimulation [41] could not be produced from the glycolytically metabolized glucose (**Figure 5**). Hence, this additional lactate had to be originated from other sources i.e., the surrounding tissue or glycogen stores [50]. As was shown by Hu and Wilson [41](**Figure 3**), a larger amount of lactate was consumed during each consecutive stimulation, while a smaller amount of glucose was

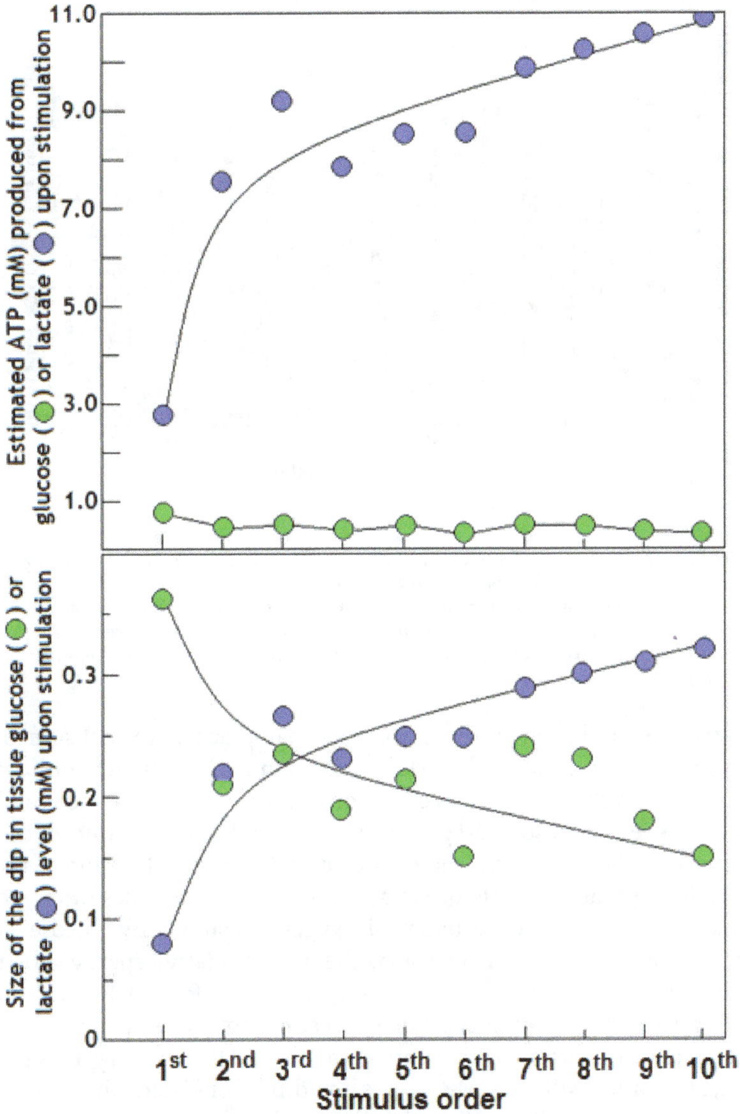

Figure4.
*Time course of changes in the amplitude of the dip in tissue glucose and lactate levels in the rat hippocampal dentate gyrus after each of the 10 electrical stimulations applied to the perforant pathway at intervals of 2 min (bottom panel). The amplitude of each dip (in mM) was calculated from the Hu and Wilson [41] as reproduced in **Figure** 3. The upper panel represents the estimated ATP amount produced based on the size of the dip (in mM) in tissue glucose and lactate levels as shown in the bottom panel. The estimated ATP levels were calculated as follows: the glucose measured dip (in mM) was multiplied by 2, the net production of 2 moles ATP from each mole of glucose metabolized glycolytically; the lactate measure dip (in mM) was multiplied by 34, the net formation of 34 moles of ATP for every 2 moles of lactate (form glycolytically from 1 mole of glucose) metabolized via the mitochondrial TCA cycle and the oxidative phosphorylation chain.*

consumed. Moreover, following each stimulation, except the first one, the lactate level measured exceeded the level expected from the amount of glucose consumed glycolytically i.e., two moles of lactate from one mole of glucose. Following the second stimulation, the tissue ratio of lactate to glucose was 3.95 and by the 10th stimulation this ratio increased to 8.33 (**Figure 5**). Meanwhile, oxygen tissue levels dipped and rose as expected during and after each stimulation, respectively, signaling that the electrical stimulation evoked an oxidative consumption of substrate. Initially, glucose and lactate were oxidatively consumed at equal amounts however,

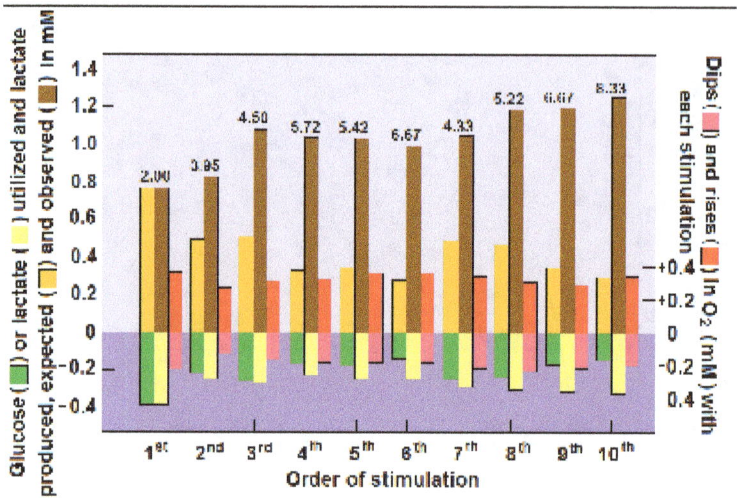

Figure5.
*The local extracellular glucose, lactate and O_2 levels in a rat hippocampal dentate gyrus during a series of 5 s electrical stimulations of the perforant pathway at 2 min rest intervals and the dynamic relationship between them. Glucose, lactate and O_2 concentrations were calculated from their dips and rises as measured by Hu and Wilson [41] using rapid response sensors. The numerical values posted above the columns representing the rises in glucose and lactate post-stimulation are the calculated ratios between the two. For additional details see **Figure 3** and [36].*

from the second stimulus onward more lactate than glucose was consumed (**Figure 5**). The oxygen level as measured by Hu and Wilson [41] fluctuated with a dip upon stimulation and a sharp rise upon its cessation (**Figure 3**). The fast rise can be interpreted as evidence that ample oxygen was available if and when needed. This rise also indicates that the tissue was well oxygenated during the duration of the experiment. Considering that one mole of lactate consumes three moles of oxygen for its full oxidation as compared to six moles of oxygen consumed by glucose for its full oxidation, if lactate, rather than glucose, is the main oxidative energy substrate during neural tissue activation, the expected ratio CMR_{O_2}: $CMR_{lactate}$ should not exceed 3:1. Therefore, it is reasonable to presume that during neural activation, when lactate oxidation is a major supplier of the ATP necessary to support said activation, the ratio CMR_{O_2}: $CMR_{glucose}$ should be considerably lower than 6:1. Obviously, most, if not all, studies aimed at measuring cerebral metabolic rates postulate that the ratio CMR_{O_2}:$CMR_{glucose}$ measured or calculated should approach 6 [23]. The conclu-sions of Fox et al. [21] are in complete disagreement with the measurements and calculations of Hyder et al. [23]. While the former concludes an almost complete uncoupling between glucose and oxygen consumption by activated neural tissue, the latter asserts the maintenance of full coupling between glucose and oxygen consumption during neural activation. In both studies [21, 23] the interpretation of the results is based entirely on the original, dogmatic paradigm of glycolysis accord-ing to which aerobic glycolysis ends with pyruvate, the assumed substrate of the mitochondrial TCA cycle.

The *in vivo* measurements performed by Hu and Wilson [41] of both glucose and lactate concentrations before, during and post electrical stimulation provide much support to the proposal that lactate plays a major role in oxidative energy metabo-lism of the activated neural tissue. The relatively small dips and rises in O_2 levels in response to the electrical stimulations, as measured polarographically, should be given additional consideration. First, the spatial and temporal resolutions provided by polarographic measurement of O_2 compared to BOLD fMRI measurements allow for a better characterization of the time-course of oxygen responses [40].

Hyder et al. [23] used BOLD fMRI and the measurements used by Fox et al. [21] were even more cumbersome, involving the use of $[^{15}O]H_2O$, $[^{15}O]O_2$ and $[^{15}O]CO_2$.

While BOLD fMRI estimates yielded a CMR_{O_2} :$CMR_{glucose}$ ratio value of 6:1, the ^{15}O measurements produced a ratio value of 0.4:1. These completely opposing outcomes make one wonder whether or not the measurements performed using these two methods and the calculated values of CMR_{O_2} they produced truly reflect the changes in the consumption of molecular oxygen upon neural activation. Could the direct measurements of CMR_{O_2}, $CMR_{glucose}$ and $CMR_{lactate}$ done by Hu and Wilson [41] along with the indirect ones made by Fox et al. [21] and Hyder et al. [23], be reconciled such that a better picture of cerebral metabolic rates of activated neural tissue can be visualized? It is widely agreed that over 90% of the normal brain's energy production originates from glucose oxidation [21, 51]. The normal glucose concentration in the brain is ~2 mM and its normal lactate concentration is about half of that of glucose. Thus, it is safe to postulate that the normal resting brain is supplied with ample amounts of oxygen to continuously oxidize more than 90% of the brain glucose. However, glucose supplies to the normal brain are limited (only 40% of normal blood glucose level). Consequently, the increased rate of CBF along with the increased consumption of glucose upon activation [21, 23, 52] should supply all the oxygen necessary to match the increased demand, in contrast to the limited supplies of glu-cose. Low resolution techniques for the measurements of oxygen concentrations are unable to detect local fluctuations accurately if at all, which could explain why Fox et al. [21] reached the conclusion regarding the very low oxygen consumption during neural activation. Nonetheless, their conclusion that the energy demands of acti-vated neural tissue are being met through glycolytic ATP production is most likely incorrect. In other words, undetectable or slightly detectable dip in tissue oxygen level upon activation is not necessarily an indication that oxygen is not consumed. The higher resolution of oxygen measurement afforded by polarography exemplifies the fact that local oxygen levels dipped upon stimulation and overshot upon its cessa-tion ([41]; **Figures 3** and **4**). Although local fluctuations in tissue oxygen levels were evident, its overall tissue concentration did not significantly change and may even have risen somewhat above its baseline level. In contrast, both glucose and lactate levels were changed significantly from their baseline levels [24, 36, 41] (**Figures 3–5**). The fluctuations between lactate and oxygen were highly synchronized, indicating that lactate is being oxidized upon tissue activation. During the 20 min following the 10th stimulation, the tissue level of both oxygen and glucose climbed above the base-line level, while the high level of lactate gradually declined ([41]; **Figure 3**). These shifts seem to indicate that upon cessation of stimulation, as the tissue is recovering from activation and high energy demands, lactate becomes the preferred oxidative energy substrate, sparing glucose. That the cerebral tissue would prefer lactate over glucose, especially when the former is abundant, is reasonable, considering the fact that lactate oxidative mitochondria. Consequently, the increased rate of CBF along with the increased consumption of glucose upon activation [21, 23, 52] should supply all the oxygen necessary to match the increased demand, in contrast to the limited supplies of glucose. Low resolution techniques for the measurements of oxygen concentrations are unable to detect local fluctuations accurately if at all, which could explain why Fox et al. [21] reached the conclusion regarding the very low oxygen consumption during neural activation. Nonetheless, their conclusion that the energy demands of activated neural tissue are being met through glycolytic ATP production is most likely incorrect. In other words, undetectable or slightly detectable dip in tissue oxygen level upon activation is not necessarily an indication that oxygen is not consumed. The higher resolution of oxygen measurement afforded by polarography exemplifies the fact that local oxygen levels dipped upon stimulation and overshot upon its cessation (**Figures 3** and **4** and [41]). Although local fluctuations in tissue oxygen levels were evident, its overall tissue concentration did not significantly change and may even have risen somewhat above its baseline level. In contrast, both glucose and lactate levels were changed significantly from their baseline levels (**Figures 3–5** and [24, 36, 41]). The fluctuations between lactate and oxygen were highly synchronized, indicating that lactate is being oxidized upon tissue activation. During the 20 min following the 10th stimulation, the tissue level of both oxygen and glucose climbed above the baseline level, while the high level of lactate gradually declined (**Figure 3** and [41]). These shifts seem to indicate that upon cessation of stimulation, as the tissue is recovering from activation and high energy demands, lactate becomes the preferred oxidative energy substrate, sparing glucose. That the cerebral tissue would prefer lactate over glucose, especially when the former is abun-dant, is reasonable, considering the fact that lactate oxidative utilization, in contrast to glucose, does not involve ATP investment prior to its utilization by mitochondria.

4. CMRs measurements and their possible implications in brain disorders

Energy metabolic interruptions are at the basis of several brain disorders and measuring CMRs of patients inflicted by such brain disorders can offer a potentially better diagnosis and treatment. Measurement of CMR_{O_2} and $CMR_{glucose}$ have been performed regularly in numerous studies of cerebral ischemia in an effort to better understand the mechanisms of neuronal ischemic damage. $CMR_{glucose}$ measure-ment has been used in studying obsessive-compulsive disorder, mood disorder and depression, where the main aim is to follow changes in glucose metabolism in specific brain regions believed to be involved in these disorders. Other brain disorders where glucose metabolic rate has been measured in include amyotrophic lateral sclerosis, Alzheimer's disease, epilepsy, Parkinson's disease and Huntington's disease. The purpose behind the measurement of glucose cerebral metabolic rate when investigating diseases and disorders is usually to identify brain regions that are involved in a given disorder or disease, not to investigate how energy metabo-lism is being affected by the disease or the disorder. Also, the energy metabolic rates of brain tumors have received great attention due to the unique energy require-ments of these tumors. Nevertheless, cerebral ischemia and traumatic brain injury (TBI) are the two disorders for which measurements of CMR_{O_2} and $CMR_{glucose}$ are most abundant. The results of these measurements prompted proposals both for treatments and mechanisms of neuronal damage due to these insults. The most heralded hypothesis attempting to explain delayed neuronal cerebral ischemic damage [20] known as the lactic acidosis hypothesis, postulated the accumulation of lactic acid as the cause of that damage. Consequently, physicians dealing with stroke patients were encouraged to control blood glucose levels in these patients, based on the assumption that the higher the glucose level during cerebral ischemia, the higher the level of lactic acid produced and the damage it causes. The lactic acidosis hypothesis was discarded, although the practice of controlling the blood glucose level of stroke patients remained. To this end, lactate was shown to support neuronal recovery post-ischemia *in vitro* [53–55]. Moreover, higher glucose level pre-ischemia (hypoxia) appear to improve neuronal recovery post-ischemia *in vitro* [56] and any exacerbation of neuronal damage due to pre-ischemic hyperglycemia was shown to be induced by increased levels of stress hormone [57]. Similarly, experimental [58] and clinical studies [59–66] over the past two decades indicate that lactate supplementation after TBI improves post-injury outcome. Measurement of $CMR_{lactate}$ could greatly enhance our understanding of the role that this monocar-boxylate plays in these two brain disorders.

5. Conclusions

A paradigm shift of a scientific model should, naturally, result in reconsideration of hypotheses and concepts that have been formulated on its foundation prior to its shift. The understanding of cerebral metabolic rates of energy substrates during rest and activation of neural tissue, the use of the method best suited for the measurement of these rates and the interpretation of the results have always relied on two fundamental assumptions. First, cerebral energy metabolism includes the obligatory glycolytic breakdown of glucose to pyruvate and the utilization of the latter by the mitochondrial TCA cycle and the electron transport chain with oxygen as its final receptor. Second, the activation of cerebral tissue is sustained by an increase in ATP production and therefore an increase in the consumption of glucose and oxygen. Two seminal papers that were published almost simultaneously [16, 21] have forced biochemists, and especially neuroscientists, to reassess these two basic postulates. The paper by Fox et al. [21] has perplexed many with its conclusion that the energy requirements of activated neural tissue are minimal and can be fulfilled by the glycolytic pathway alone (glucose → lactate + 2ATP). The paper by Schurr et al. [16] provoked great skepticism upon demonstrating that neural tissue can function and be activated when lactate is its sole oxidative energy substrate (lactate + $3O_2$ + mitochondria → pyruvate → TCA cycle → $3CO_2$ + $3H_2O$ + 17ATP). While the proposal that lactate is a suitable oxidative energy substrate had faced strong skepticism for many years, it has gained greater sup-port over the past three decades. The proposal that glycolysis could be served as the sole supplier of energy for the activated neural tissue still divides scientists working in this field. By accepting the proposed paradigm shift of glycolysis [24] and its application in the interpretation of the results obtained by Fox et al. [21], Hyder et al. [23, 67], Hu and Wilson [41] and many others, a scenario can be drawn where lactate is supplementing most if not all the energy requirements of activated neural tissue. The data and the line of reasoning presented here strongly argue against the conclusion that these energy requirements are solely fulfilled by glycolysis. Future studies of activated cerebral metabolic rates should include, along with the measurements of CMR_{O_2} and $CMR_{glucose}$, the measurement of $CMR_{lactate}$. Resolving the existing debated issues of cerebral energy metabolism is paramount for our better understanding the many brain diseases and disorders. Hopefully, this chapter provides a possible resolution of some of these issues.

Author details

Avital Schurr
Department of Anesthesiology and Perioperative Medicine, University of Louisville, School of Medicine, Louisville, KY, USA

*Address all correspondence to: avital.schurr@gmail.com

References

[1] Hill L, Nabarro DN. On the exchange of blood-gases in brain and muscle during states of rest and activity. The Journal of Physiology. 1895;**18**:218-229

[2] Tashiro S. Carbon dioxide production from nerve fibres when resting and when stimulated; a contribution to the chemical basis of irritability. The American Journal of Physiology. 1913;**32**:107-136

[3] Warburg O, Negelein E, Posener K. Versuche an Überlebendem Carcinomgewebe. Klinische Wochenschrift. 1924;**3**:1062-1064. DOI: 10.1007/BF01736087

[4] Gerard RA, Hill AV, Zotterman Y. The effect of frequency of stimulation on the heat production of nerve. The Journal of Physiology. 1927;**63**:130-143

[5] Holmes EG. The metabolism of brain and nerve. Annual Review of Biochemistry. 1932;**1**:487-506. DOI: 10.1146/annurev.bi.01.070132.002415

[6] Schurr A. Cerebral glycolysis: A century of persistent misunderstanding and misconception. Frontiers in Neuroscience. 2014;**8**:360. DOI: 10.3389/fnins.2014.00360

[7] Rogatzki MJ, Ferguson BS, Goodwin ML, Gladden LB. Lactate is always the end product of glycolysis. Frontiers in Neuroscience. 2015;**9**:22. DOI: 10.3389/fnins.2015.00022

[8] Holmes BE, Holmes EG. Contributions to the study of brain metabolism. I. Carbohydrate metabolism. Preliminary paper. The Biochemical Journal. 1925;**19**:492-499

[9] Holmes EG, Holmes BE. Contributions to the study of brain metabolism. II. Carbohydrate metabolism. The Biochemical Journal. 1925;**19**:836-839

[10] Holmes EG, Holmes BE. Contributions to the study of brain metabolism. III. Carbohydrate metabolism relationship of glycogen and lactic acid. The Biochemical Journal. 1926;**20**:1196-1203

[11] Holmes EG, Holmes BE. Contributions to the study of brain metabolism. IV. Carbohydrate metabolism of the brain tissue of depancreatised cats. The Biochemical Journal. 1927;**21**:412-418

[12] Holmes EG. Oxidations in central and peripheral nervous tissue. The Biochemical Journal. 1930;**24**:914-925

[13] Holmes EG, Ashford CA. Lactic acid oxidation in brain with reference to the Meyerhof cycle. Biochemical Journal. 1930;**24**(4):1119-1127

[14] Ashford CA, Holmes EG. Further observations on the oxidation of lactic acid by brain tissue. The Biochemical Journal. 1931;**25**:2028-2049

[15] Schurr A. Lactate: The ultimate cerebral oxidative energy substrate? Journal of Cerebral Blood Flow and Metabolism. 2006;**26**:142-152. DOI: 10.1038/sj.jcbfm.9600174

[16] Schurr A, West CA, Rigor BM. Lactate-supported synaptic function in the rat hippocampal slice preparation. Science. 1988;**240**:1326-1328. DOI: 10.1126/science.3375817

[17] Brooks GA. Lactate: Glycolytic product and oxidative substrate during sustained exercise in mammals—'the lactate shuttle'. In: Gilles R, editor. Comparative Physiology and Biochemistry—Current Topics and Trends Vol A. Respiration-Metabolism-Circulation. Berlin: Springer-Verlag; 1985. pp. 208-218

[18] Clarke DD, Sokoloff L. Circulation and energy metabolism of the brain

(Chap. 31). In: Siegle GJ, Agranoff BW, Albers RW, Molinoff PB, editors. Basic Neurochemistry. 5th ed. New York, NY: Raven Press; 1994. pp. 645-680

[19] Krebs HA, Johnson WA. The role of citric acid in intermediary metabolism in animal tissue. Enzymologia. 1937;**4**:148-156

[20] Siesjö BK. Cell-damage in the brain —A speculative synthesis. Journal of Cerebral Blood Flow and Metabolism. 1981;**1**:155-185. DOI: 10.1038/jcbfm.1981.18

[21] Fox PT, Raichle ME, Mintun MA, Dence C. Nonoxidative glucose consumption during focal physiologic neural activity. Science. 1988;**241**: 462-464. DOI: 10.1126/science. 3260686

[22] Schurr A, Gozal E. Glycolysis at 75: Is it time to tweak the first elucidated metabolic pathway in history? Frontiers in Neuroscience. 2014;**9**:170. DOI: 10.3389/fnins.2015.00170

[23] Hyder F, Rothman DL, Mason GM, Rangarajan A, Behar KL, Shulman RG. Oxidative glucose metabolism in rat brain during single forepaw stimulation: A spatially localized $^{1}H[^{13}C]$ nuclear magnetic resonance study. Journal of Cerebral Blood Flow and Metabolism. 1997;**17**:1040-1047. DOI: 10.1097/00004647-199710000-00005

[24] Schurr A. Glycolysis paradigm shift dictates a reevaluation of glucose and oxygen metabolic rates of activated neural tissue. Frontiers in Neuroscience. 2018;**12**:700. DOI: 10.3389/fnins.2018.00700

[25] Kane DA. Lactate oxidation at the mitochondria: A lactate-malate-aspartate shuttle at work. Frontiers in Neuroscience. 2014;**8**:366. DOI: 10.3389/fnins.2014.00366

[26] Brooks GA, Brown MA, Butz CE, Sicurello JP, Dubouchaud H. Cardiac and skeletal muscle mitochondria have a monocarboxylate transporter MCT1. Journal of Applied Physiology. 1999;**87**:1713-1718

[27] Valenti D, de Bari L, Atlante A, Passarella S. L-lactate transport into rat heart mitochondria and reconstruction of the L-lactate/pyruvate shuttle. The Biochemical Journal. 2002;**364**:101-104

[28] Valenti D, de Bari L, De Filippis B, Henrion-Caude A, Vacca RA. Mitochondrial dysfunction as a central actor in intellectual disability-related diseases: An overview of Down syndrome, autism, Fragile X and Rett syndrome. Neuroscience & Biobehavioral Reviews. 2014;**46**:202-217. DOI: 10.1016/j.neubiorev.2014.01.012

[29] Hashimoto T, Brooks GA. Mitochondrial lactate oxidation complex and an adaptive role for lactate production. Medicine & Science in Sports & Exercise. 2008;**40**:486-494. DOI: 10.1249/MSS.0b013e31815fcb04

[30] Hashimoto T, Hussien R, Cho H-S, Kaufer D, Brooks GA. Evidence for a mitochondrial lactate oxidation complex in rat neurons: A crucial component for a brain lactate shuttle. PLoS One. 2008;**3**:e2915. DOI: 10.1371/journal. pone.0002915

[31] Pizzuto R, Paventi G, Porcile C, Sarnataro D, Daniele A, Passarella S. L-Lactate metabolism in HEP G2 cell mitochondria due to the L-lactate dehydrogenase determines the occurrence of the lactate/pyruvate shuttle and the appearance of oxaloacetate, malate and citrate outside mitochondria. Biochimica et Biophysica Acta. 2012;**1817**:1679-1690. DOI: 10.1016/j.bbabio.2012.05.010

[32] Elustondo PA, White AE, Hughes ME, Brebner K, Pavlov E, Kane DA. Physical and functional association of lactate dehydrogenase (LDH) with skeletal muscle mitochondria.

The Journal of Biological Chemistry. 2013;**288**:25309-25317. DOI: 10.1074/jbc. M113.476648

[33] Brooks GA, Dubouchaud H, Brown M, Sicurello JP, Butz CE. Role of mitochondrial lactate dehydrogenase and lactate oxidation in the intracellular lactate shuttle. Proceedings of the National Academy of Sciences of the United States of America. 1999;**96**: 1129-1134. DOI: 10.1073/pnas.96.3.1129

[34] Havel RJ, Watkins E, Gullixson KS. The metabolism of lactate and pyruvate in children with congenital heart disease. Circulation. 1950;**2**:536-544

[35] Schurr A, Payne RS. Lactate, not pyruvate, is neuronal aerobic glycolysis end product: An in vitro electrophysiological study. Neuroscience. 2007;**147**:613-619. DOI: 10.1016/j.neuroscience.2007.05.002

[36] Schurr A, Gozal E. Aerobic production and utilization of lactate satisfy increased energy demands upon neuronal activation in hippocampal slices and provide neuroprotection against oxidative stress. Frontiers in Pharmacology. 2011;**2**:96. DOI: 10.3389/fphar.2011.00096

[37] Sokoloff L, Reivich M, Kennedy C, Des Rosiers MH, Patlak CS, Pettigrew KD, et al. The [^{14}C]deoxyglucose method for the measurement of local cerebral glucose utilization: Theory, procedure, and normal values in the conscious and anesthetized albino rat. Journal of Neurochemistry. 1977;**28**: 897-916. DOI: 10.1111/j.1471-4159.1977. tb10649.x

[38] Raichle ME. A paradigm shift in functional brain imaging. The Journal of Neuroscience. 2009;**29**:12729-12734. DOI: 10.1523/JNEUROSCI.4366-09.2009

[39] Ogawa S, Lee TM, Kay AR, Tank DW. Brain magnetic resonance imaging with contrast dependent on blood oxygenation. Proceedings of the National Academy of Sciences of the United States of America. 1990;**87**:9868-9872. DOI: 10.1073/pnas.87.24.9868

[40] Bentley WJ. Oxygen Polarography in the Awake Macaque: Bridging BOLD fMRI and Electrophysiology. 2014. Available from: http://openscholarship.wustl.edu/etd/1218

[41] Hu Y, Wilson GS. A temporary local energy pool coupled to neuronal activity: fluctuations of extracellular lactate levels in rat brain monitored with rapid-response enzyme-based sensor. Journal of Neurochemistry. 1997;**69**:1484-1490. DOI: 10.1046/j.1471-4159.1997.69041484.x

[42] Pellerin L, Magistretti PJ. Food for thought: Challenging the dogmas. Journal of Cerebral Blood Flow and Metabolism. 2003;**23**:1282-1286. DOI: 10.1097/01.WCB.0000096064.12129.3D

[43] Kasischke KA, Vishwasrao HD, Fisher PJ, Zipfel WR, Webb WW. Neural activity triggers neuronal oxidative metabolism followed by astrocytic glycolysis. Science. 2004;**305**:99-103. DOI: 10.1126/science.1096485

[44] Aubert A, Costalat R, Magistretti PJ, Pellerin L. Brain lactate kinetics: Modeling evidence for neuronal lactate uptake upon activation. Proceedings of the National Academy of Sciences of the United States of America. 2005;**102**:16448-16453. DOI: 10.1073/pnas.0505427102

[45] Medina JM, Tabernero A. Lactate utilization by brain cells and its role in CNS development. Journal of Neuroscience Research. 2005;**179**:2-10. DOI: 10.1002/jnr.20336

[46] Serres S, Bezancon E, Franconi J-M, Merle M. Ex vivo NMR study of lactate metabolism in rat brain under various depressed states. Journal of

Neuroscience Research. 2005;**179**:19-25. DOI: 10.1002/jnr.20277

[47] Dienel GA, Hertz L. Astrocytic contributions to bioenergetics of cerebral ischemia. Glia. 2005;**50**: 362-388. DOI: 10.1002/glia.20157

[48] Fillenz M. The role of lactate in brain metabolism. Neurochemistry International. 2005;**47**:413-417. DOI: 10.1016/j.neuint.2005.05.011

[49] Korf J. Is brain lactate metabolized immediately after neuronal activity through the oxidative pathway? Journal of Cerebral Blood Flow and Metabolism. 2006;**26**:1584-1586. DOI: 10.1038/sj.jcbfm.9600321

[50] Chambers TW, Daly TP, Hockley A, Brown AM. Contribution of glycogen in supporting axon conduction in the peripheral and central nervous systems: The role of lactate. Frontiers in Neuroscience. 2014;**8**:378. DOI: 10.3389/fnins.2014.00378

[51] Siesjo BK. Brain Energy Metabolism. New York: John Wiley and Sons; 1978

[52] Ueki M, Linn F, Hossmann K-A. Functional activation of cerebral blood flow and metabolism before and after global ischemia of rat brain. Journal of Cerebral Blood Flow and Metabolism. 1988;**8**:486-494

[53] Schurr A, Dong W-Q , Reid KH, West CA, Rigor BM. Lactic acidosis and recovery of neuronal function following cerebral hypoxia in vitro. Brain Research. 1988:311-314

[54] Schurr A, Rigor BM. Brain anaerobic lactate production: A suicide note or a survival kit? Developmental Neuroscience. 1998;**20**:348-357. DOI: 10.1159/000017330

[55] Schurr A, Payne RS, Miller JJ, Tseng MT, Rigor BM. Blockade of lactate transport exacerbates delayed neuronal damage in a rat model of cerebral ischemia. Brain Research. 2001;**895**:268-272

[56] Schurr A, West CA, Reid KH, Tseng MT, Reiss SJ, Rigor BM. Increased glucose improves recovery of neuronal function after cerebral hypoxia in vitro. Brain Research. 1987;**421**:135-139

[57] Schurr A. Energy metabolism, stress hormones and neural recovery from cerebral ischemia/hypoxia. Neurochemistry International. 2002;**41**:1-8

[58] Rice AC, Zsoldos R, Chen T, Wilson MS, Alessandri B, Hamm RJ, et al. Lactate administration attenuates cognitive deficits following traumatic brain injury. Brain Research. 2002;**928**:156-159. DOI: 10.1016/S0006-8993(01)03299-1

[59] Holloway R, Zhou Z, Harvey HB, Levasseur JE, Rice AC, Sun D, et al. Effect of lactate therapy upon cognitive deficits after traumatic brain injury in the rat. Acta Neurochirurgica. 2007;**149**:919-927. DOI: 10.1007/s00701- 007-1241-y. discussion: 927

[60] Ichai C, Armando G, Orban JC, Berthier F, Rami L, Samat-Long C, et al. Sodium lactate versus mannitol in the treatment of intracranial hypertensive episodes in severe traumatic brain-injured patients. Intensive Care Medicine. 2009;**35**:471-479. DOI: 10.1007/s00134-008-1283-5

[61] Ichai C, Payen JF, Orban JC, Quintard H, Roth H, Legrand R, et al. Half-molar sodium lactate infusion to prevent intracranial hypertensive episodes in severe traumatic brain injured patients: A randomized controlled trial. Intensive Care Medicine. 2013;**39**:1413-1422. DOI: 10.1007/s00134-013-2978-9

[62] Jalloh I, Helmy A, Shannon RJ, Gallagher CN, Menon DK, Carpenter

KL, et al. Lactate uptake by the injured human brain: Evidence from an arteriovenous gradient and cerebral microdialysis study. Journal of Neurotrauma. 2013;**30**:2031-2037. DOI: 10.1089/neu.2013.2947

[63] Bouzat P, Sala N, Suys T, Zerlauth JB, Marques-Vidal P, Feihl F, et al. Cerebral metabolic effects of exogenous lactate supplementation on the injured human brain. Intensive Care Medicine. 2014;**40**:412-421. DOI: 10.1007/s00134-013-3203-6

[64] Brooks GA, Martin NA. Cerebral metabolism following traumatic brain injury: New discoveries with implications for treatment. Frontiers in Neuroscience. 2015;**8**:408. DOI: 10.3389/fnins.2014.00408

[65] Glenn TC, Martin NA, Hovda DA, Vespa P, Johnson ML, Horning MA, et al. Lactate; brain fuel following traumatic brain injury. Journal of Neurotrauma. 2015;**32**:820-832. DOI: 10.1089/neu.2014.3483

[66] Carpenter KLH, Jalloh I, Hutchinson PJ. Glycolysis and the significance of lactate in traumatic brain injury. Frontiers in Neuroscience. 2015;**9**:112. DOI: 10.3389/fnins.2015.00112

[67] Hyder F, Herman P, Bailey CJ, Møller A, Globinsky R, Fulbright RK, et al. Uniform distributions of glucose oxidation and oxygen extraction in gray matter of normal human brain: No evidence of regional differences of aerobic glycolysis. Journal of Cerebral Blood Flow & Metabolism. 2016;**36**:903-916. DOI: 10.1177/0271678X15625349

3

Sports and McArdle Disease (Glycogen Storage Disease Type V): Danger or Therapy?

Georg Bollig

Abstract

McArdle disease (glycogen storage disease type V) is an inborn error of energy metabolism in the muscle. The effects of McArdle disease on physical performance have similarities with the metabolic state of marathon runners after glycogen depletion and can therefore be seen as a nature's experiment in the field of sports medicine. Many patients with McArdle disease avoid sports in general because physical activity usually leads to muscle pain and muscle cramps. Often patients therefore regard physical activity as both painful and possibly dangerous. This chapter is about the advantages and possible risks of sports for patients with McArdle disease. The scientific literature will be discussed highlighting both endurance and muscle strength exercise. It will discuss the differences of aerobic and anaerobic exercise in individuals suffering from McArdle disease. Complications as rhabdomyolysis, myoglobinuria, kidney failure, and malignant hyperthermia will be discussed. The chapter will summarize the current knowledge about the possible dangers versus possible benefits of sports for patients with McArdle disease. A summary of recommendations for physical exercise and training for McArdle patients will be provided.

Keywords: McArdle disease, glycogen storage disease type V, rhabdomyolysis, endurance exercise, muscle strength exercise, aerobic exercise, anaerobic exercise, sports medicine

1. Introduction and background

1.1 No sports?

Many people with McArdle disease do not like sports or physical activity and say therefore no to participation in sports activities or regular physical exercise. The aim of this chapter is to address patients and health care providers' queries about vMcArdle disease and sports as well as to provide guidance on physical activity for those who have to live with McArdle disease.

McArdle disease (glycogen storage disease type V) is an inborn error of energy metabolism in the muscle [1–5]. It hampers physical exercise in affected patients due to the restriction of the availability of glucose as energy source for muscular work. It can be seen as a nature's experiment in the field of sports medicine as the underlying defect of the myophosphorylase enzyme leads to metabolic effects that are similar to the effects of glycogen depletion in marathon runners [2]. Many patients with McArdle disease (McAd) avoid sports because physical activity usually leads to muscle pain and muscle cramps. Often patients with McAd therefore regard physical activity as both painful and dangerous. On the other hand, physical activity is of great importance to manage daily life. Living is based on regular motion, and muscles have to be used in order to be healthy. Not using our muscles will in the long run lead to weakness, immobility, and frailty. Therefore people affected by McArdle disease do benefit of keeping a certain degree of fitness. Regular physical exercise might play a key role in delaying progressive muscle wasting, weakness, and frailty in later life of people affected by McAd.

- But how much physical activity is beneficial and what might be dangerous?
- Are certain types of physical exercise better than others?

These and many other questions arise when talking about physical activity and sports with patients with McAd. People with McAd do therefore need guidance on physical activity based on scientific and evidence-based facts. Due to unpleasant experience with sports in the form of pain, cramps, weakness, or myoglobinuria, many patients with McAd show a tendency to avoid sports and physical activity because they are afraid that sports may be not only painful but also harmful. Unfortunately this behavior can decrease their physical activity and physical capacity further. This chapter will shed light on McArdle disease and sports in general and try to answer the above mentioned questions.

2. Method

This chapter is based on a review of the existing publications on McArdle disease (glycogenosis type V) and sports and the authors' personal experience with the topic based on his German PhD thesis on the subject [2] and practical experience with patients with McArdle disease from sports medicine and anesthesiology. A literature search with the Medical Subject Heading (MeSH) words "McArdle disease" and "sports" was performed using the search engines PubMed and Medline. Publications that had McArdle disease and sports as a topic were included. In addition reference lists of books and other sources were assessed by hand search. An overview of the existing literature on this topic is provided.

3. McArdle disease: a nature's experiment

McArdle disease was first described in 1951 by Brian McArdle, a British neurologist. It is known by different synonyms as myophosphorylase insufficiency, glycogen storage disease type V, or myophosphorylase deficiency [1–5]. **Table 1** provides an overview of the different types of glycogen storage diseases [6]. McArdle disease is caused by a lack of myophosphorylase (alpha-1,4-glucan orthophosphate glycosyl transferase) that normally initiates muscle glycogen breakdown during exercise by removing 1,4-gly-cosyl groups from the glycogen molecule leading to the release of glucose-1-phosphate [1–5] and thus providing fuel for muscular work. Patients with McAd are unable to use the glycogen storage in the muscles as an energy source to enable physical activity.

In about 50% of patients with McAd, a positive family history can be found. For most of the patients, the diagnosis is first established between the age of 10 and 30 [3]. The disease is described to be autosomal recessive, although transmission

Type of glycogen storage disease	Enzyme defect	Inheritance	Organs involved	Clinical symptoms
Type 0	Glycogen synthase deficiency		Liver	Fasting hypoglycemia, tiredness, pallor, vomiting, muscle cramps
Type I **Von Gierke disease**	Glucose-6-phosphatase deficiency	Autosomal recessive	Liver, kidney	Growth retardation, hypoglycemia
Type II **Pompe disease**	Acid maltase deficiency	Autosomal recessive	Muscle, heart, liver	Hypotonia, muscle weakness (progressive), affected: proximal and respiratory muscle, cardiac enlargement and failure
Type III **Cori disease**	Debrancher enzyme deficiency	Autosomal recessive	Liver, muscle, heart	Growth retardation, muscle weakness (liver cirrhosis can occur)
Type IV **Andersen disease**	Branching enzyme deficiency	Autosomal recessive	Liver, kidney, heart, muscle	Mild hypoglycemia
Type V **McArdle disease**	Myophosphorylase deficiency	Autosomal recessive	Skeletal muscle	Exercise intolerance, muscle cramps and pain, myoglobinuria on strenuous exercise
Type VI **Hers disease**	Liver phosphorylase deficiency	Autosomal recessive	Liver	Mild hypoglycemia
Type VII **Tarui disease**	Phosphofructokinase deficiency	Autosomal recessive	Skeletal muscle	Muscle pain and fatigue on exercise. Muscle cramps and tenderness
Type VIII	Phosphorylase b kinase deficiency	X-linked recessive	Liver, brain	Ataxia, spasms, brain degeneration
Type IX	Phosphoglycerate kinase deficiency	X-linked recessive	Liver	Mild hypoglycemia
Type X	Phosphoglycerate mutase deficiency	Autosomal recessive	Liver, muscle	Exercise intolerance, muscle pain

Table 1.
Overview of glycogen storage diseases (modified from [6]).

clinically appears to be autosomal dominant in some affected families [2–4, 7]. The gene for myophosphorylase lies on chromosome 11q13. There are a number of different mutations described in the scientific literature; the most frequent mutation is named R50X [5].

The prevalence of McArdle disease is not known exactly due to the relative benign course of the disease and the often mild and frequently misinterpreted clinical symptoms. The clinical symptoms are summarized in **Table 2**. Haller has estimated the prevalence of McArdle disease in the Dallas-Fort Worth region as 1 in 100,000 [8].

As the pathophysiological effects of McAd are similar to the state of glycogen depletion in marathon runners, it is of special interest from the view of sports medicine and has been called a nature experiment [2].

• Muscle pain (myalgia) • Fatigue • Cramps • Exercise intolerance • Intermittent claudication   ○ (Muscle pain on mild exertion in the calf muscle, usually attributed to peripheral artery disease) • Second wind phenomenon • (Exercise becomes easier after a period of moderate and tolerable exercise) • Stiffness • Muscle swelling after exercise • Myoglobinuria • Muscular atrophy • (Mostly proximal muscles affected and in elderly patients)

Table 2.
Clinical symptoms and signs of McArdle disease [1–4].

4. Typical clinical picture and diagnosis of McArdle disease

4.1 Typical clinical picture

Typical clinical symptoms of McArdle disease are muscle pain and fatigue during exercise. Pain is often located in the knee, calf, and legs. Normally pain vanishes after a few minutes rest. Clinical symptoms and signs are shown in **Table 2** [1–4].

4.2 Diagnosis of McArdle disease

Diagnosis is based on the clinical picture, muscle biopsy, biochemical tests, exercise testing, and genetic testing [2–4, 7, 9] as listed in **Table 3**.

The absence of rising lactate is a diagnostic criterion in McAd. Usually this has been tested using the forearm ischemic exercise test [2–4]. Both Bollig [2] and Vissing and Haller [9] have suggested to use cycle ergometry testing instead of the forearm ischemic exercise test. The ischemic forearm exercise is more painful for the patients and might put them at risk for severe rhabdomyolysis and myoglobinuria. The cycle ergometry as diagnostic tool from sports medicine is useful in the diagnosis of McAd and may be used to assess the actual state of cardiopulmonary fitness and may help to give patients guidance for further training. **Figure 1** shows the results of cycle ergometry testing in a typical patient with McArdle disease [2].

• Clinical picture
• Elevated creatine kinase (CK) levels in the blood
• Absence of increased venous lactate during forearm ischemic exercise test or cycle ergometry
• Low or absent myophosphorylase activity on histochemical or biochemical examination of muscle biopsy
• Genetic testing (the muscle phosphorylase gene is located on chromosome 11q13, and several mutations have been described—the most common mutation is called R50X)

Table 3.
Diagnosis of McArdle disease [2–4, 7, 9, 10].

Workload (watt)	Heartrate / min.	Bloodpressure (mmHg)	Lactate (mmol/l)
0	76	120/85	0,8
30	116	130/-	0,8
70	164	170/-	1
110	204	180/-	0,8

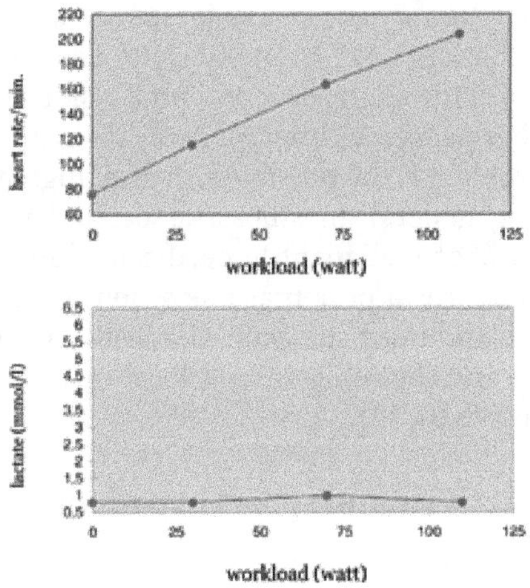

Figure 1.
Cycle ergometry of patient F.M., born 1967, 183 cm, 80 kg; modified from [2].

The prognosis of McArdle disease is usually good, and life expectancy is normal although severe cases with muscle wasting and extreme weakness and death in childhood have been described [1–4, 10].

5. Treatment options

At present there are no causal treatment options available. Some symptomatic treatment options may reduce symptoms or enhance the amount physical activity that can be tolerated. These treatment options include oral sucrose before exercise [11], a low dose of oral creatine [12], vitamin B6 [2, 10], and coenzyme Q10 [2, 10]. Our study about the use of clenbuterol over a 12-month period leads to a subjective improvement of exercise tolerance in three patients. Some relatives of the patients noted an improved exercise tolerance after clenbuterol intake over some weeks [2]. One patient from our study has used a low-dose clenbuterol (0.005–0.02 mg once daily) to enhance exercise tolerance for more than 10 years. This patient used clenbuterol for some months with regular breaks of weeks up to months between the therapy cycles. A systematic Cochrane review on pharmacological and nutritional treatment options has been published by Quinlivian et al. [12]. There do exist animal models for McArdle disease in sheep, cows, mice, and rats that may be used to test potential therapies in future studies [10]. Gene therapy of McArdle disease might be a future option, but its dangers outweigh the possible advantages at present [10].

6. Health problems and possible risks associated with McArdle disease

Patients with McArdle disease are at risk of developing myoglobinuria and even kidney failure due to rhabdomyolysis after exercise or anesthesia [2, 3, 12, 13]. Therefore, patients affected by McAd should learn how to accomplish daily activity with McAd and how to avoid major muscle damage and the risk for massive rhabdo-myolysis and acute kidney failure.

Some cases with insulin resistance and a diabetes type II-like clinical picture in patients with McArdle have been described, but there is no known connection between type I diabetes and McArdle disease [10]. Increased glycogen storage in the muscle of McArdle patients has been suggested as a probable cause of insulin resistance in McArdle patients [10]. Overweight has been observed in many patients with McAd [14]. This might be a potential risk factor for developing type II diabetes as it is for other people without McArdle disease.

Another potential problem is the possible risk of malignant hyperthermia which is a complication during general anesthesia associated with different muscular diseases. Although no case of malignant hyperthermia during anesthesia has been described in McAd so far, it is a potential risk when patients with McAd have to undergo operations with the need for general anesthesia. Therefore, precautions have to be taken by the anesthesiologist, and local or regional anesthesia may be preferred whenever feasible [6, 13].

7. Effects of physical activity and sports in patients with McArdle disease

As described above patients with McAd might suffer from exercise intoler-ance and pain, fatigue, and cramps during exercise, typical clinical symptoms of McArdle disease. Up to 50% of the patients with McAd show myoglobinuria, and unfortunately acute renal failure has been described in 27% following rhabdomyolysis as a result of vigorous or strenuous exercise [3]. Cases with extreme rhabdomyolysis and myoglobinuria have, e.g., been reported after a swimming competition, an asthma attack, after carrying a TV, and after the diagnostic use of the ischemic work test using a tourniquet [15–19]. The variation of creatine kinase (CK) levels in the blood has been investigated in a male with McAd over a period of several months by the author [2]. **Figure 2** shows the results from this German doctor thesis from the year 2000. The results indicated that anaerobic exercise and physical activity demanding great strength or strenuous exercise lead to huge increases in creatine kinase activity, whereas aerobic exercise did not increase blood creatine kinase levels to a great extent. Aerobic exercise was shown to be associated with lower creatine kinase levels after physical activity in a number of instances during the study period [2, 20]. This finding was later proofed by other researchers [10]. The oxygen uptake and physical ability of patients affected by McAd is usually limited to about 50% of comparable healthy individuals [2].

7.1 The second wind phenomenon

The second wind phenomenon is defined as "a period of less painful and more effective exercise associated with a decrease in heart rate after the initial period of cramping and/or weakness." [21]. Many patients with McArdle disease do experience this phenomenon that was first described by Pearson et al. [22]. During exercise this phenomenon can lead to better endurance because patients are able to exercise for a longer period and experience physical activity as less painful in the long run.

7.2 Aerobic exercise (endurance exercise)

Aerobic exercise is endurance exercise where oxygen is needed in energy production. Aerobic energy production takes some minutes to start but can help to

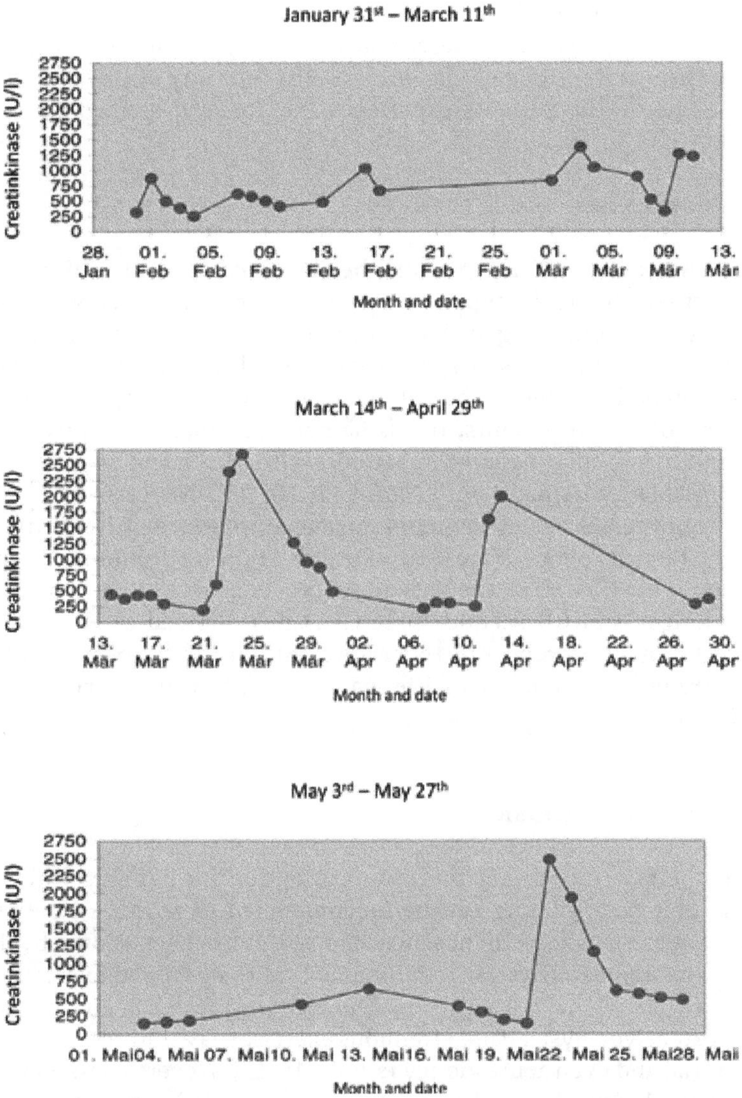

Figure 2.
Creatine kinase levels from a long-term follow-up over 5 months [modified from [2]].

supply energy for muscular activity over a longer period of time (several minutes up to several hours). Due to the lack of glucose that cannot be released from the glycogen deposits in the muscle, patients with McArdle disease rely on fatty acids, amino acids, and glucose from the liver as energy source during exercise [2, 10]. These mechanisms are based on aerobic metabolism. Patients with McAd can therefore tolerate longer periods of physical activity well if it is aerobic exercise of mild-to-moderate intensity. The work intensity that patients with McAd do tolerate can show big variations between different patients.

During the study period, different types of physical activity were recorded in a diary by the patient. Aerobic exercise as cycling and walking/hiking did not lead to CK elevation, whereas anaerobic exercise leads to CK elevation.

Different researchers have recommended aerobic training and aerobic conditioning in order to improve physical activity, oxygen uptake, cardiovascular fitness, and energy supply via the blood in McAd [2, 10, 23–26]. Especially walking and cycling with mild or moderate intensity can be recommended for all McAd patients to improve their physical capacity [2, 10, 25–27]. Aerobic metabolism usually starts after 7–10 min of exercising. Therefore, patients with McAd should warm up with low intensity and may increase the intensity of physical work after 7–10 min. Some of the patients experience the above described second wind phenomenon.

7.3 Anaerobic exercise

During anaerobic exercise (within the first seconds and minutes or using great strength), energy is supplied by anaerobic mechanisms as anaerobic glycogenolysis without oxygen. Short periods of activity with high intensity such as running, walking upstairs, and carrying or lifting heavy weights require anaerobic metabolism. Due to the deficiency of the myophosphorylase enzyme in the muscle of McAd patients, this is hampered. Anaerobic physical activity can thus lead to muscular damage in patients with McAd and should be avoided as far as possible by patients with McAd [2, 10, 25]. Nevertheless supervised resistance training has been shown to improve muscle strength in patients with McAd [28]. Pietrusz et al. state that strength training for McArdle patients is safe when it is tailored to the patient as "short bursts of resistance activity lasting no longer than 10 seconds preceded and followed by 30 seconds to 3 minutes rest." In a case report of two McAd patients, an improvement of both muscular strength and quality of life was observed after a period with resistance training [29].

8. Discussion and conclusions

Sport is the most important therapeutical option for patients with McArdle disease. Aerobic conditioning can be recommended to all McAd patients, but anaerobic exercise may lead to muscular damage. It has been shown by different researchers that regular physical activity may lead to improved exercise capacity [2, 10, 23–26]. As we have learned from practical experience and the scientific literature, extensive physical and strenuous exercise may lead to muscle damage, myoglobinuria, and even acute kidney failure [15–19]. Nevertheless Santalla et al. and Pietrusz et al. have shown that resistance training under expert supervision is feasible and improves muscle strength in McArdle patients. But it is important that this type of training is performed under supervision in order to avoid muscle damage [28, 29]. On the other hand, a case study with a long-term follow-up of one patient with McAd has shown that mostly aerobic activity did not lead to an increase in the creatine kinase level. Instead, moderate cycling or hiking led to a decrease in the creatine kinase [2]. In the same patient, anaerobic exercise lead to increased CK levels suggesting muscle damage after carrying heavy weights [2]. In order to avoid muscle damage by vigorous exercise or in a risky way, all patients with McAd should receive sport medical advice on an individualized training plan that meets their individual training needs. In order to enhance patient compliance, common aims and routines for physical activity and sports should be established.

In conclusion, regular activity and sport are paramount for patients with McArdle disease. Patients benefit from regular physical activity. Sport should be based on aerobic conditioning such as walking and cycling, whereas anaerobic exercise of high intensity over short periods should be avoided in general. Physical activity must be individualized to the patients' capacity and needs. Some case

- Do not be afraid of physical activity.
- Individualize your personal training goals.
- Compete with yourself and not with others.
- Aerobic conditioning (walking or cycling) is the preferable training method for patients with McAd.
- Keep on doing physical activity on a regular basis three to five times a week using aerobic exercise, such as walking or cycling for about 30–40 min on each occasion.
- Regular training of mild-to-moderate intensity will improve physical capacity and may postpone weakness and muscle wasting in elderly patients.
- Preexercise nutrition may enhance physical performance.
- Resistance training should only be used under competent supervision of physicians and/or physiotherapists with experience in treating McArdle patients in order to avoid muscle damage.

Table 4.
Recommendations for physical activity for patients with McArdle disease.

reports suggest that even resistance training might be feasible, effective, and safe for patients with McAd. Obviously there is individual variation of the intensity that is appropriate for different patients. Therefore, a cooperation with a doctor expe-rienced on sports medicine, trainer, and physiotherapist can help to establish an individualized training plan in order to maintain and possibly to improve physical capacity without increasing the danger for undesirable effects of too much physical activity.

Probably a self-monitoring of the CK blood level (like measuring blood-glucose in diabetes patients) could help to guide training and individual response to exercise in the future. More research on specially designed training programs for McAd is needed.

The following general recommendations shown in **Table 4** are based on personal experience of the author and the current state of the scientific literature [2, 10, 20–30].

As shown above, sport has therapeutic potential for people with McArdle disease. Sport is used with reason and is therefore not a danger but a powerful medicine.

Author details

Georg Bollig[1,2]

1 Palliative Care Team, Medical Department Sønderborg/Tønder, South Jutland Hospital, Sønderborg, Denmark

2 Medical Research Unit, Institute of Regional Health Research, University of Southern Denmark, Odense, Denmark

*Address all correspondence to: georg.bollig@rsyd.dk

References

[1] McArdle B. Myopathy due to a defect in muscle glycogen breakdown. Clinical Science. 1951;**24**:13-35

[2] Bollig G. Das McArdle-Syndrom (Glykogenose Typ 5) unter sportmedizinischen Gesichtspunkten. Köln: Inaugural-Dissertation Universität zu Köln; 2000

[3] DiMauro S, Tsujino S. Nonlysosomal glycogenoses. In: Engel AG, Franzini-Armstrong C, editors. Myology. New York: McGraw-Hill; 1994. pp. 1554-1576

[4] Bartram C, Edwards RHT, Beynon RJ. McArdle's disease-muscle glycogen phosphorylase deficiency. Biochimica et Biophysica Acta. 1995;**1272**:1-13

[5] Di Mauro S. Muscle glycogenoses: An overview. Acta Myologica. 2007;**XXVI**:35-41

[6] Bollig G. McArdle's disease (glycogen storage disease type V) and anesthesia—A case report and review of the literature. Pediatric Anesthesia. 2013;**23**(9):817-823

[7] Kubisch C, Wicklein EM, Jentsch TJ. Molecular diagnosis of McArdle disease: Revised genomic structure of the myophosphorylase gene and identification of a novel mutation. Human Mutation. 1998;**12**:27-32

[8] Haller RG. Treatment of McArdle disease. Archives of Neurology. 2000;**57**:923-924

[9] Vissing J, Haller RG. A diagnostic cycle test for McArdle's disease. Annals of Neurology. 2003;**54**:539-542

[10] Birch KE. Das McArdle-Handbuch. Deutschland: Selbsthilfegruppe Glykogenose; 2011. Available from: https://www.glykogenose.de/download/McArdle_Handbuch_DEU_Final_20_04_18.pdf

[11] Vissing J, Haller RG. The effect of oral sucrose on exercise tolerance in patients with McArdle's disease. The New England Journal of Medicine. 2003;**349**:2503-2509

[12] Quinlivan R, Beynon RJ, Quinlivan R, et al. Pharmacological and nutritional treatment for McArdle's disease (glycogen storage disease type V). Cochrane Database of Systematic Reviews. 2004;**3**:CD003458

[13] Bollig G, Mohr S, Ræder J. McArdle's disease and anaesthesia. Case reports, review of potential problems and association with malignant hyperthermia. Acta Anaesthesiologica Scandinavica. 2005;**49**:1077-1083

[14] Quinlivian R, Buckley J, James M, Twist A, Ball S, Duno M, et al. McArdle disease: A clinical review. Journal of Neurology, Neurosurgery, and Psychiatry. 2010;**81**(11):1182-1188

[15] Stamos GE, Crouch TT, Wood WG, et al. Acute renal failure in McArdle's disease. Southern Medical Journal. 1979;**72**:77-79

[16] McMillan MA, Hallworth MJ, Doyle D, et al. Acute renal failure due to McArdle's disease. Renal Failure. 1989;**11**:23-25

[17] Tsushima K, Koyama S, Ueno M, et al. Rhabdomyolysis triggered by an asthmatic attack in a patient with McArdle disease. Internal Medicine. 2001;**40**:131-134

[18] Bundschuh HD, Pfeiffer J, Schlote W. Akutes Nierenversagen bei McArdle-Syndrom. Verhandlungen der Deutschen Gesellschaft für Innere Medizin. 1977;**83**:1280-1293

[19] Meinck HM, Goebel HH, Rumpf KW, et al. The forearm ischaemic work test-hazardous to McArdle patients? Journal of Neurology,

Neurosurgery, and Psychiatry. 1982;**45**:1144-1146

[20] Bollig G. Improvement of exercise tolerance in McArdles disease—Lessons learned from sports medicine. In: Poster Presentation International GSD Conference 2013 (IGSD13). Germany: Heidelberg; 2013

[21] Scalco RS, Chatfield S, Godfrey R, Pattni J, Ellerton C, Beggs A, et al. From exercise intolerance to functional improvement: The second wind phenomenon in the identification of McArdle disease. Arquivos de Neuro-Psiquiatria. 2014;**72**(7):538-541

[22] Pearson CM, Rimer DG, Mommaerts WF. A metabolic myopathy due to absence of muscle phosphorylase. The American Journal of Medicine. 1961;**30**:502-517

[23] Perez M, Moran M, Cardona C, et al. Can patients with McArdle's disease run? British Journal of Sports Medicine. 2007;**41**:53-54

[24] Haller RG, Wyrick P, Taivassolo T, et al. Aerobic conditioning: An effective therapy in McArdle's disease. Annals of Neurology. 2006;**59**:922-928

[25] National Commissioning Group. McArdle Disease Service. Exercise and McArdle Disease. London: University College London Hospitals; 2016. Available from: https://www.uclh.nhs.uk/PandV/PIL/Patient%20information%20leaflets/Exercise%20and%20McArdle%20Disease.pdf

[26] Quinlivan R, Vissing J, Hilton-Jones D, Buckley J. Physical training for McArdle disease. Cochrane Database of Systematic Reviews. 2011;**12**:CD007931

[27] Reason SL. One Step at a Time—Walking with McArdle Disease. Droxford: Association for Glycogen Storage Disease; 2013

[28] Santalla A, Munguía-Izquierdo D, Brea-Alejo L, Pagola-Aldazábal I, Díez-Bermejo J, Fleck SJ, et al. Feasibility of resistance training in adult McArdle patients: Clinical outcomes and muscle strength and mass benefits. Frontiers in Aging Neuroscience. 2014;**6**:334

[29] Pietrusz A, Scalco RS, Quinlivan R. Case report resistance exercise training in McArdle disease: Myth or reality? Case Reports in Neurological Medicine. 2018;**2018**:9658251

[30] Nogales-Gadea G, Santalla A, Ballester-Lopez A, Arenas J, Martín MA, Godfrey R, et al. Exercise and preexercise nutrition as treatment for McArdle disease. Medicine and Science in Sports and Exercise. 2016;**48**(4):673-679

4

Metabolic Syndrome: Impact of Dietary Therapy

Suzanne Fouad Soliman

Abstract

Metabolic syndrome refers to the coexistence of insulin resistance (IR) with several risk factors, including abdominal obesity, atherogenic dyslipidemia, and hypertension, which is usually complicated by cardiovascular and/or cerebrovascular diseases. This clustering of risk factors suggests that they are interrelated and not independent of one another and that they share underlying mechanisms, mediators, and pathways. Its prevalence exceeds 40% of those over 40, and it has recently been diagnosed in adolescents and even children. Metabolic syndrome is a proinflammatory prothrombotic state with determination of elevated level of cytokines, acute phase reactants, fibrinogen, and plasminogen activator inhibitor-1. A comprehensive definition of metabolic syndrome and its pathogenesis would facilitate research into its causes and disease pathophysiology linking the components of metabolic syndrome with the increased risk of cardiovascular diseases. The management to mitigate these underlying risk factors constitutes a first-line intervention; dietary therapy of metabolic syndrome includes lifestyle modification, hypocaloric diet, and consumption of functional food. Healthy food quantity and time of consumption help restore the normal metabolic profiles. Hopefully, this will lead to new insights into facilitating epidemiological and clinical studies of pharmacological, lifestyle, and preventive treatment approaches.

Keywords: metabolic syndrome, circadian rhythm, insulin resistance, sleep, functional foods, lifestyle modification, hypocaloric diets

1. Introduction

The prevalence of metabolic syndrome (MetS) worldwide varies between 10 and 84% depending on the age, gender, and race of the population [1]. The International Diabetes Federation estimates that one-quarter of the world's population has MetS [2]. Although the prevalence of MetS varies throughout the world and depends on the particular health organization used for its definition, it is clear that the number of people complaining about this syndrome has globally risen [3], due to a more sedentary life, the increase in the number of smokers, unhealthy dietary habits, and the increase in stress [4].

MetS is a cluster of metabolic disturbances that tend to coexist. Different health organizations have suggested diverse definitions for MetS. However, it is clear that MetS is a clinical entity of substantial heterogeneity, commonly represented by the combination of central abdominal obesity, hyperglycemia, dyslipidemia, and/or hypertension [5, 6].

In 1994, the World Health Organization [7] defined MetS as the presence of one of the following: type II diabetes mellitus, insulin resistance, or impaired glucose tolerance, plus at least two of the following—triglycerides (TG) ≥150 mg/dL (≥1.7 mmol/L) and/or HDL cholesterol <40 mg/dL (≤0.9 mmol/L) (males) and <50 mg/dL (≤1.0mmol/L) (females); urine albumin excretion >20 g/min or albumin/creatinine ratio >30 mg/g; systolic blood pressure ≥140 mmHg or diastolic blood pressure ≥90 mmHg or a prescription of hypertension; and central obesity, waist circumference (WC) ≥102 cm (males) and ≥ 88 cm (females), waist/hip ratio > 0.90 (males) and >0.85 (females), and/or body mass index (BMI) >30 kg/m^2 [8].The International Diabetes Federation defined MetS as encompassing at least three of the following five conditions: (1) male waist circumference ≥ 94 cm and female waist circumference ≥ 80 cm; (2) fasting glucose level ≥ 100 mg/dL; (3) SBP ≥130 and/or DBP ≥85 mmHg or under medical treatment for hypertension; (4) TG ≥150 mg/dL or ≥ 1.7 mmol/L or prescribed pharmacological treatment for hypertri-glyceridemia; or (5) male HDL cholesterol <40 mg/dL and female HDL cholesterol <50 mg/dL [9].

Other markers used for diagnosis of metabolic syndrome include high-sensitivity C-reactive protein concentrations (hs-CRP) ≥3 mg/dL [10], uric acid [11], fibrinogen [12], plasminogen activator inhibitor-1 [13, 14], increased homocysteine [15, 16], and decreased adiponectin [17]. Fibrinogen, plasminogen activator inhibitor-1, cytokines, and hs-CRP protein are often elevated in patients with MetS, resulting in a prothrombotic and pro-inflammatory state. These biomarkers are not routinely evaluated in clinical practice. hs-CRP ≥3 mg/dL indicates a state of inflammation and a higher risk of atherosclerotic cardiovascular disease (ASCVD).

2. Characterization and risk assessment

MetS is a link between visceral adiposity, insulin resistance, inflammation, and endothelial dysfunctions [18, 19].

2.1 Visceral adiposity

Visceral adiposity (abdominal fats close to the visceral organs) leads to an imbalance in the secretion of pro-inflammatory and anti-inflammatory factors. Adipocytes produce pro-inflammatory factors in excess, such as IL-6, IL-10, TNF-α, and hs-CRP. These cytokines block the intracellular insulin signaling pathways. Moreover, the adipose tissue inhibits adiponectin [20, 21].

2.2 Insulin resistance

Insulin resistance (IR) rises with increased waist circumference or body fat and is not related to the body mass index. IR results in the diversion of excess non-estrified fatty acids from lipid-overloaded IR muscles to the hepatic cells, resulting in nonalcoholic fatty liver disease and atherogenic dyslipidemia. IR predisposes to glucose intolerance, which may be aggravated by increased hepatic gluconeogenesis with an IR liver [22].

2.3 Pro-inflammatory condition

Inflammation is a response of the immune system to injury. It is a mechanism in the pathogenesis and progression of obesity-related medical disorders and the link between adiposity, IR, MetS, and cardiovascular disease [23]. Oxidative stress is a condition in which there is an imbalance between the prooxidants and antioxidants in the body [24]. It plays a key role in the pathogenesis of atherosclerosis through different mechanisms such as the oxidation of LDL-c particles [24, 25] or the impairment of HDL-c functions [26].

2.4 High blood pressure

Hypertension is defined as a resting SBP ≥140 mmHg and/or DBP ≥90 mmHg or the use of medical prescription to lower blood pressure [27]. Activation of the sympathetic nervous system together with the activation of the renin-angiotensin system by the hyperglycemia and antinatriuretic effect of insulin, in addition to some obese people being salt sensitive, contributes to the development of hypertension. It is identified as a major cardiovascular and renal risk factor related to heart disease, myocardial infarction, cerebrovascular stroke, and vascular diseases [28].

2.5 Circadian rhythm disruption

Over the last two decades, a deeper comprehension of the molecular mechanisms that control the biological clock and circadian rhythm has been achieved. The 2017 Nobel Prize in Physiology was awarded to Jeffrey C. Hall, Michael Rosbash, and Michael Young for their work on circadian rhythms [29, 30]. There are interconnections between the circadian rhythm and endocrine homeostasis. Biological clock disruption was in fact associated to an increased incidence of metabolic and endocrine diseases. Irregular rhythms have been linked to vari-ous chronic health conditions, such as sleep disorders, obesity, type II diabetes mellitus, hypertension, gastrointestinal motility diseases, depression, cardiac diseases, cognitive dysfunction, bipolar disorder, and seasonal affective disorder [31]. The master clock in mammals that controls circadian rhythms consists of a group of nerve cells in the brain called the suprachiasmatic nucleus (SCN) which contains about 20,000 nerve cells. It is located in the hypothalamus superior to the optic chiasm and receives direct light information from the retina via the retinohypothalamic tract. The master clock in the brain coordinates all of the body's peripheral clocks through different neural and humoral signals, so they are in synchrony [32].

The nonstop 24/7 lifestyle has given rise to human clock gene mutation, for which the human biological clock does not have a suitable design for physiologi-cal adaptation; this leads to an imbalance of autonomic nervous system function toward the parasympathetic predominance in the abdominal compartment and the sympathetic predominance in the muscular and thoracic compartments, resulting in the impairment of insulin secretion, glucose uptake by the muscle, and increase in blood pressure, intra-abdominal fat, and fatty liver [33].

The timing and nutritional composition of meals can regulate circadian rhythms, particularly in the peripheral tissue. The translation of these findings to human physiology now represents an important goal. In healthy humans, glucose tolerance lowers throughout the day, leading to the term "afternoon diabetes"; a daily rhythm in pancreatic insulin production is observed, with increased insulin production in the morning compared to the evening. Moreover, lipid metabolism exhibits circadian regulation, with elevated plasma concentrations of triacylglycerol during the biological night and an elevated postprandial response following a nighttime meal compared with the same meal consumed during the day [34]. In 2006, Staels stated "When the clock stops ticking, metabolic syndrome explodes" [35]. When to eat is as important as what to eat.

2.5.1 Sleep in relation to melatonin and growth hormone secretions

Melatonin is synthesized by multiple tissues in the body, but the pineal gland is the major contributor to circulating melatonin concentration. The rhythmic activity of the SCN determines the release of melatonin, which directly correlates with day length, and it may in turn modulate other physiological functions through paracrine signaling. Pancreatic hormone insulin, gastric hormone ghrelin, growth hormone (GH) from the pituitary gland, and leptin hormone represent as key factors in metabolic regulation, and now several circadian factors are recognized as regulators of their secretion and activity. Sleep deprivation results in circadian misalignment and increases markers of inflammation and IR. A better understanding of such relationships might prove useful and supportive in designing new therapeutic approaches for metabolic syndrome [36].

Growth hormone is a peptide hormone secreted by the anterior pituitary under control of the hypothalamic-pituitary axis. Actions of GH depend on the age of the individual. The primary role of GH in children and adolescents is to promote growth until the final adult height is achieved, while in adults, the role of GH is the regulation of protein, carbohydrate, and lipid metabolism, insulin resistance, and metabolic homeostasis. Sleep restriction and disturbance have also been shown to negatively impact hormone secretion. In sleep deprivation cases, GH secretion decreases, the appetite-suppressing hormone leptin decreases, and the appetite-stimulating hormone ghrelin increases. The end result is the consumption of additional food and calories. Plasma GH peak appeared with the onset of deep sleep and did not correlate with the changes in plasma glucose and insulin. GH secretion decreased if the onset of sleep was delayed [37].

2.6 Risk assessment

Metabolic syndrome directly promotes the development of atherosclerotic cardiovascular disease (ASCVD) [38] and type II diabetes mellitus [39]. Atherogenic dyslipidemia, high circulating levels of prothrombotic factors, and the increase of inflammatory markers carry a higher risk for acute cardiovascular syndromes [40]. Previous studies on middle-aged persons with metabolic syndrome had concluded that many middle-aged people are at increased absolute risk for cardiovascular diseases in the 10-year risk. Furthermore, in young adults who develop the syndrome, long-term risk for ASCVD is increased even when the 10-year risk is not high. The Framingham risk scoring is used to classify metabolic syndrome patients into three risk categories based on a 10-year risk for coronary heart disease, cerebrovascular disease, peripheral artery disease, and heart failure risk: *low-risk* (10-year risk <10%), *moderate-risk* (10-year risk 10–20%), and *high-risk patients* (10-year risk >20%) [41, 42].

3. Metabolic syndrome, mental and cognitive functions

Obesity and dementia frequently coexist. They share many common pathways such as insulin resistance as well as inflammation and oxidative stress. Midlife obesity is directly related to the risk of developing dementia later in life. Dementia is a syndrome characterized by gradual decline in cognitive functions (memory, attention, behavior, language, mood, and learning). In 2008, Dr. Suzanne de la Monte and Dr. Jack Wands proposed that Alzheimer's disease could be termed type III diabetes, based on the fact that IR within the brain revealed to be a characteristic of Alzheimer's disease [43]. There is no definite cure for dementia, and prevention is the only option. Prevention may be achieved by early treatment of the risk factors resulting in cognitive impairments (obesity, DM, HTN, smoking, dyslipidemia, anxiety, and depression). Several biomarkers could be involved in the pathogenesis of mild cognitive impairment or early neurodegenerative diseases. Assessment of these biomarkers in obese middle-aged persons could serve as a basis for early management to alleviate the burden of cognitive impairment and dementia in the future.

3.1 Cognitive and mental evaluations

The Mini-Mental State Examination (MMSE) was performed for clinical evalua-tion of mental and cognitive status. The 30-point questionnaires take between 5 and 10 minutes and examination functions including registration, attention, calcula-tion, recall, language, commands, and orientation [44]. It is a sensitive, valid, and reliable method that is used extensively in clinical and research settings to measure cognitive impairment and to estimate the severity and progression of cognitive impairment. It is also used to follow the course of cognitive changes in an individual over time, thus making it an effective way to document an individual's response to treatment. Sleep quality and the number of sleep hours and their pattern were evaluated. Exposure to sun and length of time, duration, and clothing were recorded. General subjective life stresses, life pattern to evaluate general activity, and history of exercising were recorded.

C-peptide was detected by C-peptide enzyme immunoassay [45]. Modified homeostatic model assessment of insulin resistance (M.HOMA-IR) was calcu-lated by Eq. (1), in which insulin was replaced by C-peptide so as to be applied on diabetic patients using exogenous insulin [46]:

$$\text{M.HOMA} - \text{IR} = 1.5 + \text{Fasting blood glucose} \times \text{Fasting c} - \text{peptide}/2800 \quad (1)$$

Blood sampling and biochemical markers of cognitive function impairments were performed as ceramide kinase enzyme, alpha-synuclein, serum clusterin, amyloid beta, and inflammatory markers (hs-CRP, IL-6, and TNF-α).

Ceramides are agents involved in the pathogenesis of mild cognitive impairment or early neurodegenerative diseases. They are mediated by insulin resistance and inflammatory states. Ceramides are lipid soluble and therefore likely to readily cross the blood-brain barrier resulting in cytotoxic effects in the central nervous system with various dementia-associated diseases, including Alzheimer's disease. It is important to prevent its elevation in the human body via increasing the activity of the ceramide kinase enzyme (CERK enzyme) that converts the ceramide to ceramide-1-phosphate (C1P) via its phosphorylation. C1P is a sphingolipid metabo-lite that has been implicated in the membrane fusion of brain synaptic vesicles and neutrophil phagolysosome formation. C1P is a key regulator of cell growth and survival, it stimulates DNA synthesis and cell division, and it is a potent inhibitor of apoptosis. Many studies have shown that C1P is important for membrane biology and for the regulation of membrane-bound proteins and the CERK enzyme has appeared to be tightly regulated in order to control both ceramide levels and production of C1P [47]. Improvement of the metabolic profiles including C-peptide concentration and the M.HOMA-IR values was associated with the improvement of the serum level of the enzyme CERK and of cognitive functions [48].

Alpha-synuclein (α-Syn) protein was originally isolated from Alzheimer's disease plaques and was thought to be a presynaptic nerve terminal protein. The α-Syn is a member of the synuclein's family of cytoplasmic, predominantly neuron-specific proteins. Synucleins are small, prone to aggregate proteins associated with several neurodegenerative diseases. The α-Syn has been found in body fluids, including blood and cerebrospinal fluid, peripheral tissues, and the central nervous system. Previous reports suggest that α-Syn is widely expressed peripherally, including the macrophages. The expression of α-Syn is enhanced in activated macrophages, suggesting that α-Syn may modulate macrophage function and thereby inflammatory processes. Diet-induced obesity may be an environmental risk factor for the development alpha-synucleinopathies [49]. There was a significant positive correlation between serum IL-6 and serum α-Syn. Fighting obesity, dyslipidemia, and the associated complications especially the inflammatory processes improved the deleterious effects on the cognitive functions in obese persons [50].

Clusterin (apolipoprotein J) is a heterodimeric glycoprotein in which α and β chains are interconnected via five disulfide bonds. There is a strong associa-tion between the single-nucleotide polymorphisms in the clusterin gene and Alzheimer's disease. Moreover, plasma clusterin is considered a potential peripheral biomarker of cognitive dysfunction and AD. Clusterin may be related to AD pathogenesis through various mechanisms. It could bind amyloid extracellularly and inhibit the aggregation of amyloid beta monomers into toxic oligomers. In addition, the neurotoxicity of the amyloid might be reduced by the interaction of clusterin with the molecules involved in signal transduction, DNA repair, cell cycle, and apoptosis. Brain apolipoprotein clusterin plays an important role in cholesterol transport and neuronal lipid metabolism. Furthermore, it has a role in the inhibition of neuroinflammation which is thought to be a major factor in AD pathogenesis and identified as a key component in cerebrovascular diseases. In addition, it is known that neuroinflammation plays an important role in dementia pathogenesis and neurodegenerative diseases [51, 52]. The improvement of the C-peptide concentration and the M.HOMA values were parallel with improvement of oral cognitive tests and clusterin value; clusterin was presented as a cognitive function parameter [52].

Plasma amyloid beta (Aβ) can be applied as trait, risk, or state biomarker for AD and denote a neuropathologic condition. A previous study has reported the presence of a stronger correlation between plasma Aβ and positron emission tomography or Pittsburgh Compound-B-C11 as radiotracers illustrate fibrillar brain amyloid deposits which is a reliable method to measure brain amyloid plaque accumulation. Reduction of body weight and improving the metabolic profile reduced the level of serum Aβ protein, an effective role in improving the cognitive function. At the same time, previous studies stated that plasma Aβ measures possibly aid in clinical investigations as markers for the pharmacological impacts of medications that influence amyloid protein transformation [53].

Serum hs-CRP level may be used as a marker for cognitive functions in obese middleaged persons. Peripheral inflammatory markers are elevated in obese patients. Improvement in cognitive functions is recorded after dietary therapy, with decrease in hs-CRP serum levels. Significant inverse correlation is found between cognitive functions and hs-CRP levels, insulin resistance, minimal waist circumference, and BMI [54].

4. Management of metabolic syndrome

As the presence of MetS carries a risk for cardiovascular diseases and diabetes mellitus, the primary goal of clinical management to individuals with metabolic syndrome is directed towards decreasing the major metabolic risk factors: high LDL-C, hypertension, obesity by losing fat percentage and not muscle mass, decreasing insulin resistance, blood glucose, and maintaining normal HDL and triglyceride levels through lifestyle changes. Moreover, medical drug therapy might be considered in highrisk patients to modify cardiovascular disease risk factors [7]. Bariatric surgery has been indicated to treat morbidly obese persons. The safety and effectiveness of bariatric surgery in patients with metabolic syndrome have been studied and encouraged [55, 56].

Chronotherapy means the timing of drug and dietary treatment to obtain maximum therapeutic effect. It can be achieved through the timing of light exposure, exercise, food consumption, medication uptake, and sleep, with the goal of optimizing any treatment by taking into account the circadian rhythms of the body [57].

4.1 Therapeutic lifestyle modifications

Physical exercise, diet, and adequate sleep are the way to reach the target. Gradual permanent change in the patient's lifestyle can lead to better and easily maintain normal parameters than major food deprivation introduced all at once. Implementing the following changes increases the chances of success: changing a sedentary life through regular sustained physical exercise and eating several small portions of different foods varieties; consuming complex carbohydrates such as barley, oat, corn, quinoa, and brown rice while decreasing the consumption of simple carbohydrates such as white sugar and sweets; eating fresh seasonal fruit and vegetables daily as they are good sources of fiber, vitamins, and minerals noting that dietary fiber helps prevent gastrointestinal problems such as flatulence and constipation; eating foods rich in unsaturated fatty acids such as salmon, mackerel, sardines, tuna, raw nuts, and flaxseed oil; increasing the consumption of black beans, kidney beans, green peas, and lentils; avoiding processed and red meats; limiting the intake of sugar, salt, and carbonated drinks; replacing salt with spices; boiling, baking, or steaming food instead of frying; paying attention to portion size by utilizing smaller plates; and drinking plenty of fluids, of which 30 ml/kg/day water is the best [7].

Avoid light exposure in late evening or at night as artificial light disrupts the circadian rhythm and the production of melatonin and therefore has a negative effect on sleep quality, mood, cognition, and hormonal functions [58, 59]. Sleep early and get at least 6 hours of sleep per night [60], and stop smoking.

4.2 Energy-restricted diets

For our patients, the Harris-Benedict equation (Eq. (2)) was used to calculate the caloric requirements for each individual in order to estimate the caloric needs or basal energy expenditure (BEE) as follows [61]:

$$\begin{gathered}\text{Males: BEE} = 66.5 + 13.8\,(\text{weight in Kg}) + 5\,(\text{height in cm}) - 6.8\,(\text{age}).\\ \text{Females: BEE} = 65.5 + 9.6\,(\text{weight in Kg}) + 1.7\,(\text{height in cm}) - 4.7\,(\text{age}).\end{gathered} \tag{2}$$

These equations yield basal energy expenditure that is frequently multiplied by various activities and/or stress factors to generate the patient's estimated resting energy expenditure, and then we subtracted 500 calories per week from the calculated energy requirement, to produce weight loss of 0.5–1 kg per week. A hypocaloric diet goal is to reduce body weight by about 10% over the first 6 months. The menu varied according to the participant's age and eating habits. It was low in fat (20–25%), high in complex carbohydrate (50–60%), and sufficient in protein (25–30%) from both animal and vegetarian sources [54]. Carbohydrates in the form of complex CHO, which have low glycemic loads and indexes such as whole grains, oats, fresh seasonal fruits, and vegetables, were consumed to increase high fiber content (eight servings of fresh vegetables and fruits/day). On the other hand, they were instructed to decrease foods rich in saturated fats such as fried foods and packaged meats [62].

4.3 The Mediterranean diet

The Mediterranean diet (MedDiet) is based on scientific evidence that inhabitants of Mediterranean countries (Greece, Italy, and Spain), around the Mediterranean Sea adhering to the principles of nutrition traditional for their region, have better health indicators than people in other areas. Originally, it was developed to prevent and treat the symptoms of hypertension and/or heart disease. In this MedDiet, the emphasis is on the use of plant-based foods as the main source of nourishment: fruits, vegetables, cereals, nuts, legumes, seeds, and whole grains. It is also recommended to replace butter and hydrogenated fats with extra-virgin olive oil and salts with spices and herbs. Red meat and sweets should be eaten once a week, and the main sources of protein should be eggs, yoghurt, fish, and poultry. Red wine could be taken in limited amounts (one glass per day for women; two glasses per day for men) [63–65].

4.4 Role of functional foods

There is no universal definition for functional foods, but the easiest one is from the International Food Information Council which described functional foods as "dietary or food components that provide a health benefit beyond providing basic nutrition" [55]. Functional foods have an important role in the management of metabolic syndrome. Functional foods can help in weight loss, regulation of blood pressure, and control of blood glucose and lipid profile levels. Previous studies supported the beneficial effects of using functional foods containing bioactive components in the management of metabolic syndrome. This includes the intake of foods which have a thermogenesis effect (process in which body heat production increases, e.g., ginger, caffeine, red pepper, and green tea); functional foods with anti-inflammatory properties such as curcumin; and foods with the ability to increase insulin sensitivity, e.g., cinnamon, and increase intake of foods rich in plant estrogens such as soymilk, soybeans, flaxseed, sesame, green beans, fennel, peanut, and pumpkin seeds. In addition, soy products are rich in amino acids and can preserve lean body mass, decrease body fat percentage, and increase basal metabolic rate [48, 50, 53, 66–70].

4.4.1 Functional food classification

- Natural functional foods: food products rich in biological active compounds (food rich in dietary fiber such as oat, barley, psyllium, vegetables, and fruits) [66]

- Foods with approved scientific heath nutrients: apple (pectin), garlic (*Allium sativum*), and fish (omega 3 fatty acids) [71]

- Fortified foods with specific nutrients: iron fortified chocolate for iron deficiency anemia, calcium fortified juices, and zinc fortified products

- Fermented products rich in probiotics or synbiotics [72]

4.4.2 Eat to beat metabolic syndrome

The use of functional and prepared designed food would help fight MetS. Three thousand years ago, Hippocrates announced the tenet "Let food be thy medicine and medicine be thy food." Dietary supplements proved to improve the MetS criteria which include Syrian bread prepared mainly from barley flour and wheat germ mixed with either turmeric or ginger; doum biscuits; barley biscuits; bread pre pared with soybean flour and enriched with curcumin or ginger; and cookies prepared with whole wheat flour and fennel or chia seeds. The higher fiber content of these products has a satiety effect so as to decrease food consumption. Moreover, fibers decrease digestion and absorption of dietary carbohydrates. Soy flour is a very rich source of essential amino acids except methionine. Preparation of bakery prod-ucts using bioactive ingredients has positive beneficial effects for obesity, diabetes, and dyslipidemia management. These compounds have different benefits such as anti-atherogenic, antioxidant, and anti-inflammatory effects [48, 50, 53, 67–70].

4.5 Pharmacological management

There are no specific recommendations regarding the pharmacological management of metabolic syndrome; instead, the focus is on the management of the risk factors which need to be aggressively treated in order to prevent cardiovascular disease and type II diabetes mellitus if lifestyle changes aren't enough to reduce the risks. The available evidence of the pharmacological prescriptions is related to their cardiovascular benefits in clinical practice. Pharmacological treatments include statins and/or fibrates for dyslipidemia, angiotensin-converting enzyme inhibitors or renin-angiotensin-aldosterone system inhibitors for hypertension, metformin or sodium/glucose cotransporter 2 inhibitors or glucagon-like peptide 1 receptor agonists (GLP-1RAs) for glucose intolerance, and low-dose aspirin for prevention of arterial thrombosis [73]. Melatonin provides an innovative approach in the management of MetS through its useful effects on circadian rhythmicity, insulin resistance, dyslipidemia, high blood pressure, weight loss, and improving the antioxidative status (melatonin tablet 1 mg/kg) [74, 75]. Melatonin reduces the toxicity of many pharmaceutical agents and has a high safety profile [76].

5. Conclusion

Prevention is the best form of treatment of metabolic syndrome, and it starts from childhood by following a healthy lifestyle based on proper nutrition, exercise, and adequate early night sleep. Timing of food consumption, food quality, food quantity, light exposure, medication intake, and sleep are likely to play an important role in human health. Eat small portion sizes of food, and follow the rhythm of the day to maintain the synchronicity of the biological clock. Consuming natural effective supplements rich in bioactive substances leads toward the optimization of biochemical parameters of patients with metabolic syndrome in favor of a healthy outcome. Bariatric surgery should be considered in individuals with morbid obesity or obesity associated with comorbidity.

Financial support

The practical part of the work was financially supported by the National Research Centre, Egypt, but no fund supports for publication fees.

Abbreviations

ASCVD	atherosclerotic cardiovascular disease
α-Syn	alpha-synuclein
AD	Alzheimer's disease
Aβ	amyloid beta
BEE	basal energy expenditure
BMI	body mass index
C1P	cermide-1-phosphate
CERK	ceramide kinase enzyme
CRP	C-reactive protein
DBP	diastolic blood pressure
DM	diabetes mellitus
GLP-1RAs	glucagon-like peptide 1 receptor agonists
GH	growth hormone
HDL-c	high-density lipoprotein cholesterol
hs-CRP	high-sensitivity C-reactive protein
IL-6	interleukin-6
IL-10	interleukin-10
IR	insulin resistance
LDL-c	low-density lipoprotein cholesterol
MedDiet	Mediterranean diet
MetS	metabolic syndrome
M.HOMA-IR	modified homeostatic model assessment of insulin resistance
MMSE	mini-mental state examination
SBP	systolic blood pressure
SCN	suprachiasmatic nucleus
TG	triglycerides
TNF-α	tumor necrosis factor-α
WC	waist circumference

Author details

Suzanne Fouad Soliman[1,2]

1 Nutrition and Food Science Department, National Research Centre, Dokki, Giza, Egypt

2 Critical Care Medicine, Faculty of Medicine, Cairo University, Egypt

*Address all correspondence to: suzannefouad6161@yahoo.com

References

[1] Kaur JA. Comprehensive review on metabolic syndrome. Cardiology Research and Practice. 2014:943162. DOI: 10.1155/2014/943162

[2] O'Neill S, O'Driscoll L. Metabolic syndrome: A closer look at the growing epidemic and its associated pathologies. Obesity Reviews. 2015;**16**:1-12

[3] Oresic M, Vidal-Puig A. A Systems Biology Approach to Study Metabolic Syndrome. Heidelberg, Germany: Springer; 2014

[4] Lee EG, Choi JH, Kim KE, Kim JH. Effects of a walking program on self-management and risk factors of metabolic syndrome in older Korean adults. Journal of Physical Therapy Science. 2014;**26**:105-109

[5] Sarafidis PA, Nilsson PM. The metabolic syndrome: A glance at its history. Journal of Hypertension. 2006;**24**:621-626

[6] Bernabe GJ, Zafrilla RP, Mulero CJ, Gomez JP, Leal HM, Abellan AJ. Biochemical and nutritional markers and antioxidant activity in metabolic syndrome. Endocrinología y Nutrici6n. 2013;**61**:302-308

[7] Grundy SM, Cleeman JI, Daniels SR, Donato KA, Eckel RH, Franklin BA, et al. Diagnosis and management of the metabolic syndrome: An American Heart Association/National Heart, Lung, and Blood Institute scientific statement. Circulation. 2005;**112**:2735-2752

[8] Alberti K, Zimmet P. Definition, diagnosis, and classification of diabetes mellitus and its complications. Part 1: Diagnosis and Classification of Diabetes Mellitus Provisional Report of a WHO Consultation. Diabetic Medicine. 1998;**15**:539-553. DOI: 10.1002/

[9] Alberti KG, Zimmet P, Shaw J. The metabolic syndrome—A new worldwide definition. Lancet. 2005;**366**:1059-1062

[10] Grigore C, Vintila AM, Stoian I, Grigore O, Dawkins L, Isacoff D, et al. New markers of inflammation in metabolic syndrome. Journal of Hypertension. 2015;**33**:e244. DOI: 10.1097/01.hjh.0000468096.77160.dc

[11] Srikanthan K, Feyh A, Visweshwar H, Shapiro JI, Sodhi K. Systematic review of metabolic syndrome biomarkers: A panel for early detection, management, and risk stratification in the west Virginian population. International Journal of Medical Sciences. 2016;**13**(1):25-38

[12] Ding L, Zhang C, Zhang G, Zhang T, Zhao M, Ji X, et al. A new insight into the role of plasma fibrinogen in the development of metabolic syndrome from a prospective cohort study in urban Han Chinese population. Diabetology & Metabolic Syndrome. 2015;**7**, Article number: 110

[13] Kodaman N, Aldrich MC, Sobota R, Asselbergs FW, Brown NJ, Moore JH, et al. Plasminogen activator Inhibitor-1 and diagnosis of the metabolic syndrome in a West African population. Journal of the American Heart Association. 2016;**5**(10):e003867. DOI: 10.1161/JAHA.116.003867

[14] Marques P, Collado A, Martinez-Hervás S, Elena Domingo E, Benito E, Piqueras L, et al. Systemic inflammation in metabolic syndrome: Increased platelet and leukocyte activation, and key role of CX3CL1/CX3CR1 and CCL2/CCR2 axes in arterial platelet-proinflammatory monocyte adhesion. Journal of Clinical Medicine. 2019;**8**(5):708

[15] Sreckovic B, Sreckovic VD, Soldatovic I, Colak E,

Sumarac-Dumanovic M, Janeski H, et al. Homocysteine is a marker for metabolic syndrome and atherosclerosis. Diabetes and Metabolic Syndrome: Clinical Research and Reviews. 2017;**11**(3):179-182

[16] Srećković B, Soldatovic I, Colak E, Mrdovic I, Sumarac-Dumanovic M, Janeski H, et al. Homocysteine is the confounding factor of metabolic syndrome-confirmed by siMS score. Drug Metabolism and Personalized Therapy. 2018;**33**(2):99-103

[17] Falahi E, Hossein A, Roosta S. Diabetes & metabolic syndrome: What is the best biomarker for metabolic syndrome diagnosis? Journal of Medical Clinical Research & Reviews. 2015;**9**(4):366-372

[18] Soares R, Costa C. Oxidative Stress, Inflammation and Angiogenesis in the Metabolic Syndrome. Heidelberg, Germany: Springer; 2009

[19] Eckel RH, Grundy SM, Zimmet PZ. The metabolic syndrome. The Lancet. 2005;**365**:1415-1428

[20] Poitou C, Clément K. Le tissu adipeux: Un acteur majeur du syndrome inflammatoire de l'obésité? CAH Nutrition Diet. 2007;**42**:90-96

[21] Ridker PM, Buring JE, Cook NR, Rifai N. C-reactive proteion, the metabolic syndrome, and the risk of incident cardiovascular events: An 8-year follow-up of 14 719 initially healthy American women. Circulation. 2003;**107**:391-397

[22] Petrović G, Bjelaković G, Benedeto-Stojanov D, Nagorni A, Brzački V, Marković-Živković B. Obesity and metabolic syndrome as risk factors for the development of non-alcoholic fatty liver disease as diagnosed by ultrasound. Vojnosanitetski Pregled. 2016;**73**(10):910-920. DOI: 10.2298/VSP150514093P

[23] Ferri N, Ruscica M. Proprotein convertase subtilisin/kexin type 9 (PCSK9) and metabolic syndrome: Insights on insulin resistance, inflammation, and atherogenic dyslipidemia. Endocrine. 2016;**54**(3):588-601. DOI: 10.1007/s12020-016-0939-0

[24] Rahal A, Kumar A, Singh V, Yadav B, Tiwari R, Chakraborty S, et al. Oxidative stress, prooxidants, and antioxidants: The interplay. BioMed Research International. 2014;**761264**:19. DOI: 10.1155/2014/761264

[25] Parthasarathy S, Litvinov D, Selvarajan K, Garelnabi M. Lipid peroxidation and decomposition—Conflicting roles in plaque vulnerability and stability. Biochimica et Biophysica Acta. 2008;**1781**:221-231

[26] McGrowder D, Riley C, Morrison EY, Gordon L. The role of high-density lipoproteins in reducing the risk of vascular diseases, neurogenerative disorders, and cancer. Cholesterol. 2011;**496925**:9. DOI: 10.1155/2011/49692

[27] Zanchetti A. Challenges in hypertension: Prevalence, definition, mechanisms and management. Journal of Hypertension. 2014;**32**:451-453

[28] Thomas G, Shishehbor M, Brill D, Nally JV. New hypertension guidelines: One size fits most? Cleveland Clinic Journal of Medicine. 2014;**81**:178-188

[29] Hall J, Rosbash M, Young M. The Nobel prize in physiology or medicine 2017. Nobelprize.org. August 2, 2018 [Accessed: 03 August 2018]

[30] Hauri P. What is circadian rhythm? National Sleep Foundation. Available from: www.sleepfoundation.org [Accessed: 03 August 2018]

[31] Bixler E. Sleep and society: An epidemiological perspective. Sleep Medicine. 2009;**10**(Suppl 1):S3-S6

[32] Foster RG, Wulff K. The rhythm of rest and excess. Nature Reviews. Neuroscience. 2005;**6**(5):407-414

[33] Kalsbeek A, Scheer FA, Perreau-lenz S, et al. Circadian disruption and SCN control of energy metabolism. FEBS Letters. 2011;**585**:1412-1426

[34] Buijs RM, Scheer FA, Kreier F, Yi C, Bos N, Goncharuk VD, et al. Organization of circadian functions: Interaction with the body Chapter 20: (The link between metabolic syndrome and autonomic circadian disorganization). Progress in Brain Research. 2006;**153**:341-360

[35] Staels B. When the clock stops ticking, metabolic syndrome explodes. Nature Medicine. 2006;**12**:54

[36] Johnston JD, Skene DJ. Regulation of mammalian neuroendocrine physiology and rhythms by melatonin. Journal of Endocrinology. 2015;**226**(2):T187–T198. DOI: 10.1530/JOE-15-0119

[37] Nishad R, Mukhi D, Menon RK, Pasupulati AK. Growth hormone and metabolic homeostasis. EMJ Diabetes. 2018;**6**(1):78-87

[38] Willerson J. Third Report of the National Cholesterol Education Program (NCEP) expert panel on detection, evaluation, and treatment of high blood cholesterol in adults (adult treatment panel III) final report. Circulation. 2002;**106**:3143-3421

[39] Hanson RL, Imperatore G, Bennett PH, Knowler WC. Components of the "metabolic syndrome" and incidence of type 2 diabetes. Diabetes. 2002;**51**:3120-3127

[40] Malik S, Wong ND, Franklin SS, et al. Impact of the metabolic syndrome on mortality from coronary heart disease, cardiovascular disease, and all causes in United States adults. Circulation. 2004;**110**:1245-1250

[41] Damen JA, Pajouheshnia R, Heus P, Moons KM, Reitsma JB, Scholten RJ, et al. Performance of the Framingham risk models and pooled cohort equations for predicting 10-year risk of cardiovascular disease: A systematic review and meta-analysis. BMC Medicine. 2019;**17**:1340-1347. https://doi.org/10.1186/s12916-019-1340-7. Article number: 109

[42] Wilson PW. Estimating cardiovascular disease risk and the metabolic syndrome: A Framingham view. Endocrinology and Metabolism Clinics. 2004;**33**:467-481

[43] Ahmed S, Mahmood Z, Zahid S. Linking insulin with Alzheimer's disease: Emergence as type III diabetes. Neurological Sciences. 2015;**36**(10):1763-1769. DOI: 10.1007/s10072-015-2352-5

[44] Dean PM, Feldman DM, Morere D, Morton D. Clinical evaluation of the mini-mental state exam with culturally deaf senior citizens. Archives of Clinical Neuropsychology. 2009;**24**(8):753-760

[45] Bonger A, Garcia-Webb P. C-peptide measurement: Methods and clinical utility. CRC Critical Reviews in Clinical Laboratory Sciences. 1984;**19**:297

[46] Li X, Zhou ZG, Qi HY, et al. Replacement of insulin by fasting C-peptide in modified homeostasis model assessment to evaluate insulin resistance and islet beta cell function. Chinese. 2004;**29**(4):419-423

[47] De la Monte SM, Tong M, Nguyen V, Setshedi M, Longato L, Wands JR. Ceramide-mediated insulin resistance and impairment of cognitive-motor functions. Journal of Alzheimer's Disease. 2010;**21**(3):967-984

[48] Moaty MI, Fouad S, El Shebini S, Kazem YM, Ahmed NH, Mohamed MS, et al. Serum ceramide kinase as a biomarker of cognitive functions, and

the effect of using two slimming dietary therapies in obese middle aged females. Macedonian Journal of Medical Sciences. 2015;**3**(1):18-25

[49] Rotermund C, Truckenmüller FM, Schell H, Kahle PJ. Diet-induced obesity accelerates the onset of terminal phenotypes in α-synuclein transgenic mice. Journal of Neurochemistry. 2014;**131**(6):848-858

[50] El Shebini S, Moaty MI, Kazem YM, Ahmed NH, Fouad S, Arafa AM, et al. Human alpha Synuclein, inflammation and metabolic disorders in obesity, before and after dietary intervention. International Journal of Health Sciences & Research. 2015;**5**(11):304-312

[51] Yu JT, Tan L. The role of clusterin in Alzheimer's disease: Pathways, pathogenesis, and therapy. Molecular Neurobiology. 2012;**6**:314-326. DOI: 10.1007/s12035-012-8237-1

[52] El Shebini SM, Moaty MI, Fouad S, Kazem Y, Ahmed NH, Mohamed MS, et al. Association between serum clusterin and cognitive functions in obese Egyptian women; Potential effects of dietary therapy. Der Pharma Chemica. 2016;**8**(6):205-213

[53] El Shebini S, Moaty MI, Kazem YM, Ahmed NH, Fouad S, Mohamed MS, et al. Relation between obesity, cognition and serum amyloid β protein level and potential role of Foeniculum vulgare in reducing weight and improving cognitive functions. Journal of Biological Sciences. 2017;**17**(5):202-212

[54] Kazem Y, El Shebini SM, Moaty MA, Fouad S, Ahmed NH, Tapozada ST. High sensitivity C-reactive protein may be used as a marker for cognitive impairment in obese Egyptian middle age females. Asian Journal of Clinical Nutrition. 2017;**9**(1):17-23. DOI: 10.3923/ajcn.2016

[55] Brolin RE. Bariatric surgery and long-term control of morbid obesity. Journal of the American Medical Association. 2002;**288**:2793-2796

[56] Lee WJ, Huang MT, Wang W, Lin CM, Chen TC, Lai IR. Effects of obesity surgery on the metabolic syndrome. Archives of Surgery. 2004;**139**:1088-1092

[57] Lévi F, Okyar A. Circadian clocks and drug delivery systems: Impact and opportunities in chronotherapeutics. Expert Opinion on Drug Delivery. 2011;**8**(12):1535-1541. DOI: 10.1517/17425247.2011.618184

[58] Froy O. The relationship between nutrition and circadian rhythms in mammals. Frontiers in Neuroendocrinology. 2007;**28**(2-3):61-71

[59] Jones TM, Durrant J, Michaelides EB, Green MP. Melatonin: A possible link between the presence of artificial light at night and reductions in biological fitness. Philosophical Transactions of the Royal Society of London. Series B, Biological Sciences. 2015;**370**(1667):20140122. DOI: 10.1098/rstb.2014.0122

[60] Kazem Y, El Shebini S, Moaty MI, Fouad S, Tapozada ST. Sleep deficiency is a modifiable risk factor for obesity and cognitive impairment and associated with elevated Visfatin. Open Access Macedonian Journal of Medical Sciences. 2015;**3**(2):315-321

[61] Harris JA, Benedict FG. Standard basal metabolism constants for physiologists and clinicians: A biometric study of basal metabolism in man. In: A Biometric Study of Basal Metabolism in Man: Carnegie Institution of Washington, Pub. No. 279. Philadelphia: JB Lippincott; 1919. pp. 223-250

[62] Abete I, Astrup A, Martínez JA, Thorsdottir I, Zulet MA. Certain dietary

patterns are beneficial for the metabolic syndrome: Reviewing the evidence. Nutrition research. 2014;**34**(7):559-568. DOI: 10.1016/j.nutres.2014.06.012

[63] Davis C, Bryan J, Hodgson J, Murphy K. Definition of the Mediterranean diet: A literature review. Nutrients. 2015;7:9139-9153

[64] Sofi F, Macchi C, Abbate R, Gensini GF, Casini A. Mediterranean diet and health status: An updated meta-analysis and a proposal for a literature-based adherence score. Public Health Nutrition. 2014;**17**:2769-2782

[65] Esposito K, Marfella R, Ciotola M, Di Palo C, Giugliano F, Giugliano G, et al. Effect of a Mediterranean-style diet on endothelial dysfunction and markers of vascular inflammation in metabolic syndrome: A randomized trial. Journal of the American Medical Association. 2004;**292**:1440-1446

[66] Wildman REC. Handbook of Nutraceuticals and Functional Foods. 2nd ed. Boca Raton: CRC Press; 2006

[67] Moaty MI, Fouad S, El Shebini S, Kazem YM, Tapozada ST. Biochemical assessment of bone health in working obese Egyptian females with metabolic syndrome: The effect of weight loss by natural dietary therapies. Open Access Macedonian Journal of Medical Sciences. 2015;**3**(4):582-589

[68] El Shebini S, Moaty MI, Fouad S, Tapozada ST. Serum retinol binding protein-4 to probe insulin resistance and atherogenic dyslipidemia in metabolic syndrome patients following a diet therapy. Research Journal of Pharmaceutical, Biological and Chemical Sciences. 2016;**7**(1):1903-1909

[69] Fouad S, El Shebini S, Moaty MI, Ahmed NH, Tapozada ST. Effect of soya beans bread fortified with turmeric or ginger on Diabesity. Der Pharma Chemica. 2016;**8**(18):398-405

[70] El Shebini S, Moaty MI, Fouad S, Ahmed NH, Tapozada ST. Obesity related metabolic disorders and risk of renal disease: Impact of hypocaloric diet and Avena sativa supplement. Open Access Macedonian Journal of Medical Sciences. 2018;**6**(8):1376-1381

[71] Van SC, Angerer P, Kothny W, Theisen K, Mudra H. The effect of dietary omega-3 fatty acids on coronary atherosclerosis. A randomized, double-blind, placebo-controlled trial. Annals of Internal Medicine. 1999;**130**(7):554-562

[72] Goldin BR. Health benefits of probiotics. The British Journal of Nutrition. 1999;**80**:S203-S207

[73] Rask Larsen J, Dima L, Correll CU, Manu P. The pharmacological management of metabolic syndrome. Expert Review of Clinical Pharmacology. 2018;**11**(4):397-410. DOI: 10.1080/17512433.2018.1429910

[74] Cardinali DP, Cano P, Jiménez-Ortega V, Esquifino AI. Melatonin and the metabolic syndrome. Physiopathologic and therapeutical implications. Neuroendocrinology. 2011;**93**(3):133-142. Available from: http://bibliotecadigital.uca.edu.ar/repositorio/investigacion/melatonin-metabolic-syndrome-physiopathologic.pdf

[75] Kozirog M, Poliwczak AR, Duchnowicz P, Koter-Michalak M, Sikora J, Broncel M. Melatonin treatment improves blood pressure, lipid profile, and parameters of oxidative stress in patients with metabolic syndrome. Journal of Pineal Research. 2011;**50**(3):261-266. DOI: 10.1111/j.1600-079X.2010.00835.x

[76] Reiter RJ, Tan DX, Sainz RM, Mayo JC, Lopez-Burillo S. Melatonin: Reducing the toxicity and increasing the efficacy of drugs. The Journal of Pharmacy and Pharmacology. 2002;**54**:1299-1321

5

Dipeptidyl Peptidase-4 Inhibitor-Associated Bullous Pemphigoid

Ágnes Kinyó

Abstract

Bullous pemphigoid (BP) is the most common type of autoimmune bullous diseases; drug-induced bullous pemphigoid is a rare variant of it. In the last decade, there is an increasing prevalence of BP, especially dipeptidyl peptidase-4 inhibitor-associated BP (DPP-4i-BP), with the higher prevalence of BP in diabetic patients. Recently, several clinical phenotypes of BP were detected, but there is a tendency in BP cases related to DPP-4 inhibitors to show an atypical noninflammatory form with less distributed skin symptoms, mild erythema, decreased eosinophilic infil-tration in the periblister area, and normal or slightly elevated peripheral eosinophil count. Anti-NC16A BP180 autoantibodies are less frequently detected by ELISA, but they response to other epitopes of BP180. Clinical outcome is similar such as in classical non-DPP-4 BP patients, regardless of withdrawal of DPP-4 inhibitors.

Keywords: bullous pemphigoid, noninflammatory, dipeptidyl peptidase-4 inhibitor, gliptins, eosinophil

1. Introduction

Bullous pemphigoid (BP) is a rare autoimmune blistering disease of elderly patients, but in the last decades, it shows increasing incidence [1–8]. Higher prevalence of BP may be according to the increasing global life expectancy of the population, increasing incidence of predisposing neurological diseases, growing numbers of provoking drugs, and improving awareness of newly recognized atypical clinical phenotypes and better diagnostic methods [1]. The role of culprit drugs such as neuroleptics, diuretics, and antidiabetics is reported in several studies [1, 2, 9]. BP is typically present in elderly with a higher predominance in female patients [4, 7]. The classical clinical features of BP are generalized bullous skin eruptions with sur-rounding erythema and itching; peripheral eosinophilia is also common. Mucosal involvement was observed in 10–30% of patients [2]. Atypical clinical variants may be present up to 20% in BP, including the more common prurigo-like or urticarial type, eczema-like type, dyshidrosiform type, erosive type, and erythema annulare centrifugum-like phenotype [1, 7]. The diagnosis is based on the histological findings, including direct and indirect immunofluorescence microscopy and anti-BP180/BP230 enzyme-linked immunosorbent assays (ELISA) [2]. The gold standard for the treatment of the disease according to guidelines is corticosteroid, in topical or systemic administration and in severe cases with adjuvant immuno-suppressive medications, such as azathioprine, methotrexate, or mycophenolate mofetil [2]. In the case of drug-induced BP, the most important therapeutic step is the withdrawal of the culprit drug [9].

2. Bullous pemphigoid and diabetes

BP is a chronic, relapsing disease in patients with several comorbidities and significant morbidity. Investigating the prevalence of diabetes mellitus (DM) in BP, a higher frequency of DM has been found in the last decade [10, 11]. In accordance to several case reports [12–18], case series, case-control studies, pharmacovigilance reports, and retrospective investigations [19–37], the growing number of DM in BP patients is in association with the increasing use of an antidiabetic drug, the dipeptidyl peptidase-4 inhibitor (DPP-4i). DPP-4i, also called gliptins, was approved in 2006 to treat type 2 DM. Sitagliptin, vildagliptin, linagliptin, saxagliptin, and alogliptin have been approved by FDA or EMA; anagliptin, trelagliptin, omarigliptin, and teneligliptin have approval only in Japan. Gliptins are used in monotherapy or in combination with metformin. However, there is a clear evidence of provoking role of DPP-4 inhibitors in BP; the pathomechanism of it is still not understood.

3. Demographics of DPP-4 inhibitor-related BP

DPP-4 inhibitor intake was associated with a threefold increased risk for BP [27, 31, 38]. According to the former investigations [14, 23, 26, 27, 29, 31, 38, 39], vildagliptin has the strongest association with BP; the risk was tenfold (ranged between 7.23 and 11.8) in a systemic review and meta-analysis by Kridin et al. [38]. A higher, sixfold risk was also observed with linagliptin [27, 38]. Higher risk of DPP-4i-BP was also found with sitagliptin by Lee et al. [40], and they also found in a larger sample size (patients with DPP-4i-BP n = 260) that the risks associated with specific DPP-4 inhibitors were lower than the previous studies [27, 38], 1.81 for vildagliptin, 1.64 for linagliptin, and 1.7 for sitagliptin. However, a significant association was detected with vildagliptin, linagliptin, and sitagliptin in age- and sex-matched controlled population; the associa-tion with saxagliptin in BP was not significant [27, 31]. Saxagliptin, anagliptin- and alogliptin-induced BP were presented only in a few sporadic cases [29, 31, 32, 40, 41]. DPP-4 inhibitors are often given in combination with metformin; the two recent publications reported that the association of BP and gliptins is independent of met-formin exposure [27, 30]. Despite the BP is more common in females, a multicenter investigation and EudraVigilance data showed that DPP-4 inhibitor-associated BP tends to be more common in men [14, 23], similarly to Kridin et al. [27] and Lee et al. [40] and in contrast to Varpuluoma et al., who found a higher risk for BP in women taking DPP-4 inhibitors [30]. The mean age of DPP-4i-related BP patients ranged between 76.6 and 79.1 years [23, 27, 39]. Kridin et al. presented the strongest associa-tion in patients younger than 70 years [27], while Benzaquen et al. found stronger association in patients older than 80 years [23], while Lee et al. observed 1.76-fold risk in patients younger than 75 years and 1.5-fold risk in patients older than 75 years [40]. The mean latency period between the initiation of gliptin and the appearance of BP is ranging from 6 to 26.4 months [23, 27, 29, 30, 33]. That means DPP-4 inhibitors can be suspected in the pathogenesis of BP if the drug has been initiated at least 6 months, but it also has to be considered if the drug intake last for more than 2 years prior to the onset of BP.

4. Clinical features of DPP-4 inhibitors-related noninflammatory BP

The classical clinical picture of BP is generalized as bullous skin lesions, tense blisters with severe urticarial erythema. Recent publications characterized a noninflammatory form of BP with limited distribution, smaller blisters, and scant erythema

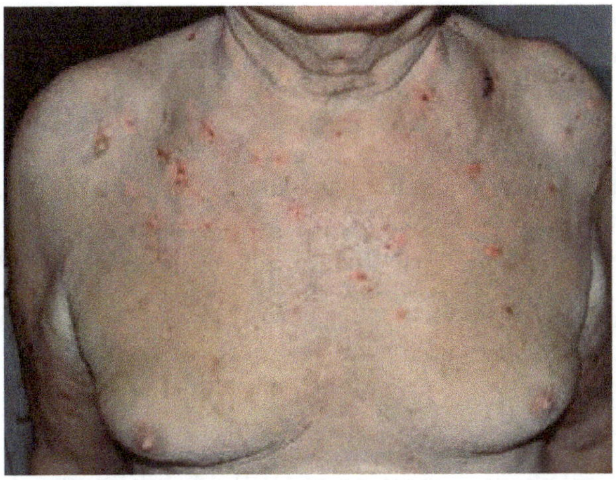

Figure 1.
DPP-4i-related bullous pemphigoid. Mild skin involvement localized to the upper part of the trunk with small blisters and erosions without erythema in a male patient.

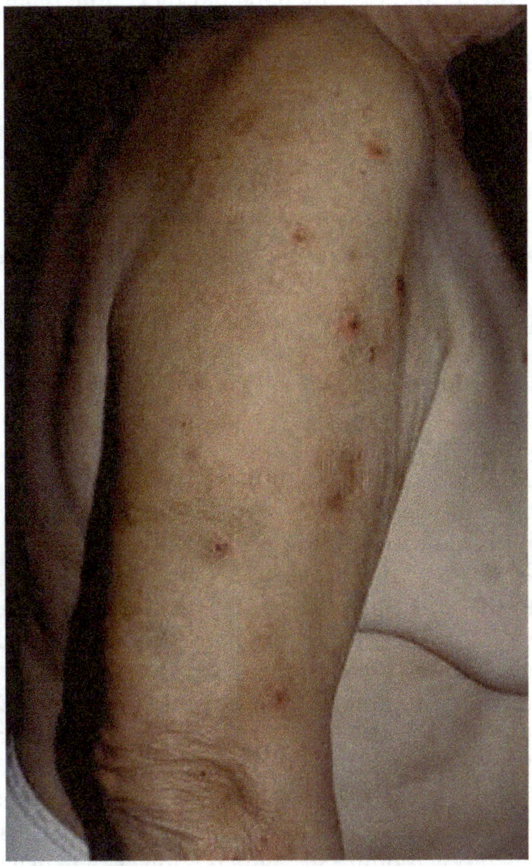

Figure 2.
DPP-4i-related bullous pemphigoid in a female patient with small, round erosions without erythema on the upper extremities.

(**Figures 1** and **2**) [37, 42, 43]. These noninflammatory phenotypes do not react with the NC16A domain of BP, show better clinical outcome, and has a higher prevalence in DPP-4 inhibitor taking patients [22, 32, 34–37, 42, 43]. Noninflammatory BP can also be found in non-DPP-4-related cases but in a significantly lower manner. Higher frequency of mucosal involvement was reported in gliptin-associated BP in two studies (Kridin et al., 22.2%, n = 36, and Chijiwa et al., 78%, n = 9) [27, 33], but this observation was not supported by Plaquevent (n = 108) [31]. Interestingly, in gliptin-associated mucous membrane pemphigoid (MMP) there is a significantly lower buc-cal and more common cutaneous involvement [44]. Previous studies demonstrated that eosinophil count is in correlation with the severity of BP [45–47] and with BPDAI score [46]. Comparing the Bullous Pemphigoid Disease Area Index (BPDAI) [48], BPDAI scores for urticaria/erythema (U/E) were significantly lower in noninflam-matory phenotypes [33, 36, 42, 43], and lower BPDAI U/E was in correlation with decreased eosinophil count in the perilesional skin [33, 42]. Significantly decreased peripheral eosinophil count was detected in patients of Kridin et al. [27].

5. Immunological characterization of DPP-4 inhibitor-related BP

In BP, there are two targets for autoantibodies, the hemidesmosomal BP180 and BP230 and the juxtamembranous extracellular non-collagenous 16A (NC16A), both of them can be easily detected by commercially available ELISA tests [2, 49, 50]. The domain of BP180 (also called COL17) is a major target epitope in 80–90% of cases [49]. In several investigations, noninflammatory BP patients did not show reactivity against the NC16A domain of BP180, but they were positive for full-length BP180 and its ectodomain midportion with ELISA [34, 36, 42]. The midportion of BP180 is more likely to be recognized than the NC16A domain in DPP-4i-associated noninflammatory BP patients; they are presented with localized symptoms and mild erythema [32, 35, 36, 42, 51]. Although there was a positive reaction to anti-NC16A in DPP-4i-BP cases, but they were mainly presented with prominent erythema and inflammation, concurring with classical phenotype of BP [31, 35, 42]. Kawaguchi et al. showed that the rate of ELISA positivity and antibody titers for anti-BP180 NC16A was significantly lower in DPP-4i-BP than the non-DPP-4 group [22], and this lower titer was also observed by García-Díez et al. [52]. Kawaguchi has also emphasized that patients with DPP-4i-BP tended to have noninflammatory phenotype of BP and presented with negative ELISA for BP180 NC16A domain [22]. Some studies reported noninflammatory DPP-4-induced BP patients who were negative for anti-NC16A domain initially but responded to the full-length BP180 and became positive for NC16A during the course of the disease. This epitope spreading was observed in several cases, after the prolonged use of DPP-4 inhibitors after the onset of BP [17, 37, 52]. Other investigations also demonstrated that multiple epitopes of BP180 are targeted in DPP-4i-BP (midportion, C terminus, and LAD1) [35, 52], and it may suggest that epitope spreading is more common in DPP-4i-BP than in classical BP cases [35]. García-Díez et al. suggested the major role of the midportion of BP180 in DPP-4i-BP, while other BP180 regions are involved later by epitope spreading [52]. Indirect IF positivity was also detected in DPP-4i-BP patients [13, 27, 34], and anti-BP230 autoantibodies were present [34, 35], but the sensitivity of the test was only 38%, which is lower than usually reported [51].

6. Clinical outcome

Based on several investigations, withdrawal of the DPP-4 inhibitors was the first therapeutic step in most cases in the treatment of the disease [23, 34]. Regarding to these data, discontinuation of DPP-4i treatment seems to have a favorable impact on the clinical outcome of BP [23, 27, 34], but Plaquevent et al. have found that there is no difference in outcome in patients who have further got the gliptin treatment after the diagnosis of BP [31]. Regardless of DPP-4 inhibitor withdrawal, standard treatment protocol was applied in most cases, topical potent corticosteroid treatment in localized form and gold standard systemic corticosteroid treatment with adjuvant immunosuppressive therapy, such as azathioprine, mycophenolate mofetil, or methotrexate in severe cases [27, 31]. Relapse rates were similar in DPP- 4i-BP patients such it was previously reported [31].

7. Conclusions

Dipeptidyl peptidase-4 inhibitors (also called gliptins) are widely used drug in the treatment of type 2 diabetes mellitus. There is an increased risk of BP in patients during DPP-4 inhibitor treatment. The exact pathomechanism of DPP-4i-associated BP is still unclear. Dipeptidyl peptidase 4 (also called CD26) is a 110 kDa transmembrane glycoprotein, which is expressed on the surface of several cells, such as T cells [53, 54]. DPP-4 has antihyperglycemic effect and enzymatic activity; it plays a major role in glucose metabolism by blocking incretin [54]. DPP-4 is a plasminogen receptor that activates plasminogen resulted in plasmin formation [55, 56]. Plasmin, a serine protease, which has a high level in lesional skin and in blister fluid in BP [57], cleaves BP180 within the NC16A domain [58]. Cleavage of BP180 in the NC16A region can induce neoepitopes with altered antigenicities [42, 59]. The antifibrotic effect of DPP-4 inhibitors in the skin also supports the role of DPP-4 in collagen metabolism [56]. DPP-4 is involved in immune cell activation, and its inhibition can modify the immune response, which may increase the activation of eosinophil recruitment into the dermis, which is considered to be essential in blister formation in BP [60]. In contrary to these findings, in patients with gliptin-associated noninflammatory BP, both peripheral and perilesional skin eosinophil counts are significantly lower than in classical BP [27, 33], so the exact pathognostic role of eosinophils in DPP-4-related BP needs further investigations. It is also not elucidated why vildagliptin has the strongest association with BP, but it is known that vildagliptin has the lowest selectivity among gliptins with strong inhibition of DDP8 and DPP9 isozymes [22, 61]. Some re s ults suggest that DPP-4 inhibitor has immunomodifier effect mainly in

	DPP4i-related BP	Classical type BP
Distribution	Limited	Generalized
Bullous eruption	Smaller blisters	Tense bullas
Erythema	Absent or sparse	Pronounced
Perilesional eosinophilic infiltration	Absent or mild	Pronounced
Serum eosinophil count	Normal or slightly elevated	Elevated

Table 1.
Comparison of clinical features in DPP-4i-related and classical type bullous pemphigoid.

genetically susceptible individuals, and they have detected that HLA-DQB1*03:01 allele has higher prevalence in DPP-4i-BP patients [22, 62].

In conclusion, DPP-4 inhibitor-related BP tends to be presented with noninflammatory phenotype of BP, with limited extension, smaller blisters, scant perilesional erythema and eosinophilic infiltration, and normal or slightly elevated peripheral eosinophil count (**Table 1**). Anti-NC16A BP180 positivity is less common, and ELISA titers are slightly elevated, similarly to anti-BP230, but positivity for full-length BP180 or other epitopes of BP180 may be detected. Response to therapy is similar such as in classical non-DPP-4i-BP patients, regardless of withdrawal of DPP-4 inhibitors.

Acronyms and abbreviations

BP	bullous pemphigoid
BPDAI	bullous pemphigoid disease area index
COL17	collagen XVII
DM	diabetes mellitus
DPP-4	dipeptidyl peptidase-4
DPP-4i-BP	dipeptidyl peptidase-4 inhibitor-associated bullous pemphigoid
ELISA	enzyme-linked immunosorbent assays
EMA	European Medicines Agency
FDA	Food and Drug Administration
NC16A	non-collagenous 16A
MMP	mucous membrane pemphigoid
BPDAI	BP disease area index

Author details

Ágnes Kinyó
Department of Dermatology, Venereology and Oncodermatology, University of Pécs, Pécs, Hungary

*Address all correspondence to: kinyo.agnes@pte.hu

References

[1] Kridin K, Ludwig RJ. The growing incidence of bullous Pemphigoid: Overview and potential explanations. Frontiers in Medicine (Lausanne). 2018;**5**:220. DOI: 10.3389/fmed.2018.00220

[2] Bernard P, Antonicelli F. Bullous pemphigoid: A review of its diagnosis, associations and treatment. American Journal of Clinical Dermatology. 2017;**18**:513-528. DOI: 10.1007/s40257-017-0264-2

[3] Schmidt E, Spindler V, Eming R, Amagai M, Antonicelli F, Baines JF, et al. Meeting report of the pathogenesis of pemphigus and Pemphigoid meeting in Munich, September 2016. The Journal of Investigative Dermatology. 2017;**137**:1199-1203. DOI: 10.1016/j.jid.2017.01.028

[4] Langan SM, Smeeth L, Hubbard R, Fleming KM, Smith CJP, West J. Bullous pemphigoid and pemphigus vulgaris—Incidence and mortality in the UK: Population based cohort study. BMJ. 2008;**337**:160-163. DOI: 10.1136/bmj.a180

[5] Hübner F, Recke A, Zillikens D, Linder R, Schmidt E. Prevalence and age distribution of pemphigus and pemphigoid diseases in Germany. The Journal of Investigative Dermatology. 2016;**136**:2495-2498. DOI: 10.1016/j.jid.2016.07.013

[6] Bernard P, Vaillant L, Labeille B, Bedane C, Arbeille B, Denoeux JP, et al. Incidence and distribution of subepidermal autoimmune bullous skin diseases in three French regions. Bullous diseases French study group. Archives of Dermatology. 1995;**131**:48-52

[7] Joly P, Baricault S, Sparsa A, Bernard P, Bédane C, Duvert-Lehembre S, et al. Incidence and mortality of bullous pemphigoid in France. The Journal of Investigative Dermatology. 2012;**132**:1998-2004. DOI: 10.1038/jid.2012.35

[8] Försti AK, Jokelainen J, Timonen M, Tasanen K. Increasing incidence of bullous pemphigoid in Northern Finland: A retrospective database study in Oulu University Hospital. The British Journal of Dermatology. 2014;**171**:1223-1226. DOI: 10.1111/bjd.13189

[9] Stavropoulos PG, Soura E, Antoniou C. Drug-induced pemphigoid: A review of the literature. Journal of the European Academy of Dermatology and Venereology. 2014;**28**:1133-1140. DOI: 10.1111/jdv.12366

[10] Fania L, Di Zenzo G, Didona B, Pilla MA, Sobrino L, Panebianco A, et al. Increased prevalence of diabetes mellitus in bullous pemphigoid patients during the last decade. Journal of the European Academy of Dermatology and Venereology. 2018;**32**:e153-e154. DOI: 10.1111/jdv.14649

[11] Gravani A, Gaitanis G, Tsironi T, Tigas S, Bassukas ID. Changing prevalence of diabetes mellitus in bullous pemphigoid: It is the dipeptidyl peptidase-4 inhibitors. Journal of the European Academy of Dermatology and Venereology. 2018;**32**:e438-e439. DOI: 10.1111/jdv.14957

[12] Aouidad I, Fite C, Marinho E, Deschamps L, Crickx B, Descamps V. A case report of bullous pemphigoid induced by dipeptidyl peptidase-4 inhibitors. JAMA Dermatology. 2013;**149**:243-245. DOI: 10.1001/jamadermatol.2013.1073

[13] Attaway A, Mersfelder TL, Vaishnav S, Baker JK. Bullous pemphigoid associated with dipeptidyl peptidase IV inhibitors. A case report and review of literature. Journal of Dermatological Case

Reports. 2014;**8**:24-28. DOI: 10.3315/jdcr.2014.1166

[14] García M, Aranburu MA, Palacios-Zabalza I, Lertxundi U, Aguirre C. Dipeptidyl peptidase-IV inhibitors induced bullous pemphigoid: A case report and analysis of cases reported in the European pharmacovigilance database. Journal of Clinical Pharmacy and Therapeutics. 2016;**41**:368-370. DOI: 10.1111/jcpt.12397

[15] Harada M, Yoneda A, Haruyama S, Yabuki K, Honma Y, Hiura M, et al. Bullous Pemphigoid associated with the dipeptidyl peptidase-4 inhibitor sitagliptin in a patient with liver cirrhosis complicated with rapidly progressive hepatocellular carcinoma. Internal Medicine. 2017;**56**:2471-2474. DOI: 10.2169/internalmedicine.8703-16

[16] Maki N, Nishie W, Takazawa M, Kakurai M, Yamada T, Umemoto N, et al. Dipeptidyl peptidase-4 inhibitor-associated bullous pemphigoid in a patient with acquired reactive perforating collagenosis. The Journal of Dermatology. 2018;**45**:600-602. DOI: 10.1111/1346-8138.14254

[17] Takama H, Yoshida M, Izumi K, Yanagishita T, Muto J, Ohshima Y, et al. Dipeptidyl peptidase-4 inhibitor-associated bullous pemphigoid: Recurrence with epitope spreading. Acta Dermato-Venereologica. 2018;**98**: 983-984. DOI: 10.2340/00015555-3010

[18] Oya K, Fujii M, Taguchi S, Nishie W, Izumi K, Shimizu H. Localized bullous pemphigoid associated with dipeptidyl peptidase-4 inhibitor treatment. European Journal of Dermatology. 2018;**28**:250-251. DOI: 10.1684/ejd.2018.3230

[19] Pasmatzi E, Monastirli A, Habeos J, Georgiou S, Tsambaos D. Dipeptidyl peptidase-4 inhibitors cause bullous pemphigoid in diabetic patients: Report of two cases. Diabetes Care. 2011;**34**:e133. DOI: 10.2337/dc11-0804

[20] Mendonça FM, Martín-Gutierrez FJ, Ríos-Martín JJ, Camacho-Martinez F. Three cases of bullous pemphigoid associated with dipeptidyl peptidase-4 inhibitors—One due to Linagliptin. Dermatology. 2016;**232**:249-253. DOI: 10.1159/000443330

[21] Haber R, Fayad AM, Stephan F, Obeid G, Tomb R. Bullous pemphigoid associated with linagliptin treatment. JAMA Dermatology. 2016;**152**:224-226. DOI: 10.1001/jamadermatol.2015.2939

[22] Kawaguchi Y, Shimauchi R, Nishibori N, Kawashima K, Oshitani S, Fujiya A, et al. Dipeptidyl peptidase-4 inhibitors-associated bullous pemphigoid: A retrospective study of 168 pemphigoid and 9,304 diabetes mellitus patients. Journal of Diabetes Investigation. 2018. DOI: 10.1111/jdi.12877. [Epub ahead of print]

[23] Benzaquen M, Borradori L, Berbis P, Cazzaniga S, Valero R, Richard MA, et al. Dipeptidyl peptidase IV inhibitors, a risk factor for bullous pemphigoid: Retrospective multicenter case-control study from France and Switzerland. Journal of the American Academy of Dermatology. 2018;**78**:1090-1096. DOI: 10.1016/j.jaad.2017.12.038

[24] Béné J, Jacobsoone A, Coupe P, Auffret M, Babai S, Hillaire-Buys D, et al. Bullous pemphigoid induced by vildagliptin: A report of three cases. Fundamental & Clinical Pharmacology. 2015;**29**:112-114. DOI: 10.1111/bjd.14601

[25] Skandalis K, Spirova M, Gaitanis G, Tsartsarakis A, Bassukas ID. Drug-induced bullous pemphigoid in diabetes mellitus patients receiving dipeptidyl peptidase-IV inhibitors plus metformin. Journal of the European Academy of Dermatology and Venereology. 2012;**26**:249-253. DOI: 10.1111/j.1468-3083.2011.04062.x

[26] Schaffer C, Buclin T, Jornayvaz FR, Cazzaniga S, Borradori L, Gilliet M,

et al. Use of Dipeptidyl-peptidase IV inhibitors and bullous pemphigoid. Dermatology. 2017;**233**:401-403. DOI: 10.1159/000480498

[27] Kridin K, Bergman R. Association of bullous pemphigoid with dipeptidyl-peptidase 4 inhibitors in patients with diabetes: Estimating the risk of the new agents and characterizing the patients. JAMA Dermatology. 2018;**154**:1152-1158. DOI: 10.1001/jamadermatol.2018.2352

[28] Arai M, Shirakawa J, Konishi H, Sagawa N, Terauchi Y. Bullous pemphigoid and dipeptidyl peptidase 4 inhibitors: A disproportionality analysis based on the Japanese adverse drug event report database. Diabetes Care. 2018;**41**:e130-e132. DOI: 10.2337/dc18-0210

[29] Béné J, Moulis G, Bennani I, Auffret M, Coupe P, Babai S, et al. French Association of Regional PharmacoVigilance Centres: Bullous pemphigoid and dipeptidyl peptidase IV inhibitors: A case-noncase study in the French Pharmacovigilance database. The British Journal of Dermatology. 2016;**175**:296-301. DOI: 10.1111/fcp.12083

[30] Varpuluoma O, Försti AK, Jokelainen J, Turpeinen M, Timonen M, Tasanen K, et al. Oral diabetes medications other than dipeptidyl peptidase 4 inhibitors are not associated with bullous pemphigoid: A Finnish nationwide case-control study. Journal of the American Academy of Dermatology. 2018;**79**:1034-1038. DOI: 10.1016/j.jaad.2018.05.030

[31] Plaquevent M, Tétart F, Fardet L, et al. Higher frequency of dipeptidyl peptidase IV inhibitor intake in bullous pemphigoid patients than in the French general population. The Journal of Investigative Dermatology. 2018;**18**:32923-3. DOI: 10.1016/j.jid.2018

[32] Yoshiji S, Murakami T, Harashima SI, Ko R, Kashima R, Yabe D, et al. Bullous pemphigoid associated with dipeptidyl peptidase-4 inhibitors: A report of five cases. Journal of Diabetes Investigation. 2018;**9**:445-447. DOI: 10.1111/jdi.12695

[33] Chijiwa C, Takeoka S, Kamata M, Tateishi M, Fukaya S, Hayashi K, et al. Decrease in eosinophils infiltrating into the skin of patients with dipeptidyl peptidase-4 inhibitor-related bullous pemphigoid. The Journal of Dermatology. 2018;**45**:596-599. DOI: 10.1111/1346-8138.14245

[34] García-Díez I, Ivars-Lleó M, López-Aventín D, Ishii N, Hashimoto T, Iranzo P, et al. Bullous pemphigoid induced by dipeptidyl peptidase-4 inhibitors. Eight cases with clinical and immunological characterization. International Journal of Dermatology. 2018;**57**:810-816. DOI: 10.1111/ijd.14005

[35] Fania L, Salemme A, Provini A, Pagnanelli G, Collina MC, Abeni D, et al. Detection and characterization of IgG, IgE, and IgA autoantibodies in patients with bullous pemphigoid associated with dipeptidyl peptidase-4 inhibitors. Journal of the American Academy of Dermatology. 2018;**78**: 592-595. DOI: 10.1016/j.jaad.2017.09.051

[36] Horikawa H, Kurihara Y, Funakoshi T, Umegaki-Arao N, Takahashi H, Kubo A, et al. Unique clinical and serological features of bullous pemphigoid associated with dipeptidyl peptidase-4 inhibitors. The British Journal of Dermatology. 2018;**178**:1462-1463. DOI: 10.1111/bjd.16479

[37] Mai Y, Nishie W, Izumi K, Yoshimoto N, Morita Y, Watanabe M, et al. Detection of anti-BP180 NC16A autoantibodies after the onset of dipeptidyl peptidase-IV inhibitor-associated bullous pemphigoid: A report of three patients. The British Journal of

Dermatology. 2018;**179**:790-791. DOI: 10.1111/bjd.16656

[38] Kridin K, Cohen AD. Dipeptidyl-peptidase IV inhibitor-associated bullous pemphigoid: A systematic review and meta-analysis. J Am Acad Dermatol. 2018;**18**:32660-32664. DOI: 10.1016/j.jaad.2018.09.048

[39] Varpuluoma O, Försti AK, Jokelainen J, Turpeinen M, Timonen M, Huilaja L, et al. Vildagliptin significantly increases the risk of bullous pemphigoid: A Finnish Nationwide registry study. The Journal of Investigative Dermatology. 2018;**138**:1659-1661. DOI: 10.1016/j.jid.2018.01.027

[40] Lee SG, Lee HJ, Yoon MS, Kim DH. Association of dipeptidyl peptidase 4 inhibitor use with risk of bullous pemphigoid in patients with diabetes. JAMA Dermatology. 2019;**155**:172-177. DOI: 10.1001/jamadermatol.2018.4556

[41] Oya K, Fujii M, Taguchi S, Nishie W, Izumi K, Shimizu H. Bullous pemphigoid following the replacement of vildagliptin with anagliptin. The Journal of Dermatology. 2017;**44**:e238-e239. DOI: 10.1111/1346-8138.13941

[42] Izumi K, Nishie W, Mai Y, Wada M, Natsuga K, Ujiie H, et al. Autoantibody profile differentiates between inflammatory and noninflammatory bullous Pemphigoid. The Journal of Investigative Dermatology. 2016;**136**:2201-2210. DOI: 10.1016/j.jid.2016.06.622

[43] Nakama K, Koga H, Ishii N, Ohata C, Hashimoto T, Nakama T. Clinical and immunological profiles of 14 patients with bullous Pemphigoid without IgG autoantibodies to the BP180 NC16A domain. JAMA Dermatology. 2018;**154**:347-350. DOI: 10.1001/jamadermatol.2017.5465

[44] Gaudin O, Seta V, Alexandre M, Bohelay G, Aucouturier F, Mignot-Grootenboer S, et al. Gliptin accountability in mucous membrane Pemphigoid induction in 24 out of 313 patients. Frontiers in Immunology. 2018;**9**:1030. DOI: 10.3389/fimmu.2018.01030

[45] Bernard P, Venot J, Constant F, Bonnetblanc JM. Blood eosinophilia as a severity marker for bullous pemphigoid. Journal of the American Academy of Dermatology. 1987;**16**:879-881

[46] Messingham KN, Holahan HM, Frydman AS, Fullenkamp C, Srikantha R, Fairley JA. Human eosinophils express the high affinity IgE receptor, FcεRI, in bullous pemphigoid. PLoS One. 2014;**9**:e107725. DOI: 10.1371/journal. pone.0107725

[47] Kridin K. Peripheral eosinophilia in bullous pemphigoid: Prevalence and influence on the clinical manifestation. The British Journal of Dermatology. 2018;**179**:1141-1147. DOI: 10.1111/bjd.16679

[48] Murrell DF, Daniel BS, Joly P, Borradori L, Amagai M, Hashimoto T, et al. Definitions and outcome measures for bullous pemphigoid: Recommendations by an international panel of experts. Journal of the American Academy of Dermatology 2012:**66**:479-485. DOI: 10.1016/j.jaad.2011.06.032

[49] Schmidt E, Zillikens D. Pemphigoid diseases. Lancet. 2013;**381**(9863):320-332. DOI: 10.1016/S0140-6736(12)61140-4

[50] Sárdy M, Kostaki D, Varga R, Peris K, Ruzicka T. Comparative study of direct and indirect immunofluorescence and of bullous pemphigoid 180 and 230 enzyme-linked immunosorbent assays for diagnosis of bullous pemphigoid. Journal of the American Academy of Dermatology. 2013;**69**:748-753. DOI: 10.1016/j.jaad.2013.07.009

[51] Sakai A, Shimomura Y, Ansai O, et al. Linagliptin-associated bullous pemphigoid that was most likely caused by IgG autoantibodies against the midportion of BP180. The British Journal of Dermatology. 2017;**176**: 541-543. DOI: 10.1111/bjd.15111

[52] García-Díez I, España A, Iranzo P. Epitope spreading phenomena in dipeptidyl peptidase-4 inhibitor-associated bullous pemphigoid. The British Journal of Dermatology. 2019. DOI: 10.1111/bjd.17690. [Epub ahead of print]

[53] Klemann C, Wagner L, Stephan M, von Hörsten S. Cut to the chase: A review of CD26/dipeptidyl peptidase-4's (DPP4) entanglement in the immune system. Clinical and Experimental Immunology. 2016;**185**:1-21. DOI: 10.1111/cei.12781

[54] Röhrborn D, Wronkowitz N, Eckel J. DPP4 in Diabetes. Frontiers in Immunology. 2015;**6**:386. DOI: 10.3389/fimmu.2015.00386

[55] Gonzalez-Gronow M, Kaczowka S, Gawdi G, Pizzo SV. Dipeptidyl peptidase IV (DPP IV/CD26) is a cell-surface plasminogen receptor. Frontiers in Bioscience. 2008;**13**:1610-1618

[56] Thielitz A, Ansorge S, Bank U, Tager M, Wrenger S, Gollnick H, et al. The ectopeptidases dipeptidyl peptidase IV (DP IV) and aminopeptidase N (APN) and their related enzymes as possible targets in the treatment of skin diseases. Frontiers in Bioscience. 2008;**13**:2364-2375

[57] Schmidt E, Wehr B, Tabengwa EM, Reimer S, Bröcker EB, Zillikens D. Elevated expression and release of tissue-type, but not urokinase-type, plasminogen activator after binding of autoantibodies to bullous pemphigoid antigen 180 in cultured human keratinocytes. Clinical and Experimental Immunology. 2004;**135**:497-504

[58] Hofmann SC, Voith U, Schönau V, et al. Plasmin plays a role in the In vitro. Generation of the linear IgA Dermatosis antigen LADB97. Journal of Investigative Dermatology. 2009;**129**:1730-1739. DOI: 10.1038/jid.2008.424

[59] Nishie W, Lamer S, Schlosser A, Licarete E, Franzke CW, Hofmann SC, et al. Ectodomain shedding generates Neoepitopes on collagen XVII, the major autoantigen for bullous pemphigoid. Journal of Immunology. 2010;**185**:4938-4947. DOI: 10.4049/jimmunol.1001524

[60] Forssmann U, Stoetzer C, Stephan M, Kruschinski C, Skripuletz T, Schade J, et al. Inhibition of CD26/dipeptidyl peptidase IV enhances CCL11/eotaxin-mediated recruitment of eosinophils in vivo. Journal of Immunology. 2008;**181**:1120-1127

[61] Filippatos TD, Athyros VG, Elisaf MS. The pharmacokinetic considerations and adverse effects of DPP-4 inhibitors [corrected]. Expert Opinion on Drug Metabolism & Toxicology. 2014;**10**:787-812. DOI: 10.1517/17425255.2014.907274

[62] Ujiie H, Muramatsu K, Mushiroda T, Ozeki T, Miyoshi H, Iwata H, et al. HLA-DQB1*03:01 as a biomarker for genetic susceptibility to bullous pemphigoid induced by DPP-4 inhibitors. The Journal of Investigative Dermatology. 2018;**138**:1201-1204. DOI: 10.1016/j. jid.2017.11.023

6

Lipid Disorders in Uremia

Valdete Topçiu-Shufta and Valdete Haxhibeqiri

Abstract

Lipoprotein has important physiologic functions within the human body. Many enzymes, enzyme activators, and protein parts, such as apolipoproteins and specific hepatic and extrahepatic receptors, are involved in their metabolism. Renal failure is associated with an increased risk of cardiovascular disease. One of the main mechanisms underlying this increased cardiovascular risk is dyslipid-emia. Abnormal lipoprotein profiles are generally a combination of abnormalities of all fractions. Uremic lipoprotein profile includes increased triglyceride-rich lipoproteins, small dense LDL particles, increased lipoprotein (a), and decreased HDL. Enhanced oxidative stress and uremic environment can strongly modify plasma lipoproteins, changing their interactions with biological functions and especially cardiovascular physiology. This profound lipoprotein disorder has led to the formulation of an accelerated atherogenesis hypothesis and has been commonly linked with their metabolic alteration associated with uremia.

Keywords: chronic uremia, lipoproteins, dyslipidemia, cardiovascular risk

1. Introduction

Urea cycle is one of the most important pathways in the human body. The continuous degradation and synthesis of cellular proteins occur in all forms of life. High rates of protein degradation occur in tissue undergoing structural rearrangements.

Approximately 75% of liberated amino acids are reutilized. Since the excess amino acids are not stored, those not immediately incorporated into new proteins are degraded rapidly. The excess nitrogen from amino acids forms urea. As a hydrosoluble compound, urea is excreted by the kidney. Uremia is a clinical syndrome marked by elevated concentrations of urea in the blood and is associated with many metabolic disorders such as acidosis, abnormalities in lipids, mineral and homocysteine metabolism, oxidative stress, chronic inflammation, insulin and erythropoietin resistance, vitamin D deficiency, and malnutrition. Uremia more commonly develops with chronic kidney disease (CKD), but it also may occur with acute kidney injury if loss of renal function is rapid. Nearly all body organs and systems are affected by the toxicity of uremic compounds retained in the course of renal dysfunction. According to the European Uremic Toxin Work Group, uremic toxins are defined as accumulated solutes, normally excreted by the kidneys, that interact negatively with biological functions [1]. This has shown the need for the search for new uremic compounds, combining them into panels of substances involved in the pathophysiological processes. As example we can mention uridine adenosine, a strong vasoconstrictor, which is considered as a new uremic toxin. It has been demonstrated that uremic patients have increased levels of uridine adenos-ine, which can influence blood pressure, proliferation rate of vascular smooth muscle cells, and vascular calcification [2]. All these effects correlated with vascular dysfunctions and development of atherosclerosis. As uremic toxins are considered some components, which concentrations are not directly associated with glomerular filtration, but interacts negatively with vascular physiology. Several acute-phase proteins, IL-1, IL-6, IL-12, α2-macroglobulin, fibrinogen, and myeloperoxidase, together with endothelium-related proteins, such as vascular cell adhesion molecule 1, vascular endothelial growth factor 1, and soluble vascular endothelial growth factor receptor, increased in CKD and play a crucial role in endothelium dysfunction promoting the development of atherosclerosis. Renal failure is associated with an increased risk of cardiovascular disease [3, 4]. One of the main mechanisms underlying this increased cardiovascular risk is dyslipidemia [2]. In uremic environment lipids are affected by oxidative stress. The end products of lipid peroxidation process affect the circulating lipoproteins, lipidic and proteinic, leading to profound alterations of their biological properties, changing their interactions with biological functions and especially cardiovascular physiology. For this reason, lipoproteins, in renal failure, can be also considered as uremic toxins.

In the human body, dietary lipids absorbed from intestine and lipids synthesized by the liver and adipose tissue must be transported between the various tissues and organs for utilization and storage. Since lipids are insoluble in water, the problem on how to transport them in aqueous blood plasma is solved by associating nonpolar lipids (triacylglycerol and cholesterol esters) with amphipathic lipids (phospho-lipids and cholesterol) and proteins, to form water-soluble particle known as lipoproteins.

The plasma lipoproteins are classified as chylomicrons and very-low-density (VLDL), intermediate-density (IDL), low-density (LDL), and high-density (HDL) lipoproteins, according to their ultracentrifugation characteristics. Chylomicrons and VLDL serve as vehicles to transport triglycerides to the sites of consumption, as myocytes and suprarenal glands or storage in adipocytes. HDL fraction serves as a vehicle to transport surplus cholesterol from peripheral tissues to the liver for disposal. Many enzymes, enzyme activators, and protein parts, such as apolipoproteins and specific hepatic and extrahepatic receptors, are involved in lipoprotein metabolism.

Apolipoproteins (Apo), the protein part of lipoproteins, are present in each lipoprotein and carry out several roles. They can be part of the structure of lipoproteins, serve as an enzyme cofactors or inhibitors, and finally act as ligands for interaction with lipoprotein receptor in tissue. Apolipoproteins of HDL are designated as A (A-I, A-II, A-IV). Apo A-I is an activator of enzyme lecithin-cholesterol acyltransferase (LCAT) and serves as a ligand for HDL binding to specific scavenger receptor B1 (SR-B1). Apo A-II is an inhibitor of enzyme lipoprotein lipase. The main apolipoprotein of LDL and VLDL is Apo B-100, while the chylomicrons contain Apo B-48. Apo B-100 acts as ligand of LDL for LDL receptors in the liver and extrahepatic tissue. Apo B-48 is part of the structure of chylomicrons. Apo E is found in chylomicrons, VLDL, and HDL, and its role is to uptake the remnant of chylomicrons by a receptor specific for apolipoprotein E, in the liver. Apo C-I, Apo C-II, and Apo C-III are transferable between several different lipoproteins. Apo C-II is activator, whereas Apo C-III is an inhibitor for enzyme lipoprotein lipase. The Apo C-I is an inhibitor for enzyme cholesteryl ester transfer protein (CETP).

The main enzymes involved in lipoprotein metabolism are lipoprotein lipase (EC 3.1.1.34), hepatic lipase (EC 3.1.1.3), lecithin-cholesterol acyltransferase (EC 2.3.1.43), and acyl-CoA cholesterol acyltransferase (ACAT) (EC 2.3.1.26).

Lipoprotein lipase is located on the walls of blood capillaries of the heart, adipose tissue, spleen, lung, renal medulla, aorta, lactating mammary gland, and diaphragm. It is abundantly produced as an inactive enzyme by myocytes, adipocytes, and several other cell types. The inactive enzyme requires sequential glycation and cleavage of a 27-amino acid peptide to become functionally active. The role of lipoprotein lipase is the hydrolysis of triglyceride-rich lipoproteins, as chylomicrons and VLDL. Apo C-II and phospholipids are cofactors for enzyme activity, while Apo A-II and Apo C-III act as inhibitors.

Hepatic lipase is bound to the surface of hepatic cells. Hepatic lipase catalyzes hydrolysis and removal of the triglyceride content of HDL and chylomicron remnant. Accordingly, hepatic lipase plays a central role in the metabolism of chylomicron remnants and HDL.

LCAT is the enzyme of HDL, which is activated by Apo A-I, the structural protein of HDL. The enzyme plays an important role in HDL-mediated cholesterol uptake from the extrahepatic tissues and, as such, serves as a main determinant of HDL maturation and plasma HDL cholesterol level.

The formation of cholesteryl esters from cholesterol and long-chain fatty-acyl-coenzyme A catalyzes the enzyme called ACAT. It is a membrane-bound protein and, at the single-cell level, serves as a regulator of intracellular cholesterol homeostasis. In addition, ACAT supplies cholesteryl esters for lipoprotein assembly in the liver and small intestine.

Cholesteryl ester transfer protein is a hydrophobic glycoprotein that is secreted mainly from the liver and circulates in the plasma, bounded mainly to HDL [5]. It mediates cholesterol ester transfer from HDL to IDL in exchange for triglycerides. CETP promotes the transfer of cholesteryl esters from anti-atherogenic HDLs to pro-atherogenic Apo B-containing lipoproteins, including VLDL, VLDL remnants, IDL, and LDL. In this way CETP transfers lipids from one lipoprotein particle to another in a process that results in equilibration of lipids between lipoprotein fractions.

2. Lipoprotein metabolism

2.1 The metabolism of chylomicrons

The exogenous pathway of lipid metabolism begins with chylomicrons. Chylomicrons are responsible for the transport of all dietary lipids into the circulation. They are produced within the enterocytes containing triglycerides, cholesterol ester, and phospholipids. Apo B-48 is essential for chylomicron formation. The nascent chylomicrons, from the small intestine, are released into the circulation via the lymphatic system. In the circulation, the nascent chylomicrons acquire Apo E and Apo C-II, which are in the surface of HDL. Apo C-II is an activator for enzyme lipoprotein lipase. The endothelium binding accommodates interaction of chylo-microns with the endothelium-bound lipoprotein lipase. Reaction with lipoprotein lipase results in the loss of approximately 90% of triglycerides in chylomicrons. The majority of fatty acids released diffuse into the adjacent myocytes for energy production or into adipocytes for energy storage. After hydrolysis chylomicron remnants are subsequently cleared by the liver and other tissues. Uptake is mediated by a receptor specific for Apo E. Both the LDL (Apo B-100 and Apo E) receptor and LDL receptor-related protein (LRP), specific for Apo E, are believed to take part. Chylomicron remnants return the borrowed Apo C- II and Apo E to HDL before their uptake by the liver and other tissues (**Figure 1**).

2.2 The metabolism of VLDL, IDL and LDL

VLDL particles are produced by the liver and are precursor of IDL and LDL. VLDL serves as the vehicle for delivery of endogenous lipids, endogenous

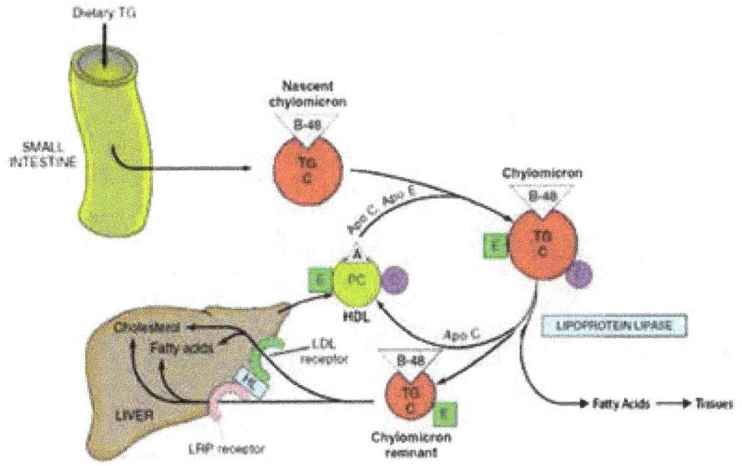

Figure 1.
Chylomicron metabolism. Chylomicrons, from the small intestine, are released into the circulation by apolipoprotein B-48 (B48). Nascent chylomicrons acquire apolipoprotein (Apo) E (green square) and C-II (purple circle), which are in the surface of HDL. Apo A-I (white triangle marked with a) is a main apolipoprotein of HDL. Apo E and Apo C-II are necessary for activation of lipoprotein lipase and for uptakes of remnant chylomicrons by an LDL receptor and LDL receptor-related protein. TG-triglycerides, C-cholesterol, P-phospholipids.

triglycerol, and cholesterol, to the peripheral tissues. Nascent VLDL is formed within the hepatocyte and Apo B-100. Those are triglyceride-rich lipid droplet, followed by the addition of Apo E, Apo A-I, and Apo A-II. The triglycerides and cholesterol ester used by hepatocytes for incorporation into VLDL are generated by the enzymes acyl-CoA diacylglycerol acyltransferase (DGAT; EC 2.3.1.20) and ACAT. Apo C-II and Apo E, borrowed from HDL, are important for subsequent metabolism of VLDL by lipoprotein lipase and the VLDL receptor. Enzyme lipoprotein lipase is activated by the apolipoprotein C-II, and this is followed from the hydrolysis of VLDL triglycerides by the activated enzyme, leading to release fatty acids, which diffuse into the adjacent myocytes or adipocytes for energy production or storage. Lipolysis of VLDL results in reduction in their triglyceride content and detachment and release of a remnant particle, known as IDL. IDL particles may undergo further lipolysis via hepatic triglyceride lipase. Apo E serves as a ligand for remnant VLDL or IDL binding to specific receptors in the liver. This leads to the extraction of nearly all remaining triglycerides from IDL by the liver and forma-tion of cholesterol-rich LDL. LDL particles are then removed via LDL (Apo B-100) receptor by the liver, as well as extrahepatic tissue (**Figure 2**).

2.3 The metabolism of HDL

HDL is synthesized and secreted from the liver and intestine. A major func-tion of HDL is to act as a repository for the Apo C-II and Apo E, for metabolism of triglyceride-rich lipoproteins, chylomicrons, and VLDL. Also, the primary function of HDL is retrieval and transport of cholesterol from the tissue to the liver which is known as reverse cholesterol transport. This cycle is very important for cellular cholesterol homeostasis. The principal apolipoprotein constituents of HDL are Apo A-I and Apo A-II. As the main structural constituent of HDL, Apo A-I is the activator of enzyme LCAT. LCAT system is involved in HDL-mediated removal of excess unesterified cholesterol from triglyceride-rich lipoproteins and tissues. Apo A-II serves as an activator of hepatic lipase, which plays a central role in the removal of HDL triglycerides by the liver. HDL-mediated removal of surplus

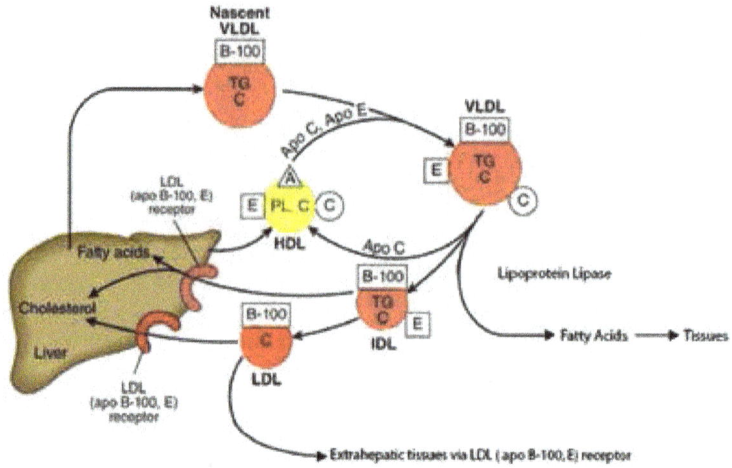

Figure 2.
Very-low-density lipoprotein metabolism. In circulation VLDL are transformed into intermediate-density lipoprotein after lipoprotein lipase activation by apolipoprotein C-II. IDL are removed by hepatic LDL receptors specific for apolipoprotein B and E. Apo E and Apo C-II are borrowed from high-density lipoprotein. A, Apolipoprotein a; B-100, Apolipoprotein B-100; C, Apolipoprotein C; E, Apolipoprotein E; LDL, low-density lipoproteins; TG, triglycerides; C, cholesterol; P, phospholipids.

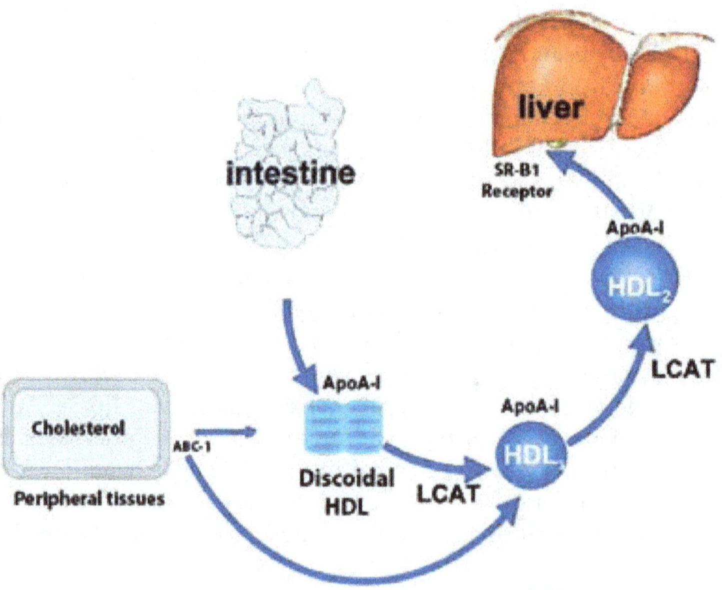

Figure 3.
High-density lipoprotein metabolism. As the main structural constituent of HDL, apolipoprotein A-I (Apo A-I), is the activator of enzyme lecithin-cholesterol acyltransferase. LCAT system is involved in HDL-mediated removal of excess unesterified cholesterol from tissues and its esterification. Scavenger receptor B1,and ATP-binding cassette transporter type I.

cholesterol from extrahepatic tissues requires attachment of nascent HDL to the ATP-binding cassette transporter type I (ABCA1). Binding to ABCA1 appears to trigger active transfer of phospholipids to nascent HDL, a step which is neces-sary for efficient translocation of free cholesterol from adjacent caveolae to the surface of HDL. Free cholesterol reaching the surface of HDL moves to the core of HDL. In this process nascent discoidal HDL is transformed into spherical HDL 3. After being accepted by HDL3, the free cholesterol is then esterified by LCAT to cholesterol esters, increasing the size of the particles to form the less dense HDL2. In the next step, HDL2, released in circulation, participates in a series of elaborate exchanges of apoproteins and lipids with the Apo B-containing lipoproteins such as chylomicrons, VLDL, and IDL, before reaching the liver. Actually, HDL in circulation receives triglycerides from Apo B-containing lipoproteins in exchange for cholesterol esters, a process catalyzed by CETP. Finally, HDL-2, via Apo A-I, binds to the scavenger receptor B1, which has been identified as a HDL receptor in the liver. The cycle is completed by the reformation of HDL 3, either after selective delivery of cholesteryl esters to the liver via SR-B1 or by hydrolysis of HDL2 phospholipids and triglycerides by hepatic lipase. Released, free Apo A-I forms preβ-HDL with the minimum amount of phospholipid and cholesterol. Preβ-HDL is considered the most potent form of HDL in inducing cholesterol efflux from the tissues to form discoidal HDL (**Figure 3**).

3. Oxidative stress, lipid peroxidation, and lipoprotein modifications

CKD is associated with increased oxidative stress, which promotes covalent modifications of lipids and lipoproteins. Oxidative stress is an imbalance in the reactive oxygen species (ROS) production and their degradation ratio. ROS include various compounds such as superoxide anions, hydroperoxide, and hydroxyl radical. These compounds are produced under physiologic conditions, during energy production in mitochondria by reducing oxygen during aerobic respiration.

But excessive ROS levels may have a harmful effect on tissue function and struc-ture, because of their interaction with different biomolecules in the human body, such as nucleic acids, proteins, and lipids. This interaction results with oxidative modifications of these biomolecules.

Under physiologic conditions, the production of ROS is balanced by antioxidant mechanisms that protect the cells from oxidative damages. The antioxidant mechanisms include enzymes; superoxide dismutase (SOD, EC 1.15.1.1) which catalyzes the dismutation of $O2^{\bullet-}$ into H_2O_2; and glutathione peroxidase (GPX, EC 1.11.1.9), which detoxifies H_2O_2 and other hydroperoxides. Reduced glutathione (GSH), as a non-enzymatic antioxidant, allows the scavenging of OH. The redox reactions are catalyzed by glutathione peroxidase. In antioxidant mechanisms also included several compounds such as HDL, albumin, tocopherols, ferritin, ceruloplasmin, transferrin, ubiquinol, flavonoids, and carotenoids.

HDL is well known for its protective antioxidant properties. Protein paraoxonase-1 (PON1, EC 3.1.8.1), bound to HDL, exhibited antioxidant effects, against lipid peroxidation. Selenium Glutathione-peroxidase 3, also known as glutathione peroxidase 3 (GPX3, EC 1.11.1.9), is another antioxidant enzyme, which is associated with HDL. Besides many functions in the human body, albumins are known for the antioxidant function too. In the first place concerning the lipid peroxida-tion, albumin can scavenge hypochlorous acid, responsible for chlorination of proteins mediated by myeloperoxidase, and through its reduced cysteine residue can scavenge hydroxyl radicals. One of the physiological functions of albumins is the transportation of insoluble components, through the blood plasma. In this way, albumins bind the long-chain fatty acids (LCFA), polyunsaturated fatty acids (PUFAs), and cholesterol and in the circulation, preventing them from oxidative modifications. Albumins bind also the ligands such as copper, iron, α-tocopherol, bilirubin, and homocysteine and prevent their antioxidant damages. Tocopherol is an important antioxidant in the human body, because of its ability to intercept intermediary radicals during the lipid peroxidation process. Most antioxidant mechanisms described above are decreased in patients with renal failure, leading to a higher sensitivity to oxidative stress. These patients have low activity and concentration of Glutathione, low concentration of HDL, PON-1 and GPX3 enzymes, albumins and antioxidant vitamins such as vitamin E, D and C. This decreased antioxidant status, enhanced oxidative stress, and affected lipids and proteins lead-ing to lipoproteins modifications and dysfunction. Lipids are one of the compounds mostly attached to oxidative stress. The peroxidation of lipids began with the reaction between a free radical with a polyunsaturated fatty acid containing more than two double bounds and formation of a lipid radical. In the next reaction, lipid radical can create lipid peroxyl radicals (LOO$^{\bullet}$) in reaction with oxygen, which can further react with other lipids forming new lipid radicals and lipid hydroperoxide (LOOH). Malondialdehyde (MDA) and 4-OH-2,3 alkenals are the end products of lipid hydroperoxide degradation. MDA covalently binds to proteins and nucleic acids, interfering with their normal biological functions. Binding to nucleic acids, MDA induce mutations and base-pair substitutions [6].

Binding to lysine amino group of protein part of lipoproteins, MDA created toxic adducts known as advanced lipoxidation end products (ALEs). In general, ALEs exhibit several pro-inflammatory effects and are involved in atherosclerosis [7]. These ALEs on Apo B result with oxidative modification of [8]. 4-OH-2,3 alkenals can also react with proteins, exactly with histidine, cysteine, and lysine residues and, create ALEs [9], which generate modified LDL. In this modified form, LDL can activate macrophages and increase the upregulation of class A scavenger receptors involved in the transformation of LDL into foam cells [10]. Another end product of lipid peroxidation is F2 α-isoprostanes. Oxidation of arachidonic acid by a cyclooxygenase-independent pathway generates F2α-isoprostanes, known for atherogenic properties, because of their implication on platelet aggregation via Thromboxane A2 receptor, vasoconstrictive effects on smooth muscle cells, and endothelial cell proliferation and endothelin-1 secretion [11]. These three end products are routinely used for in vivo evaluation of lipid peroxidation level [12].

4. Lipid disorders in chronic uremia

Renal failure is characterized by specific metabolic abnormalities of plasma lipoproteins [13]. These abnormalities involve all lipoprotein classes and show variations depending on the degree of renal impairment. Uremic lipid profile includes increased VLDL, IDL, small dense LDL particles, lipoprotein (a), and decreased HDL. Besides the changes in their concentration and structure, as stated above, uremic environment can strongly modified circulating lipoproteins leading to profound alterations of their biological properties and toxic effects in different cells and tissues. This has led to the formulation of an accelerated atherogenesis hypothesis and has been commonly linked with the lipid metabolic alteration associated with uremia.

4.1 Chylomicrons and VLDL

Hypertriglyceridemia is common a disorder in uremic patients. Several studies have shown increased concentration of triglycerides even though serum creatinine levels are within normal range [14]. The predominant mechanism responsible for increased concentration of triglyceride-rich lipoproteins, including chylomicrons, VLDL, and their remains, is delayed catabolism and increased synthesis Apo B-48, the essential for chylomicrons metabolism. There are evidences that Apo B-48 levels are increased and inversely correlated with glomerular filtration and proteinuria [15]. In circulation, triglyceride-rich lipoproteins acquire Apo E and Apo C-II, which are in the surface of HDL. In uremic patients, concentrations of Apo E and Apo C-II, which are necessary for activation of lipoprotein lipase and for uptakes of remnant chylomicrons and VLDL by a receptor specific for Apo E, are reduced. Such defect leads to a reduced release of triglycerides in peripheral tissues and to an accumulation of triglycerides. Delayed catabolism of triglyceride-rich lipoproteins occurs probably because of a decreased activity of hepatic triglyceride lipase and lipoprotein lipase. Moreover, significant evidence showed that enzyme lipoprotein lipase is lacking in renal failure [16]. There are evidences that diminished activity of enzyme is a consequence of the downregulation of the enzyme gene [17]. There is also downregulation of hepatic lipase expression [18].

The presence of lipoprotein lipase inhibitors also contributes to delayed triglyceride-rich lipoprotein catabolism. Apolipoprotein C-II is an activator, whereas apolipoprotein C-III is a direct lipoprotein lipase inhibitor. A decrease in apolipoprotein C-II/ apolipoprotein C-III ratio due to a disproportionate increase in plasma apolipoprotein C-III may be the cause of lipoprotein lipase inactivation, which further contributes to hypertriglyceridemia [19].

As it is mentioned above, triglyceride-rich lipoproteins, chylomicrons, and VLDL,need apolipoprotein C-II and apolipoprotein E for their maturation, which are delivered by HDL-2. In uremic patients HDL metabolism is impaired and HDL-3 are not maturated into HDL-2 due to a LCAT deficiency [20].

In healthy persons, VLDL and chylomicrons are transformed into IDL and chylomicron remnants after lipolysis in peripheral tissue. Chylomicron remnants are removed by the specific receptors of the liver, via LDL (Apo B-100 and Apo E)receptor and LDL receptor-related protein. It has been demonstrated that LDL receptor protein is downregulated in uremic patients [21] which leads to increasing levels of exogenous triglycerides. In physiological conditions, surplus IDL is transformed into LDL by the removal of their triglycerides by the hepatic lipase and enrichment in cholesteryl esters from HDL-2 by CETP. But the lack of HDL-2 impedes this process and leads to the accumulation of pro-atherogenic IDL [22]. There is a downregulation of hepatic lipase expression [18]; thus hepatic lipase deficiency which decreased conversion of IDL to LDL and lack of HDL work in concert to rise plasma concentration of IDL. A part of VLDL is removed by VLDL receptors, but in chronic uremia, the expression of VLDL receptors in tissues is also downregulated [23]. This makes impossible the VLDL binding with VLDL receptors in adipocytes and myocytes and their removal from the circulation (**Figure 4**). Insulin resistance is often associated with chronic uremia and seems to be responsible for a hepatic VLDL overproduction [24]. Secondary hyperparathyroidism, in renal failure, may play an additional role in triglyceride-rich lipoprotein catabolism impairment.

The predominant mechanism responsible for delayed metabolism of chylomicrons and very-low-density lipoproteins is increased synthesis apolipoprotein (Apo B-48) and low activity of lipoprotein lipase (LPL). Decrease concentration of high-density lipoproteins in renal failure results with decreased Apo E and Apo C-II, which are necessary for activation of LPL and for uptakes of remnant chylomicrons and intermediate-density lipoproteins by a receptor specific for Apo E. Such defect, together with the downregulation of LDL receptor protein and hepatic lipase (HL), leads to accumulation of chylomicron remnants and IDL, reducing the release of fatty acids into peripheral tissues. In physiological conditions, surplus IDL is transformed into LDL by the removal of their triglycerides and enrichment in cholesteryl esters from HDL-2 by CETP. But the lack of HDL-2 impedes this process and leads to the accumulation of pro-atherogenic IDL. Increased activity of CETP contrib-uted in reducing HDL concentration. The presence of lipoprotein lipase inhibitor, Apo C-III, also contributes to delayed triglyceride-rich lipoprotein metabolism.

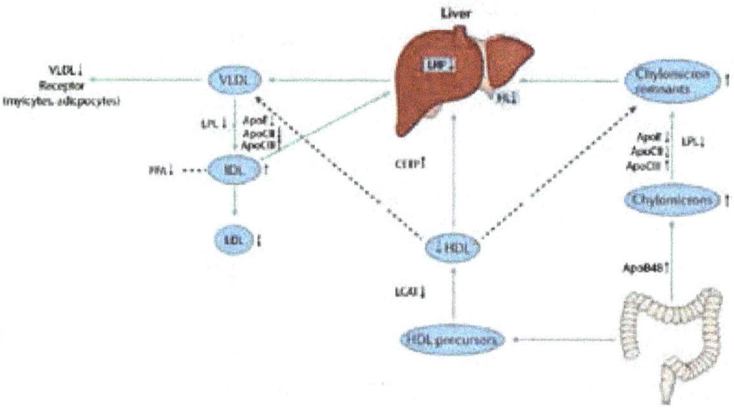

Figure 4.
Changes in chylomicrons and VLDL metabolism in renal failure.

4.2 HDL cholesterol

Uremic patients have decreased HDL in comparison with healthy population [25, 26]. Several mechanisms, working in concern, may underlie the reduction in HDL levels, which is usually indicative of impaired reverse cholesterol transport. Specifically, maturation of HDL is impaired and its composition is altered. Thus, uremic patients usually exhibit decreased levels of apolipoproteins A-I and A-II (the main protein constituents of HDL), diminished activity of LCAT, the enzyme responsible for the esterification of free cholesterol in HDL particles, as well as increased activity of CETP that facilitates the transfer of cholesterol esters from HDL to triglyceride-rich lipoproteins. One of the mechanisms for impaired HDL metabolism in uremia is the increased activity of enzyme ACAT which is responsible for intracellular cholesterol esterification. In physiological conditions, Apo A-I and Apo A-II, in the circulation, are loaded with cholesterol and phospholipids to form nascent HDL. Then, nascent HDL binds to the ABCA-1 receptor on circulating macrophages and activates cholesterol ester hydrolase allowing their loading with cholesterol. ACAT limits this reverse efflux of cholesterol from macrophages by catalyzing the esterification of intracellular cholesterol. Oxidative modification of Apo A-I can limit HDL binding on macrophages [27] and upregulation of hepatic ACAT [28] contributing in impaired cholesterol efflux. Therefore, an increase in ACAT activity can potentially limit HDL-mediated cholesterol uptake and con-tribute to the reduction in plasma HDL cholesterol and impaired maturation of HDL. Although the effect of chronic renal failure on ACAT expression and activity in the extrahepatic tissues is not known, chronic renal failure has been recently shown to markedly raise hepatic ACAT-2 mRNA and protein abundance, as well as total ACAT activity [29].

On the other hand, the activity of enzyme LCAT is decreased [30, 31]. Apo A-I is the activator of LCAT, the essential enzyme for the HDL-mediated cholesterol retrieval from extrahepatic tissues and as well as ligand for the SR-B1 and HDL-binding protein (ABCA1 transporter). Apo A-II serves as an activator of hepatic lipase, which plays a central role in the removal of HDL triglycerides by the liver. As mentioned above, in patients with impaired kidney function, Apo A-I and Apo A-II levels are decreased. This reduction contributes to diminished HDL concentration and impaired HDL maturation. Until recently, it was not clear whether the reported reduction in plasma LCAT activity is caused by the reduction in its hepatic production and plasma concentration or is a consequence of its inhibition by an unknown uremic toxin [32]. Another enzyme with diminished activity is CETP. The enzyme mediates transfer of cholesterol ester from HDL to IDL in exchange for triglycerides. Increased activity of CETP in uremic patients facilitates the transfer of cholesterol esters from HDL to triglyceride-rich lipoproteins, reducing the HDL cholesterol ester and elevation of HDL triglycerides. The mechanism responsible for the elevation of CETP is unknown, but some investigation connected its increased synthesis with proteinuria. Probably the same mechanism is responsible for the dysregulation of hepatic SR-B1. Hepatic SR-B1 is the primary pathway for the disposal of HDL-borne cholesterol ester and triglycerides, and dysregulation of this protein can impact HDL metabolism. Heavy glomerular proteinuria has been shown to significantly reduce hepatic SR-B1 protein expression in experimental animals [29]. HDL has a protective effect against inflammation, platelet adhesion, and LDL oxidation. Those protective functions of HDL can be attributed to HDL-associated enzymes on its surface. Paraoxonase-1 is considered as the main antioxidant enzyme bound to HDL. Mainly expressed in the liver and the kidney, this enzyme exhibited antioxidant properties against lipid peroxidation as it binds to HDL and in a minor part to VLDL [33]. Glutathione seleno-peroxidase 3, also known as glutathione peroxidase 3, is another antioxidant enzyme associated with HDL [34].

One of main anti-atherogenic properties of HDL is a reverse cholesterol transport from circulating macrophages. HDL also increases the production of nitric oxide (NO), through the activation of the endothelial NO synthase in endothelial cells resulting in a vasorelaxant phenotype. In CKD the production of NO by endothelial cells is significantly reduced with HDL [28]. HDL also inhibits the expression of adhesion molecules such as intercellular adhesion molecule-1 (ICAM-1) and vascular adhesion molecule-1 (VCAM-1), which prevent the attachment of circulating monocytes to endothelial cells. In uremic patients, HDL promotes an enhanced expression of VCAM-1 and ICAM-1 on endothelial cells [35, 36]. Moreover, CKD-HDL upregulates the expression of pro-inflammatory mediators such as monocyte chemoattractant protein-1 (MCP-1), interleukin-1ß (IL-1ß), and tumor necrosis factor α (TNF-α) [36, 37]. And finally normal HDL exhibit anti-apoptotic effects on endothelial cells through the downregulation of caspase-3 (a member of the cysteine-aspartic acid protease) activity [38]. All these dimin-ished protective functions of HDL can contribute to accelerated atherogenesis [39]. HDL is very sensitive in oxidative stress and posttranslational modifications. Renal failure is associated with an enhanced activity of enzyme myeloperoxidase (MPO, EC 1.11.2.2) that plays a crucial role in the generation of posttranslational modification derived products (PTMDPs). MPO catalyzed the oxidative reactions and formation of a variety of chlorinated protein and lipid adducts. MPO-modified ApoA-1 results in decreased reverse cholesterol efflux and a reduced binding with ABCA-1 receptor, which disturbed cholesterol homeostasis (**Figure 5**). 3-chlorotyrosine, an oxidation product of MPO, impairs the activity of enzymes, LCAT, and PON-1, resulting with decreased anti-inflammatory effects of HDL. And through the activation of SR-B1 in macrophages, MPO-modified HDL directly contributes in atherosclerosis (**Figure 5**).

In renal failure, decreased activity of lecithin-cholesterol acyltransferase impaired the transformation of nascent cholesterol into HDL3 and then into HDL2. Increased activity of cholesteryl ester transfer protein facilitates the transfer of cholesterol esters from HDL to triglyceride-rich lipoproteins, reducing HDL concentration. Removal of free cholesterol from macrophages proceeds by scavenger receptor 1. Nascent HDL is generated when Apo A-I interacts with ATP-binding cassette transporter type 1 (ABCA1). Than nascent HDL activates cholesterol ester hydrolase allowing their loading with cholesterol. ACAT limits this reverse efflux of cholesterol from macrophages by catalyzing the esterification of intracellular

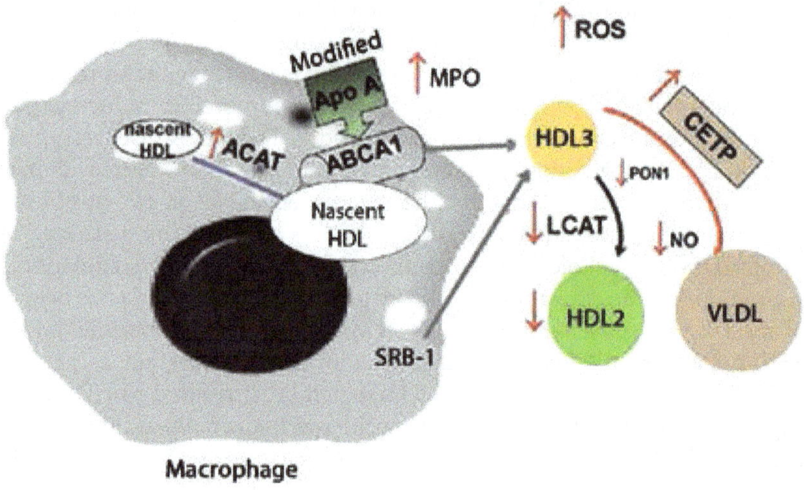

Figure 5.
HDL metabolism in renal failure.

cholesterol. Increased activity of ACAT, in renal failure, participates in impaired cholesterol efflux. Antioxidative and anti-inflammatory functions of HDL are impaired due to reduced activity of HDL enzyme PON1. HDL from patients with renal failure loses its vasoprotective properties, inhibiting nitric oxide produc-tion. Oxidative modification of Apo A-I decreases HDLs binding to macrophages. Myeloperoxidase-modified Apo A-I decrease reverse cholesterol efflux, reduce binding with ABCA1, and impair HDLs anti-apoptotic properties.

4.3 LDL cholesterol

Beyond atherogenic risk of LDL level itself, renal failure leads to various struc-tural modifications of LDL particles. The lipoproteins found in uremic patients are disproportionately modified, with LDL that is enriched in triglycerides. These modified LDL particles tend to be smaller and denser in their form. Small dense LDL is believed to be markedly pro-atherogenic, and this is attributed to its ability to infiltrate the vessel wall and its increased susceptibility to oxidative modifica-tion. Because of the significantly modified lipid subfraction turnover, residence time of lipoproteins in the circulation is prolonged. Thus, lipoproteins are at risk of posttranslational modification. LDL receptor-mediated cholesterol uptake plays an important role in cholesterol homeostasis. Modified LDL have reduced affin-ity for the classic LDL receptors and are taken up by the scavenger receptors on the surface of the macrophages. These receptors are increased in chronic uremia. High affinity for macrophages results in the accumulation of cholesterol and the formation of foam cells in the vascular walls, resulting in the development of atherosclerotic plaques [40, 41]. Heavy proteinuria alone or in combination with chronic uremic state results in acquired LDL receptor deficiency and plays a central role in the genesis of the atherosclerosis and cardiovascular diseases. Several levels of LDL oxidation can coexist in the bloodstream and lead to the activation of several pathways involved in atherosclerosis through their binding to scavenger receptors [42] and smooth muscle cell proliferation. There is an evidence that OxLDL are accumulated in uremic patients and are correlated with the intensity of peripheral arterial disease [43]. Oxidized epitopes of LDL can activate immunity and then lead to the formation of antibodies directed against OxLDL. OxLDL/antibodies against OxLDL ratio were also correlated with carotid atherosclerosis and cardiovascular events [44]. Formation of OxLDL is a consequence of oxida-tive stress. As mentioned above, the breakdown of polyunsaturated fatty acids produces highly reactive molecules, such as MDA and 4-OH-2,3 alkenals. MDA and 4-OH-2,3 alkenals can form Schiff bases and covalent Michael-type adducts, with lysine residues of Apo B-100, in LDL (**Figure 6**). The oxidized fatty acid fragments which can remain attached via ester bridges, may also contain terminal reactive phospholipids which may form adducts with Schiff base lysine residues of Apo B-100. Similarly with HDL modifications, increased levels of MPO are involved in LDL modifications. MPO can modify LDL through several mechanisms. MPO initiated the reaction between hypochlorous acid and tyrosine residues of Apo B-100, protein part of LDL, resulting with 3-chlorotyrosine formation, which is known for pro-atherogenic properties through its binding with lectin-like oxidized LDL receptor 1. MPO also generated reactive nitrogen species, converting LDL into a nitrosilated-LDL form. This reaction resulted in nitration of Apo B-100 tyrosyl residues of LDL. Carbamylated LDL (cLDL) is another modified form of LDL, initiated by MPO. In this reaction MPO catalyzed the addition of thiocyanate, derived from the decomposition of urea to the lysine residues of LDL, and leads to the formation of carbamylated LDL [45, 46]. The carbamylation occurs by spontaneous, nonenzymatic chemical modification of Apo B-100, by thiocyanate. It is a irreversibly reaction of thiocyanate with free amino groups and ε-NH2 of lysine residues in protein part of LDL (**Figure 7**). cLDL have pro-atherogenic effects such as the transformation of macrophages into foam cells [47] through their binding to the pro-atherogenic CD36 receptor [48, 49]. cLDL are associated with endothelial toxicity [50, 51] through lectin-like oxidized LDL receptor 1 [52] (**Figure 8**). cLDL levels are raised by chronic uremia [53, 54].

Modified forms of LDL; carbamylated LDL and oxidized LDL; activated lectin-like oxidized LDL receptor 1, on endothelial cells; and initiated formation of macrophages and smooth muscle cell proliferation.

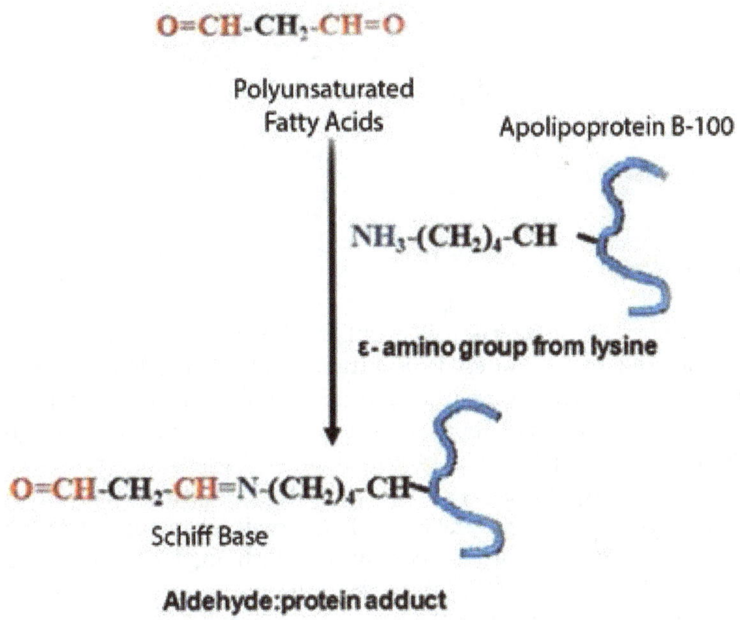

Figure 6.
Formation of oxidized LDL.

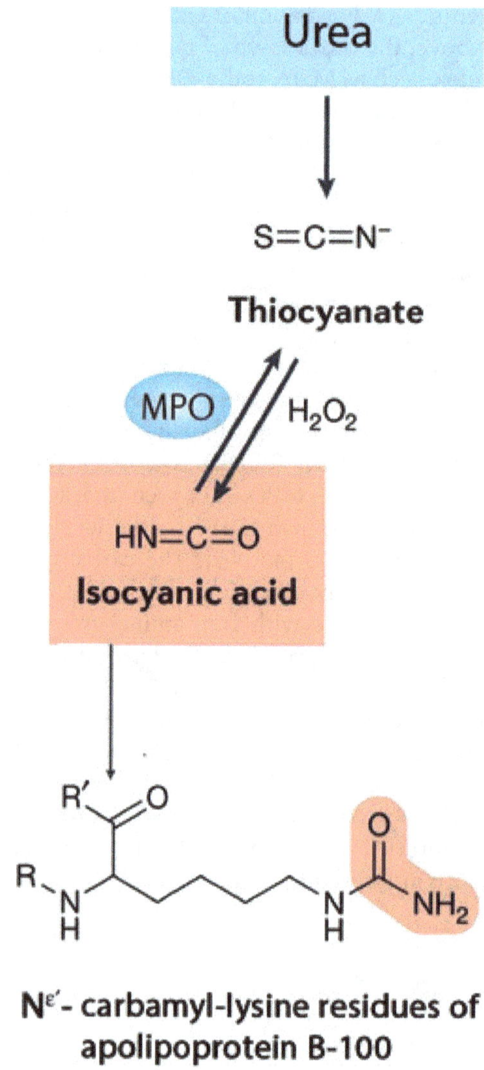

Figure 7.
Formation of carbamylated LDL.

4.4 Lipoprotein (a)

The contribution of cardiovascular events to the extraordinary high mortality in CKD has generated some interest in nontraditional atherosclerotic cardiovascular disease risk factors, which are prevalent in this population, such as Lipoprotein (a) [Lp (a)]. Lp (a) is an LDL-like lipoprotein containing a unique apolipoprotein called Apo(a). Serum levels of Lp(a) are determined largely by genetic variation in the gene encoding for Apo(a). Apo(a) is very homologous to plasminogen [55] and exhibits an extreme size polymorphism with the Apo(a) isoproteins, ranging in size from 420 to 840 kDa. Inherited in an autosomal codominant fashion, the Apo(a) isoprotein is closely correlated with serum Lp(a) concentrations, with an inverse correlation between the size of the Apo(a) isoprotein and the serum Lp(a) concentrations. Lp(a) has been implicated in the regulation of plasminogen activator inhibitor-1 expression in endothelial cells and shown to inhibit endothelial cell surface fibrinolysis to

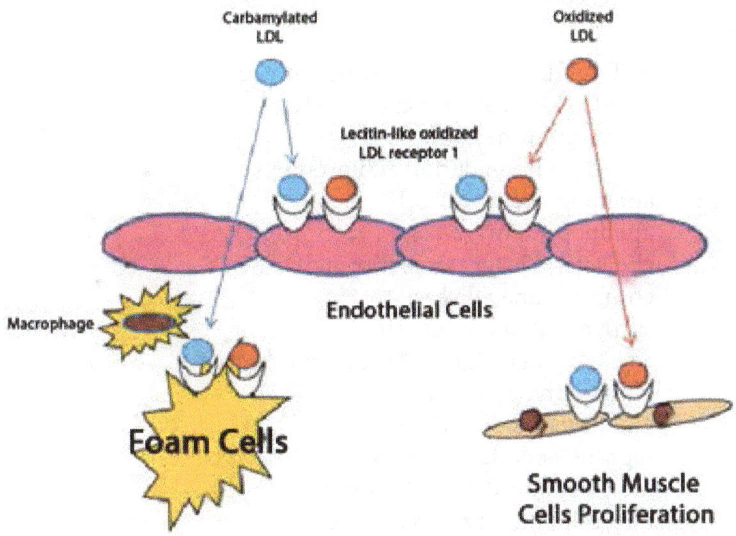

Figure 8.
Oxidized LDL and carbamylated LDL effects.

attenuate plasminogen binding to platelets and to bind to plaque matrix components. Autopsy studies in humans have documented the presence of Lp(a) in aortic and coronary atherosclerotic plaques and an apparent colocalization with fibrinogen [56]. Lp(a) levels are frequently elevated in uremic patients with CKD [57] and have been associated with a frequency distribution of apolipoprotein (a)-Lp(a) isoforms, similar to those found in general population. This indicates that elevated Lp(a) levels in these patients are not due to the genetic origin [58]. It has been suggested that kidneys have an important role in Lp(a) metabolism [59]. In CKD, Lp(a) occurs at high concentra-tions, largely because of reduced clearance or as a result of increased hepatic synthe-sis, induced by an acute-phase reaction or by protein losses from proteinuria [60]. Uremia can be considered to be a state of activated acute-phase response, and in the micro-inflammatory milieu, a number of atherogenic proteins like Lp(a) are acting as an acute-phase reactant. Based in all these properties, Lp(a) is a prototype candidate to be classified as a uremic toxin.

5. Conclusion

Chronic uremia causes profound alteration in lipoprotein metabolism, promoting the development of atherosclerosis and cardiovascular disease. Besides the changes in their concentration, enhanced oxidative stress and uremic environment can strongly modify circulating lipoproteins leading to profound alterations of their biological properties and can be considered as uremic toxins. Uremic lipoprotein profile is directly involve in glomerular capillary endothelial damage and in the progression of renal disease. This “reverse epidemiology” shows the importance of lipid control to prevent the progression of renal failure.

Conflict of interest

The authors declare no conflict of interest.

Abbreviations

ABCA1	ATP-binding cassette transporter type I
ACAT	acyl-CoA cholesterol acyltransferase
α-TNF	alpha tumor necrosis factor
ALEs	advanced lipoxidation end products
Apo A,B,C,E	apolipoprotein A,B,C,E
Apo (a)	apolipoprotein
CETP	cholesteryl ester transfer protein
CKD	chronic kidney disease
cLDL	carbamylated low-density lipoprotein
DGAT	acyl-CoA diacylglycerol acyltransferase
GPX	glutathione peroxidase
GPX3	glutathione peroxidase 3
HDL	high-density lipoproteins
IDL	intermediate-density lipoproteins
ICAM-1	intercellular adhesion molecule-1
IL-1ß	interleukin-1ß
LCAT	lecithin-cholesterol acyltransferase
LDL	low-density lipoproteins
Lp(a)	lipoprotein (a)
LRP	LDL related protein
MCP-1	monocyte chemoattractant protein-1
mRNA	messenger ribonucleic acid
MDA	malondialdehyde
MPO	myeloperoxidase
NO	nitric oxide
OxLDL	oxidized-LDL
LCFA	long chains fatty acids
$LOO^{\bullet}$	peroxyl radicals
LOOH	lipid hydroperoxide
PON1	paraoxonase 1
PTMDPs	posttranslational modification derived products
PUFAs	polyunsaturated fatty acids
ROS	reactive oxygen species
SOD	superoxide dismutase
SR-B1	scavenger receptor B1
VCAM	vascular adhesion molecule-1
VLDL	very-low-density lipoprotein

Author details

Valdete Topçiu-Shufta and Valdete Haxhibeqiri*
Faculty of Medicine, Clinic of Medical Biochemistry, University Clinical Center of Kosova, University of Prishtina, Prishtina, Republic of Kosova

*Address all correspondence to: valbera@yahoo.com

References

[1] Florens N, Calzada C, Lyasko E, Juillard L. Modified lipids and lipoproteins in chronic kidney disease: A new class of uremic toxins. Toxins (Basel). 2016;**8**(12):376. DOI: 10.3390/toxins8120376

[2] Duranton F, Cohen G, De Smet R, Rodriguez M, Jankowski J, Vanholder R, et al. On behalf of the European uremic toxin work group Normal and pathologic concentrations of uremic toxins. Journal of the American Society of Nephrology. 2012;**23**:1258-1270. DOI: 10.1681/ASN.2011121175

[3] McCullough PA, Agrawal V, Danielewicz E, Abela GS. Accelerated atherosclerotic calcification and Monckeberg's sclerosis: A continuum of advanced vascular pathology in chronic kidney disease. Clinical Journal of the American Society of Nephrology. 2008;**3**:1585-1598. DOI: 10.2215/CJN.01930408

[4] Schiffrin EL, Lipman ML, Mann JFE. Chronic kidney disease: Effects on the cardiovascular system. Circulation. 2007;**116**:85-97. DOI: 10.1161/CIRCULATIONAHA.106.678342

[5] Tall AR. Plasma cholesteryl ester transfer protein. Journal of Lipid Research. 1993;**34**:1255-1274

[6] VanderVeen LA, Hashim MF, Shyr Y, Marnett LJ. Induction of frameshift and base pair substitution mutations by the major DNA adduct of the endogenous carcinogen malondialdehyde. PNAS. 2003;**100**:14247-14252. DOI: 10.1073/pnas.2332176100

[7] Negre Salvayre A, Coatrieux C, Ingueneau C, Salvayre R. Advanced lipid peroxidation end products in oxidative damage to proteins. Potential role in diseases and therapeutic prospects for the inhibitors. British Journal of Pharmacology. 2008;**153**:6-20. DOI: 10.1038/sj.bjp.0707395

[8] Palinski W, Ord VA, Plump AS, Breslow JL, Steinberg D, Witztum JL. ApoE-deficient mice are a model of lipoprotein oxidation in atherogenesis. Demonstration of oxidation-specific epitopes in lesions and high titers of autoantibodies to malondialdehyde-lysine in serum. Arteriosclerosis, Thrombosis, and Vascular Biology. 1994;**14**:605-616. DOI: 10.1161/01.ATV.14.4.605

[9] Petersen DR, Doorn JA. Reactions of 4-hydroxynonenal with proteins and cellular targets. Free Radical Biology and Medicine. 2004;**37**:937-945. DOI: 10.1016/j.freeradbiomed.2004.06.012

[10] Hoff HF, O'Neil J, Chisolm GM, Cole TB, Quehenberger O, Esterbauer H, et al. Modification of low density lipoprotein with 4-hydroxynonenal induces uptake by macrophages. Arteriosclerosis, Thrombosis, and Vascular Biology. 1989;**9**:538-549. DOI: 10.1161/01. ATV.9.4.538

[11] Yura T, Fukunaga M, Khan R, Nassar GN, Badr KF. Free-radical-generated F2-isoprostane stimulates cell proliferation and endothelin-1 expression on endothelial cells. Kidney International. 1999;**56**:471-478. DOI: 10.1046/j.1523-1755.1999.00596.x

[12] Grotto D, Santa Maria LD, Boeira S, Valentini J, Charão MF, Moro AM, et al. Rapid quantification of malondialdehyde in plasma by high performance liquid chromatography-visible detection. Journal of Pharmaceutical and Biomedical Analysis. 2007;**43**:619-624. DOI: 10.1016/j.jpba.2006.07.030

[13] Tsimihodimos V, Dounousi E, Siamopoulos KC. Dyslipidemia in

chronic kidney disease an approach to pathogenesis and treatment. American Journal of Nephrology. 2008;**28**:958-973

[14] Sechi LA, Catena C, Zingaro L, Melis A, De Marchi S. Abnormalities of glucose metabolism in patients with early renal failure. Diabetes. 2002;**51**:1226-1232

[15] Okubo M, Hanada H, Matsui M, Hidaka Y, Masuda D, Sakata Y, et al. Serum apolipoprotein B-48 concentration is associated with a reduced estimated glomerular filtration rate and increased proteinuria. Journal of Atherosclerosis and Thrombosis. 2014;**21**:974-982. DOI: 10.5551/jat.23309

[16] Vaziri ND, Yuan J, Ni Z, Nicholas SB, Norris KC. Lipoprotein lipase deficiency in chronic kidney disease is accompanied by down-regulation of endothelial GPIHBP1 expression. Clinical and Experimental Nephrology. 2012;**16**:238-243. DOI: 10.1007/s10157-011-0549-3

[17] Vaziri ND, Liang K. Down-regulation of tissue lipoprotein lipase expression in experimental chronic renal failure. Kidney International. 1996;**50**:1928-1935

[18] Sato T, Liang K, Vaziri ND. Protein restriction and AST-120 improve lipoprotein lipase and VLDL receptor in focal glomerulosclerosis. Kidney International. 2003;**64**:1780-1786. DOI: 10.1046/j.1523-1755.2003.00281.x

[19] Chan DT, Dogra GK, Irish AB, et al. Chronic kidney disease delays VLDL apoB-100 particle catabolism: Potential role of apo C-III. Journal of Lipid Research. 2009;**50**:2524-2531

[20] Moradi H, Pahl MV, Elahimehr R, Vaziri ND. Impaired antioxidant activity of high-density lipoprotein in chronic kidney disease. Translational Research. 2009;**153**:77-85. DOI: 10.1016/j.trsl.2008.11.007

[21] Kim C, Vaziri ND. Down-regulation of hepatic LDL receptor-related protein (LRP) in chronic renal failure. Kidney International. 2005;**67**:1028-1032. DOI: 10.1111/j.1523-1755.2005.00166.x

[22] Hirowatari Y, Homma Y, Yoshizawa J, Homma K. Increase of electronegative-LDL-fraction ratio and IDL-cholesterol in chronic kidney disease patients with hemodialysis treatment. Lipids in Health and Disease. 2012;**11**(1). DOI: 10.1186/1476-511X-11-111

[23] Sato T, Liang K, Vaziri ND. Down-regulation of lipoprotein lipase and VLDL receptor in rats with focal glomerulosclerosis. Kidney International. 2002;**64**:1780-1786. DOI: 10.1046/j.1523-1755.2003.00281.x

[24] Tsimihodimos V, Mitrogianni Z, Elisaf M. Dyslipidemia associated with chronic kidney disease. Open Cardiovascular Medicine Journal. 2011;**5**:41-48

[25] Attman PO, Samuelsson O, Alaupovic P. Lipoprotein metabolism and renal failure. American Journal of Kidney Diseases. 1993;**21**:573-592

[26] Vaziri ND, Deng G, Liang K. Hepatic HDL receptor, SR-B1 and Apo A-I expression in chronic renal failure. Nephrology, Dialysis, Transplantation. 1999;**14**:1462-1466

[27] Shao B, Oda MN, Oram JF, Heinecke JW. Myeloperoxidase: An inflammatory enzyme for generating dysfunctional high density lipoprotein. Current Opinion in Cardiology. 2006;**21**:322-328. DOI: 10.1097/01.hco.0000231402.87232.aa

[28] Liang K, Vaziri ND. Upregulation of acyl-CoA: Cholesterol acyltransferase in chronic renal failure. American Journal of Physiology. Endocrinology and Metabolism. 2002;**283**:676-681. DOI: 10.1152/ajpendo.00364.2001

[29] Blaton V. Dyslipidemia at chronic renal failure. EJIFCC. 2009;**20**(1):59-66

[30] Shoji T, Nishizawa Y, Nishitani H, Yamakawa M. Impaired metabolism of high density lipoprotein in uremic patients. Kidney International. 1992;**41**:1653-1661. DOI: 10.1038/ki.1992.238

[31] Vaziri ND, Liang K, Parks JS. Down-regulation of hepatic lecithin: Cholesterol acyltransferase gene expression in chronic renal failure. Kidney International. 2001;**59**:2192-2196. DOI: 10.1046/j.1523-1755.2001.00734.x

[32] Vaziri ND. Dyslipidemia of chronic renal failure: the nature, mechanisms, and potential consequences. American Journal Of Physiology. Renal Physiology. 2006;**290**(2):F262-F272. DOI: 10.1152/ajprenal.00099.2005

[33] Litvinov D, Mahini H, Garelnabi M. Antioxidant and anti-inflammatory role of paraoxonase 1: Implication in arteriosclerosis diseases. North American Journal of Medical Sciences. 2012;**4**:523-532

[34] Maddipati KR, Marnett LJ. Characterization of the major hydroperoxide-reducing activity of human plasma. Purification and properties of a selenium-dependent glutathione peroxidase. The Journal of Biological Chemistry. 1987;**262**(36):17398-17403

[35] Shroff R, Speer T, Colin S, Charakida M, Zewinger S, Staels B, et al. HDL in children with CKD promotes endothelial dysfunction and an abnormal vascular phenotype. Journal of the American Society of Nephrology. 2014;**25**:2658-2668. DOI: 10.1681/ASN.2013111212

[36] Kaseda R, Jabs K, Hunley TE, Jones D, Bian A, Allen RM, et al. Dysfunctional high-density lipoproteins in children with chronic kidney disease. Metabolism. 2015;**64**:263-273. DOI: 10.1016/j.metabol.2014.10.020

[37] Yamamoto S, Yancey PG, Ikizler TA, Jerome WG, Kaseda R, Cox B, et al. Dysfunctional high-density lipoprotein in patients on chronic hemodialysis. Journal of the American College of Cardiology. 2012;**60**:2372-2379. DOI: 10.1016/j.jacc.2012.09.013

[38] Kimura T, Sato K, Malchinkhuu E, Tomura H, Tamama K, Kuwabara A, et al. High-density lipoprotein stimulates endothelial cell migration and survival through sphingosine 1-phosphate and its receptors. Arteriosclerosis, Thrombosis, and Vascular Biology. 2003;**23**:1283-1288. DOI: 10.1161/01.ATV.0000079011.67194.5A

[39] Kwan BCH, Kronenberg F, Beddhu S, Cheung AK. Lipoprotein metabolism and lipid management in chronic kidney disease. Journal of the American Society of Nephrology. 2007;**18**:1246-1261

[40] Shurraw S, Tonelli M. Statins for treatment of dyslipidemia in chronic kidney disease. Peritoneal Dialysis International. 2006;**26**:523-539

[41] Piecha G, Adamczak M, Ritz E. Dyslipidemia in chronic kidney disease. Polskie Archiwum Medycyny Wewnętrznej. 2009;**119**:487-492

[42] Levitan I, Volkov S. Oxidized LDL: Diversity, patterns of recognition, and pathophysiology. Antioxidants and Redox Signaling. 2010;**13**:39-75. DOI: 10.1089/ars.2009.2733

[43] Takenaka T, Takahashi K, Kobayashi T, Oshima E, Iwasaki S, Suzuki H. Oxidized low density lipoprotein (ox-LDL) as a marker of atherosclerosis in hemodialysis (HD) patients. Clinical Nephrology. 2002;**58**:33-37. DOI: 10.5414/CNP58033

[44] Pawlak K, Myśliwiec M, Pawlak D. Oxidized LDL to autoantibodies against oxLDL ratio—The new biomarker associated with carotid atherosclerosis and cardiovascular complications in dialyzed patients. Atherosclerosis. 2012;**224**:252-257. DOI: 10.1016/j.atherosclerosis.2012.07.011

[45] Shiu SWM, Xiao SM, Wong Y, Chow WS, Lam KSL, Tan KCB. Carbamylation of LDL and its relationship with myeloperoxidase in type 2 diabetes mellitus. Clinical Science. 2013;**126**:175-181. DOI: 10.1042/CS20130369

[46] Exner M, Hermann M, Hofbauer R, Hartmann B, Kapiotis S, Gmeiner B. Thiocyanate catalyzes myeloperoxidase-initiated lipid oxidation in LDL. Free Radical Biology and Medicine. 2004;**37**:146-155. DOI: 10.1016/j.freeradbiomed.2004.04.039

[47] Apostolov EO, Ok E, Burns S, Nawaz S, Savenka A, Shah SV, et al. Carbamylated-oxidized LDL: Proatherosclerotic effects on endothelial cells and macrophages. Journal of Atherosclerosis and Thrombosis. 2013;**20**:878-892. DOI: 10.5551/jat.14035

[48] Park YM. CD36, a scavenger receptor implicated in atherosclerosis. Experimental and Molecular Medicine. 2014;**46**:e99. DOI: 10.1038/emm.2014.38

[49] Chmielewski M, Bryl E, Marzec L, Aleksandrowicz E, Witkowski JM, Rutkowski B. Expression of scavenger receptor CD36 in chronic renal failure patients. Artificial Organs. 2005;**29**:608-614. DOI: 10.1111/j.1525-1594.2005.29097.x

[50] Carracedo J, Merino A, Briceno C, Soriano S, Buendia P, Calleros L, et al. Carbamylated low-density lipoprotein induces oxidative stress and accelerated senescence in human endothelial progenitor cells. The FASEB Journal. 2011;**25**:1314-1322. DOI: 10.1096/fj.10-173377

[51] Speer T, Owala FO, Holy EW, Zewinger S, Frenzel FL, Stähli BE, et al: Carbamylated low-density lipoprotein induces endothelial disfunction. Eur Heart J. 2014;**35**(43):3021-32

[52] Son JN, Lho Y, Shin S, Kwon S-H, Moon KC, Ha E. Carbamylated low-density lipoprotein increases reactive oxygen species (ROS) and apoptosis via lectin-like oxidized LDL receptor (LOX-1) mediated pathway in human umbilical vein endothelial cells. International Journal of Cardiology. 2011;**146**:428-430. DOI: 10.1016/j.ijcard.2010.10.098

[53] Apostolov EO, Ray D, Savenka AV, Shah SV, Basnakian AG. Chronic uremia stimulates LDL carbamylation and atherosclerosis. American Society of Nephrology. 2010;**21**:1852-1857. DOI: 10.1681/ASN.2010040365

[54] Apostolov EO, Basnakian AG, et al. Carbamylated low-density lipoprotein: Nontraditional risk factor for cardiovascular events in patients with chronic kidney disease. Journal of Renal Nutrition. 2012;**22**:134-138. DOI: 10.1053/j.jrn.2011.10.023

[55] Kirmizis D, Tsiandoulas A, Pangalou M, et al. Validity of plasma fibrinogen, D-dimer, and the von Willebrand factor as markers of cardiovascular morbidity in patients on chronic hemodialysis. Medical science monitor international medical journal of experimental and. Clinical Research. 2006;**12**:55-62

[56] Costa E, Rocha S, Rocha-Pereira P, et al. Cross-talk between inflammation, coagulation/fibrinolysis and vascular access in hemodialysis patients. The Journal of Vascular Access. 2008;**9**(4):248-253

[57] Shibata T, Magari Y, Kamberi P, et al. Significance of urinary fibrin/fibrinogen degradation products (FDP) D- dimer measured by a highly sensitive

ELISA method with a new monoclonal antibody (D-D E72) in various renal diseases. Clinical Nephrology. 1995;**44**(2):91-95

[58] Péquériaux NC, Fijnheer R, Gemen EF, et al. Plasma concentration of von Willebrand factor predicts mortality in patients on chronic renal replacement therapy. Nephrology, Dialysis, Transplantation. 2012;**27**(6):2452-2457

[59] Longenecker C, Klag M, Marcovina S, et al. High lipoprotein(a) levels and small Apolipoprotein(a) size prospectively predict cardiovascular events in dialysis patients. American Society of Nephrology. 2005;**16**:1794-1802

[60] Dieplinger H, Lackner C, Kronenberg F. Elevated plasma concentrations of lipoprotein(a) in patients with end stage renal disease are not related to the size polymorphism of apolipoprotein(a). The Journal of Clinical Investigation. 1993;**91**(2):397-401

7

Brown Adipose Tissue Energy Metabolism

Yuan Lu

Abstract

The brown adipose tissue (BAT) evolved as a specialized thermogenic organ in mammals. Nutrients (i.e., fatty acids and glucose) from the intracellular storage and peripheral tissues are critical to the BAT thermogenic function. The BAT converts the chemical energy stored in nutrients to thermo energy through UCP1-mediated nonshivering thermogenesis (NST). Activated BAT contributes significantly to the whole body energy substrate homeostasis. It is now well-recognized that adult humans possess BAT with functional thermoactivity. Thus, BAT energy metabolism has a significant therapeutic potential in the management of metabolic disorders, such as obesity, insulin resistance, type 2 diabetes, and lipid abnormality in humans.

Keywords: brown adipose tissue, brown adipocyte, metabolism, fatty acid, glucose, metabolic disorders

1. Introduction

Brown adipose tissue (BAT) evolved as a specialized thermogenic organ in the modern eutherian mammals, including *Homo sapiens* [1–7]. The main function of BAT is mediating adaptive thermogenesis or nonshivering thermogenesis (NST), when mammals are challenged in the cold environment. The NST function is a critical adaptation, which helps to maintain the homeothermy in mammals and gives them the evolutionary advantage to survive in the cold habitat, and to thrive from the Artic to the Antarctic region [3, 4]. During the cold challenge, BAT metabolizes nutrients (i.e., fatty acids and glucose) and converts the chemical energy to thermo energy through NST. In addition to its thermogenic function, activated BAT also contributes significantly to whole body energy substrate homeostasis. When activated, BAT presents physiologically significant benefit in fatty acids and glucose homeostasis as well as insulin sensitivity in mammals [8–24]. It is now well-recognized that adult humans possess BAT, which has functional thermoactivity [8–10, 25]. Cold challenge-activated BAT is detected mainly in the supraclavicular, paravertebral, and cervical regions in adult humans [9, 10, 25, 26]. Accumulating evidences indicate that BAT function is inversely associated with age, body mass index (BMI), and diabetic status in adult humans, which indicates that the activation of BAT has potential translational implication in the management of metabolic disorders, such as diabetes and obesity in humans [8–14, 16, 17, 21, 27].

Brown adipocyte, which is endowed with mitochondria, is the most important thermogenic functional unit of BAT [4, 28]. In the mitochondrion of brown adipocyte, energy generated from nutrients is initially stored as proton gradient membrane potential across the mitochondrial inner membrane, and then converted directly to thermo energy by the uncoupling protein 1 (UCP1)-mediated proton flow [4, 29]. In BAT, brown adipocytes are surrounded by abundant capillaries. The heat generated in the brown adipocyte mitochondria can be quickly distributed by the blood flow to maintain the steady core body temperature in mammals [4]. In addition to classical brown adipocytes, beige/brite adipocytes can be induced from the white adipocytes to conduct thermogenesis upon the cold challenge or catecholamine stimulation [26, 30, 31]. This process is termed as "browning" [26, 30, 31]. Thermogenesis in beige/brite adipocytes can also contribute to mammals' body temperature maintenance and whole body metabolism [26, 30, 31]. Beige/brite adipose tissue generates heat through both UCP1-independent thermogenesis, including calcium cycling in and out of endoplasmic reticulum, futile cycle between creatine and phosphocreatine, and UCP1-dependent thermogenesis in mitochondria [26, 30–33]. The energy metabolism is essential for the optimal UCP1-mediated mitochondrial thermogenic function of brown and beige/brite adipocytes [4, 34, 35]. In this chapter, we discuss the importance of energy metabolism in maintaining brown adipocyte thermogenic function and the proceeding of targeting metabolic disorder through BAT activation in human studies.

2. Fatty acid metabolism in brown adipocytes

Thermogenic brown adipocyte possesses a high capacity for fatty acid β-oxidation that has been reported in both rodent and humans [4, 12, 14, 36]. Fatty acids serve as the activator for UCP1, which is a fatty acid/H^+ symporter directly mediating proton flow and thermogenesis [37]. Fatty acids also serve as the main energy substrate mediating the uncoupling and thermogenic function in brown adipocytes [2, 36–41]. In addition, fatty acids can enhance brown adipocyte ther-mogenic capacity through the nuclear receptor peroxisome proliferator-activated receptors (PPARs), which are the master transcription regulators for the expression of genes involved in lipid metabolism, oxidative phosphorylation, and the key thermogenic protein UCP1 in brown adipocytes [42, 43].

Intracellular fatty acids are stored in the format of triglyceride in the heterogeneous multilocular lipid droplets in the brown adipocyte [4]. Upon the cold challenge, triglycerides stored in the brown adipocyte lipid droplet are lipolysed. Triglyceride lipolysis is a sequential process that involves different enzymes, resulting in the liberation of glycerol and fatty acids for heat production [44]. The lipid droplet is composed of triglycerides and cholesterol esters, which are surrounded by a monolayer of phospholipids [44]. Important proteins with regulatory and enzymatic functions, including perilipin and CGI58, coexist on the phospholipid monolayer to regulate the lipid trafficking and other functions of the lipid droplet [44–47]. Perilipin stabilizes the lipid droplet and prevents it from lipolysis under basal condition. Upon cold challenge, β-adrenergic stimulation leads to the activation of G-protein-coupled receptor (GPCR) and adenylate cyclase, which subsequently increases the cAMP level in brown adipocyte [48]. cAMP then activates protein kinase A (PKA), which phosphorylates perilipin at its serine residues [49–52]. The phosphorylated perilipin releases CGI-58, an adipose triglyceride lipase (ATGL)-activating protein. CGI-58, subsequently, binds and actives ATGL. Activated ATGL hydrolyzes triglycerides and generates free fatty acids and diglycerides [50, 53–55]. Upon the cold challenge, PKA also phosphorylates serine residues on another key lipolysis enzyme hormone sensitive lipase (HSL) [56]. Although HSL is capable of hydrolyzing triglycerides, diglycerides, monoglycerides, and a broad array of other lipid substrate, it is the rate-limiting enzyme for hydrolyzing the diglycerides *in vivo* [56]. The phosphorylated HSL catalyzes the diglycerides and generates free fatty acids and monoglycerides [56]. As the last step of lipolysis, monoglycerides is hydrolyzed by monoacylglycerol lipase (MGL) to form free fatty acids and glycerol [57].

Given the importance of fatty acid in brown adipocyte thermogenic function, it is reasonable to predict that the deficiency of the key lipolysis enzymes and fatty acid transportation proteins in brown adipocytes can lead to a defected BAT thermogenic function. Human and rodent *in vivo* studies using both pharmacological and genetic approaches to manipulate triglyceride lipolysis processes were reported [36, 39, 58–60]. In the initial studies, nicotinic acid (NiAc) was used to inhibit intracellular triglyceride lipolysis through acting on metabolite-sensing Gi-protein-coupled GPR109A and subsequent PKA activation [36, 58, 61]. A study in rats suggested the brown adipocyte intracellular lipid lipolysis played a major role in thermogenesis in rodents [36]. It showed that the intracellular triglyceride lipolysis contributes to 84% of thermogenesis during an acute cold challenge (10° C, 2–6 hours) and 74% of thermogenesis during a chronic cold challenge (10°C, 21 days) [36]. The importance of the brown adipocyte intracellular triglyceride-derived fatty acids was also reported in a human study [58]. In this study, administrating intracellular triglyceride lipolysis inhibitor NiAc significantly blunted BAT oxidative metabolism in cold-challenged young healthy humans (average 30 years of age with an average BMI of 24.5 kg/m^2). During a 3-hour cold challenge at 10°C, NiAc administration suppressed BAT intracellular triglyceride lipolysis by about 50% and BAT oxidative metabolism by 70%, despite of the increased blood flow in the BAT [58].

One caveat from both the aforementioned human and rodent *in vivo* studies is that NiAc-mediated lipolysis inhibition took effect in both brown and white adipocytes in addition to other tissues, which makes it hard to delineate the contribution of lipolysis from each cell type. An *in vitro* study nicely confirmed the importance of intracellular lipolysis in brown adipocytes [2]. In cultured primary mouse brown adipocytes, the inhibition of both of the key lipolysis enzymes adipose triglyceride lipase (ATGL) and hormone-sensitive lipase (HSL) led to a 97% decrease in isopropanolinduced brown adipocyte respiration, indicating the imperative requirement of intracellular lipolysis in activated brown adipocytes [2]. These studies suggested that fatty acids liberated from the intracellular triglyceride storage serve as a critical fuel resource and contribute significantly to BAT thermogenesis in brown adipocytes upon activation.

Further studies in genetically manipulated mice showed similar results. The ATGL-knockout mouse presented defective triglyceride lipolysis function with increased BAT weight (8.5-fold), enlarged lipid droplet (20-fold), and decreased BAT explant lipid hydrolysis activity (−85%) [38]. The impaired triglyceride lipolysis activity in the *ATGL-knockout* mouse led to a defective thermogenic function. Upon a 5-hour 4°C challenge, mouse body temperature dropped to a critical low point at around 25°C [38]. Another study using mice with CGI-58 deficiency in both white and brown adipocytes also showed decreased BAT thermogenic func-tion. The *adipose-CGI-58-knockout* mice were cold-sensitive, but only under fasted state [60]. The *HSL-null* mouse adipose tissue also presented defected triglyceride lipolysis function. Upon catecholamine-stimulation, *HSL-null* adipose tissue explants exhibited significantly reduced fatty acid and almost blunted glycerol release into the culture medium, parallel with diglycerides accumulation in both white and brown adipose tissue [56, 62]. However, it seems that HSL-mediated lipolysis function is not as critical as ATGL-mediated lipolysis function in brown adipocytes. An in vitro study showed that isopropanol-induced UCP1 activity was largely dependent on ATGL function (80%) compared to HSL function (35%) in cultured primary brown adipocytes, and the combined inhibition of both ATGL and HSL functions led to an almost complete block of UCP1-mediated thermogenic function (97%) [2].

These studies highlight the cardinal role of intracellular triglyceride liberation in the brown adipocyte thermogenic function, but raise the question if the brown adipocyte intracellular lipolysis is essential for BAT to maintain the adaptive thermogenesis *in vivo*. Interestingly, recent studies using *UCP1-Cre*-mediated knockout of either ATGL or CGI-58 gene in mice showed an unexpected insignificant impact of brown adipocyte intracellular lipolysis and suggested that the circulating fatty acids mobilized from peripheral tissues may play more important roles in BAT thermogenesis in mice [59, 60]. These studies demonstrated that the loss of either ATGL or CGI58-mediated triglycerides lipolysis in brown adipocytes did not compromise the cold challenge-induced thermogenic response in mice [59, 60]. The phenotypic discrepancy between the systemic and BAT-specific *ATGL* and *CGI-58* knockout mice brings very important insights to *in vivo* BAT fatty acid metabolism, and indicates that the brown adipocyte intracellular lipolysis is not the only energy resource for *in vivo* adaptive thermogenesis during cold challenge. Although it is not clear about the exact partition of intracellular and circulating fatty acids contribution to the BAT thermogenic function, it is reasonable to speculate that the core body temperature maintenance is vital for mammals to maintain their optimal function during the cold challenge, when the intracellular lipolysis function is impaired or insufficient, circulating energy substrates from other metabolic tissues, that is, white adipose tissue and liver, are mobilized to provide energy substrates for BAT-mediated adaptive thermogenesis to maintain the adequate core body tempera-ture and ensure the optimal functionality of mammals.

Long-chain fatty acids (LCFAs) are the most abundant format of fatty acid energy substrate in mammals [63–66]. The liberated LCFAs from intracellular lipid droplet are facilitated and transported to mitochondrion and nucleus by fatty acid-binding proteins (FABPs) to conduct their functions [67]. Of the six FABP isoforms, FABP4 (also termed adipocyte p2) and FABP5 are the major FABP isoforms in brown adipocytes [68–74]. Mice mutated in both *FABP4* and *FABP5* gene developed severe hypothermia during fasting after an acute cold challenge (1–3 hours), indicating that fatty acids transportation plays an indispensable role in BAT thermogenic function [74].

The LCFAs-mediated mitochondrial oxidation and UCP-1 activation function require sequential carnitine acyltransferases in order to translocate the LCFA into mitochondrial matrix. Carnitine palmitoyltransferase 1 (CPT1), located on the mitochondrial outer membrane, is the rate-limiting enzyme that mediates LCFAs inward translocation into mitochondrial matrix [63–66]. CPT1 exists in tissue-specific isoforms, and CPT1b is the major isoform expressed in brown adipocytes [75–77]. Mouse embryos-carried homozygous knockout of *Cpt1b* gene were lethal before embryonic day 9.5–11.5 and a normal percentage of $CPT1b^{+/-}$ mice was born from the $CPT1b^{+/-}$ and wild type breeding pairs [78]. However, more than 50% of the $CPT1b^{+/-}$ pups were lost before weaning [78]. The detailed experiment showed that ~7% $CPT1b^{+/-}$ mice developed fatal hypothermia following a 3-hour cold challenge (4°C) and ~52% $CPT1b^{+/-}$ mice developed fatal hypothermia following a 6-hour cold challenge (4°C), indicating the critical contribution of CPT1b-mediated LCFAs transportation and thermogenesis during the cold challenge in infant/young mice [78]. Carnitine palmitoyltransferase 2 (CPT2) is located on mitochondrial inner membrane to further medicate LCFAs translocation into mitochondrial matrix. In line with the study using the *CPT1b*-deficient mice, an *adipose-specific CPT2 knockout* mice also presented hypothermic phenotype after 3 hours cold challenge (4°C) and decreased LCFAs oxidation in isolated MEF cells [39, 79]. These studies indicate that fatty acid transportation is critical for BAT thermogenic func-tion during the cold challenge.

The intramitochondrial LCFAs contribute to the thermogenesis through UCP-1 activation and β-oxidation. A detailed study confirmed that mice with impaired fatty acid β-oxidation are cold-intolerant [80]. The acyl-CoA dehydrogenases, which catalyze the initial steps of fatty acid β-oxidation, are composed of a group of enzymes, including very long-chain acyl CoA dehydrogenase (VLCAD), long-chain acyl CoA dehydrogenase (LCAD), and short-chain acyl CoA dehydrogenase (SCAD) [81, 82]. A detailed experiment showed that fetal hypothermia was presented in all of the homozygous knockout mice ($VLCAD^{-/-}$, $LCAD^{-/-}$, and $SCAD^{-/-}$) after 1-hour 4°C challenge [80]. Although the mice with single heterozygous of VLCAD, LCAD, and SCAD genes were cold tolerate; more than 30% of mice with double heterozygous $VLCAD^{+/-}//LCAD^{+/-}$ *or* $LCAD^{+/-}//SCAD^{+/-}$ and triple heterozygous $VLCAD^{+/-}//LCAD^{+/-}//SCAD^{+/-}$ combinations developed hypothermia upon the cold challenge [80], indicating the essential role of the LCFAs-mediated thermogenic function in mitochondria.

In summary, these studies highlight the importance of the brown adipocyte lipolysis and the liberated intracellular fatty acid transportation and oxidation during thermogenesis. In addition, these studies indicate that brown adipocytes not only use the intracellular lipid storage, but also the circulating energy substrate to maintain the critical thermogenic function for the organismal survival and optimal function in mammals [2, 12, 38, 44, 56, 59, 60, 80]. In modern days, BAT's ability to metabolize fatty acids mobilized from other peripheral storages, including white adipose tissue and liver, makes it a good potential therapeutic target in humans. Studies have shown that BAT mediates significant plasma lipid clearance during the cold challenge in rodent and humans under both physiological and pathological conditions [83–85].

One study showed that the activated BAT is involved in the basal and post-prandial triglyceride metabolism in rodents [83]. In this study, compared to mice kept at 22°C, mice kept at 4°C had significantly lower triglyceride-rich lipoproteins (TRLs)-triglyceride concentration [83]. The study also showed that the activated BAT was involved in the post-prandial triglyceride metabolic process, evidenced by an oral ^{3}H-triolein tolerance test. Under cold challenge condition, the BAT ^{3}H-triolein uptake was significantly higher than that of either liver or skeletal muscle, suggest-ing the significant triglyceride/triglyceride-rich lipoprotein metabolism in activated BAT. For the triglyceride clearance, circulating triglycerides rose to a peak at 2 hours postprandial and declined subsequently in mice kept at 22°C. In contrast, the triglyceride level reminded persistently low in mice kept at 4°C, suggesting that the BAT possesses a high postprandial triglyceride clearance function [83]. Most interestingly, the cold challenge increased the uptake of radio-labeled lipoprotein in BAT and reduced the uptake in liver [83]. The cold challenge-induced lipid clearance shift suggests that BAT can be targeted for lipid metabolism *in vivo*. In a pathological setting, the genetically manipulated $Apoa5^{-/-}$ mice with severe hyperlipidemia were studied. A 4–24 hours 4°C challenge corrected $Apoa5^{-/-}$ mouse plasma lipid concentration to the values comparable to wild-type mice, indicating the significant impact of BAT on whole body lipid metabolism in the rodent under the pathological condition [83, 86, 87]. Other studies also showed that the BAT preferentially uptook plasma triglycerides through peripheral tissue lipolysis, and the selective fatty acids uptake from triglyceride-rich lipoprotein ameliorated hyperlipidemia in rodents [84, 85].

Although the studies in rodent strongly support the importance of BAT in lipid clearance, the significance of BAT in human lipid metabolism is still unclear. The contribution of activated BAT in human body was studied using a dietary radio-labeled LCFAs tracer 14(R,S)-[^{18}F]-fluoro-6-thia-heptadecanoic acid (^{18}FTHA) [88].

This study showed that a 4-hour mild cold-challenge at 18°C significantly increased dietary fatty acids distribution in BAT in humans [88]. However, given the relative small volume of BAT tissue, the dietary fatty acids clearance contribution from the BAT is less significant compared to other organs including heart, liver, white adipose tissue, or skeletal muscles, and only contributed to ~0.3% of total body dietary fatty acids clearance upon the mild cold challenge [88]. Similarly, another study using fatty acid tracer ^{18}F-fluoro-thiaheptadecanoic acid (^{18}FTHA) showed that a 3-hour cold challenge at 18°C led to four times higher radio-labeled tracer uptake and ~80% metabolic rate increase in the BAT, contributing <1% of total fatty acids clearance rate in human subjects [12]. Nevertheless, these experiments suggest that BAT not only exists, but also is functionally active and can contribute to systemic metabolism in humans. One possible reason for the relatively low BAT contribution in fatty acids metabolism in humans could be due to the short cold challenge time and mild cold challenge conditions. It is possible that during the acute cold challenge, intracellular lipid lipolysis serves as the main energy resource so that it ameliorates the clearance of circulating TRLs or fatty acids.

The circulating TRLs or fatty acids are transported into brown adipocyte by a group of proteins, including lipoprotein lipase (LPL) and fatty acid transport proteins [83, 89–93]. LPL is a multifunctional protein produced by many tissues, including the adipose tissue [94]. LPL serves as a rate-limiting enzyme mediating extracellular lipolysis [94]. It hydrolyzes triglycerides into lipid-rich proteins into fatty acids and monoacylglycerol. It also mediates the cellular uptake of triglyceride and other lipids [94]. Studies have shown that cold challenge or catecholamine-stimulation induce the expression and activity of LPL in brown adipocytes through a cAMP-mediated mechanism [89–91]. After activation, LPL is released from the brown adipocyte, transferred to the capillary endothelium lumen, and serves as the anchor between the endothelium cell surface and the TRLs [95–97]. Next, the LCFAs liberated from LPL channel into brown adipocyte for thermogenic function [98, 99]. A study indicated that the local LPL activity is required for TRLs uptake into the BAT [83]. In this study, it is shown that either LPL-specific inhibitor tetrahydrolipstatin pretreatment or releasing LPL from endothelium by heparin significantly blocked the uptake of TRL and fatty acids in BAT [83].

The liberated LCFAs in circulation can be transported into cells and activated by both transmembrane fatty acid transporter proteins (FATPs) and scavenger receptor CD36 [92, 93]. FATPs are composed by a family of six proteins mediating circulating LCFAs uptake and distribution in cells [83, 92, 93, 100]. Among the six isoforms, FATP1 (SLC27A1) is the major isoform in brown and white adipose tissue [83, 92, 93, 100]. Studies showed that postprandial lipid uptake is highly dependent on the adipose tissue FATP1 [101]. The FATP1-knockout mice have decreased lipid uptake in adipose tissue and a compensatory lipid redistribution in liver and heart, where FATP1-mediated lipid uptake function is not required under normal conditions [101]. The cold challenge can induce FATP1 expression in BAT [100]. In line with this, the isolated FATP1-null brown adipocytes showed significantly less fatty acids uptake upon catecholamine stimulation [100]. *In vivo* studies showed that in cold challenged FATP1-knockout mice, BAT triglyceride storage was decreased and circulating serum free fatty acids was increased, indicating the importance of FATP1-mediated fatty acid uptake in BAT [100]. The LCFAs uptake can also be mediated by the transmembrane class B scavenger receptor CD36 [83, 102, 103]. Upon the cold challenge, CD36 is significantly upregulated in adipocytes [83, 102]. Other studies using *CD36*$^{-/-}$ mice clearly indi-cated the importance of CD36-mediated LCFAs transportation during thermogenesis. Around 60% of *Cd36*$^{-/-}$ mice died during a 24-hour cold challenge, which is paralleled with drastically increased plasma free fatty acids concentration [83, 102]. A recent study indicated that CD36-medicated coenzyme Q (CoQ) uptake is required for the BAT mitochondrial thermogenic function [104].

In summary, fatty acid is indispensable for the brown adipocyte thermogenic function. Fatty acid mediates brown adipocyte thermogenesis through mitochondrial β-oxidation, UCP1-activation, and fatty acid-mediated thermogenic gene regulation. Both the intracellular fatty acids from brown adipocyte lipolysis and the liberated fatty acids from peripheral tissues play essential roles mediating the critical BAT adaptive thermogenic function to main the adequate core body temperature in mammals. In modern days, activating BAT thermogenic function to increase fatty acids uptake and utilization may offer new therapeutics to treat human metabolic disorders.

3. Glucose metabolism in BAT

The radio-labeled glucose analog ^{18}F-fluorodeoxyglucose (^{18}F-FDG), in combination with positron emission tomography (PET) and computed tomography (CT), provides a reliable method for the *in vivo* tissue glucose uptake study [105, 106]. Based on studies using this method, it is now well-recognized that BAT in both rodents and humans possesses a significant glucose uptake capacity upon the cold challenge [4, 9–12, 25, 26, 107–112].

Glucose uptake is regulated in brown adipocytes. *In vivo* studies showed that cold challenge significantly enhanced insulin sensitivity and subsequent glucose uptake in the BAT [9, 10, 25, 26, 107, 112, 113]. Interestingly, other studies showed that starved rats with low insulin level also had increased BAT glucose uptake during the cold challenge, indicating that the enhanced glucose uptake is not completely insulin-dependent [114]. Additional studies also indicated that the enhanced glucose uptake in brown adipocytes can be mediated by different pathways in addition to insulin stimulation, including β-adrenergic receptor activation, AMP-activated protein kinase (AMPK) activation, and mTOR activation [115–117].

The importance of glucose metabolism in BAT is a long-standing question. Glucose transporters, which facilitate glucose across the cell plasma membrane is the first rate-limiting step of the glucose metabolism [34, 118, 119]. Intracellular glucose is subsequently phosphorylated to glucose-6-phosphate (G6P) by the enzyme hexokinase (HK). Glucose-6-phosphate serves as the substrate into different pathways, including glycolysis, glycogen synthesis, and the pentose phosphate pathway (PPP) [34, 118, 119]. Glycolysis breaks down glucose to pyruvate to generate small amount of ATP and NADH [34, 118, 119]. The pyruvate is then transported into mitochondria for oxidation and energy production. Under hypoxia condition, pyruvate is disposed in the form of lactate [34, 118, 119].

Early study indicates that glucose and its metabolites contribute to the brown adipocyte thermogenesis by showing that catecholamine-induced glucose uptake was decreased when mitochondrial β-oxidation was inhibited in brown adipocytes [120]. Other studies in brown adipocyte glucose transportation also indicate the importance of the glucose in brown adipocyte metabolism [6, 112, 121, 122]. Both glucose transporter 1 (Glut1) and glucose transporter 4 (Glut4) are abundantly expressed in brown adipocytes and the insulin sensitive Glut4 is the major isoform [112, 122]. The *in vitro* study showed that knock-down of Glut1 and/or Glut4 gene in cultured brown adipocytes impaired the catecholamine-induced cell oxygen consumption by 30–50% [6, 121]. Other studies indicate the importance of glycolysis in brown adipocyte thermogenesis [115, 123, 124]. An *in vitro* study showed that the knockdown of two glycolysis enzymes, HK2 or pyruvate kinase M (PKM), decreased glucose uptake and catecholamine-induced cell oxygen consumption by 67% or

34% respectively, in brown adipocytes [6]. These studies suggested the importance of glucose metabolism in brown adipocyte function. However, another detailed study indicated that glucose oxidation does not play a major role in brown adipocyte metabolism and thermogenesis, by showing that the rate of $^{14}CO_2$ formation from the ^{14}C glucose was relatively small compared with the maximum rate of oxygen consumption in activated brown adipocytes [125]. Studies using radio-labeled glucose also suggested that glucose uptake was only sufficient to fuel maximally ~15% of the thermogenic capacity in activated rodent brown adipocytes, suggesting that the significantly upregulated glucose uptake in activated brown adipocytes is disassociated with the relative low glucose-mediated thermogenic capacity [123, 125, 126]. A more recent study indicates that the brown adipocyte energy production from glucose depends on the state of the cell: glucose and fatty acid contribute equally to brown adipocyte oxygen consumption under the basal condition; upon catecholamine-activation, oxygen consumption is mainly fueled by fatty acids [6].

The dissociation between high glucose uptake and low glucose-mediated thermogenesis in activated BAT raises an important question: what is the function of the intracellular glucose in brown adipocytes? The importance of glucose in the *de novo* lipogenesis in brown adipocytes has been reported [127–129]. Studies indicate that glucose uptake is an independent process of thermogenesis in both coldchallenged and catecholamine-activated BAT, which further supports that glucose uptake might play other role as energy substrate in activated brown adipocytes [116, 117, 130–132]. One detailed *in vitro* study using rat brown adipocytes showed that norepinephrine significantly enhanced glucose oxidation by sevenfold, while it also inhibited lipogenesis at the same time. On the other hand, insulin stimulation increased the lipogenesis by sevenfold in brown adipocytes whereas glucose oxidation remained very low. Most interestingly, the addition of insulin to the norepinephrine only potentiated the enhanced glucose oxidation, but do not enhance the lipogenesis [115]. On the other hand, another study showed that brown adipocytes converted a greater proportion of metabolized glucose into lactate and pyruvate, but only a smaller proportion into fatty acids through insulin-mediated pathway [125]. These studies suggest that glucose metabolism is involved in two different states in brown adipocytes, the thermogenic state and nonthermogenic state. It is plausible that during the thermogenic state, glucose contributes to both the energy production and lipogenesis; and during the nonthermogenic state, glucose contributes mainly to the lipogenesis and energy storage process in brown adipocytes, which explains the disassociation of glucose uptake and glucose metabolism in brown adipocytes. In addition to glucose, other energy substrates, including glycerol and amino acid (glutamate), might also contribute to BAT metabolism and thermogenesis in human and rodent brown adipocytes. However, the relative contribution and partition of these energy substrates are unclear [132–135].

4. Energy storage in BAT

The cold challenge not only enhances catabolic processes mediating energy substrate metabolism and heat generation, but also induces anabolic processes mediating fatty acid synthesis and lipogenesis, as well as glycogenesis [36, 111, 136–139].

Glucose can be stored as glycogen in brown adipocytes [36, 139]. Studies showed that glycogen synthases (GStot) and uridine diphosphate glucose pyrophosphorylase (Udgp), which mediates glycogenesis, were upregulated upon the cold challenge in BAT [36, 139]. Interestingly, glycogen hydrolysis enzyme, glycogen phosphorylase (Pygl), was also upregulated after cold challenge [36]. Although cold-challenge upregulates both glycogen synthesis and degradation, it is reported that the stored glycogen is used up shortly after the cold challenge [139]. Indicating that glycogen is not an efficient format for energy storage and a sustainable energy resource in brown adipocytes.

The glucose uptaken by brown adipocyte can also be converted to fatty acid through the *de novo* lipogenesis. Carbohydrate response elementbinding protein (ChREBP) is the major transcription factor mediating fatty acid synthesis in adipocytes. There are two isoforms of ChREBP, ChREBPα and ChREBPβ. ChREBPα is abundantly expressed in BAT from rodents and humans [137, 140, 141]. *De novo* lipogenesis involves a series of enzymes mediating the sequential reactions converting glucose-derived citrate into fatty acids, including ATP-citrate lyase (ACLY), acetyl-CoA carboxylases 1 (ACC1), fatty acid synthase (FASN), and stearoyl-CoA desaturase-1 (SCD1). ChREBPα, coordinates with another transcription factor Max-like protein (MLX), directly binds to the carbohydrate response elements (ChoRE) and upregulates these *de novo* lipogenic genes expressions. ChREBPα can also increase the expression of ChREBPβ to further activate the *de novo* lipogenesis enzymes [142, 143]. It has been reported that lipogenic genes and AKT2-ChREBP pathways are upregulated to optimize fuel storage and thermogenesis upon cold stimulation in BAT [136]. In accordance with these studies, *ChREBP*$^{-/-}$ mice presented less BAT weight [142], and adipose-specific *ChREBP-knockout* mice had decreased carbohy-drate-induced lipogenesis in BAT [144]. These studies indicate that ChREBP plays important roles in brown adipocyte's *de novo* lipogenesis and energy storage.

Sterol regulatory element-binding protein-1 (SREBP-1) is another transcrip-tion factor mediating the *de novo* lipogenesis [145]. Of the three different SREBP isoforms, SREBP1c is more abundant and SREBP1a is less abundant in the adipose tissue [137, 145, 146]. *In vitro* study showed that SREBP1c is sufficient to regulate lipogenic enzymes in cultured adipocytes [147]. In the adipose-specific *ap2-SREBP1c* transgenic mice, lipogenic enzyme Acc1, FASN and Scd1 expression as well as fatty acids synthesis rate were significantly upregulated in brown adipocytes [148]. In line with this study, *ap2-SREBP1a* transgenic mice also developed adipose tissue hypertrophy in accordance with an increased lipogenic enzyme profile and enhance *de novo* lipogenesis in the BAT [148]. However, some *in vivo* studies indicated that SREBP1c's role in the *de novo* lipogenesis in adipocytes is dispensable, as evidenced by *SREBP1-knockout* mice have normal lipogenic enzymes gene expression profile and normal lipid storage in their adipose tissue [149, 150]. These studies suggest that SREBP-1 is involved in the brown adipocyte lipogenesis and triglyceride storage when the excessive energy resources are available; however, SREBP-1's function can be compensated by other factors when it is absent.

In summary, these studies suggest that in activated brown adipocytes, glycogenesis and lipogenesis are upregulated to store/restore energy substrates, which is parallel with energy substrate metabolism and thermogenesis. These coordinated anabolic and catabolic processes are important to maintain the brown adipocyte energy homeostasis.

5. The proceeding of targeting BAT in human metabolic disorders

The understating of BAT energy homeostasis and the discovery of the functional BAT in humans lead to significant interests in targeting BAT for metabolic disorders, for example, obesity, insulin resistance, type 2 diabetes, and lipid profile abnormality. The ability of BAT metabolizing fatty acid and glucose from the intracellular storage, the peripheral tissues liberation, and teh dietary nutrition absorption makes it a good potential therapeutic target for combating metabolic disorders in humans. In addition to t he BAT, beige/brite fat, which is coexisted in white adipose tissue, can be recruited and activated (browning) in respond to cold challenge or pharmacological stimulation and serves as a target for metabolic disorders [151–154].

It has been reported that the BAT ^{18}F-FDG uptake in humans correlates inversely with aging, adiposity, diabetic status, and BMI, indicating that the manipulation of BAT function is a possible approach for combating metabolic disorders [8–14, 16, 21–24, 155, 156]. The studies in mouse models and humans provide evidences for the metabolic benefit of BAT. Mice with genetic ablation of BAT and the *UCP1-knockout* mice under thermoneutrality, both developed obesity [18–20, 85]. In addition, BAT activation reduced hypercholesterolemia and protected mice from atherosclerosis development and liver steatosis [18–20, 85]. It is well-recognized that BAT can be activated upon seasonal temperature changes or short time period (several hours) mild cold challenge (16-19°C) in humans [7, 9, 10, 12, 21, 22, 24–27, 157]. However, the significance of BAT's contribution in human metabolism has not been clearly elucidated. Human studies indicated that BAT activity is inversely correlated with body fat deposition, suggests BAT can serve as a target for obesity [9, 10, 16, 21–24]. Studies also showed that fasting glucose was lower in the human subjects with higher BAT prevalence and the BAT activity were blunted in subjects with obesity [11, 16, 155]. More interestingly, in the same patients with multiple PET scans, BAT was more detectable when fasting glucose in the subjects were lower [16]. Other studies showed that a mild cold challenge significantly increased whole body glu-cose disposal, glucose oxidation, insulin sensitivity, and whole body energy expen-diture in human subjects [13, 17, 158, 159]. Additional study showed that moderate cold challenge (18.06°C) significantly improved the peripheral glucose uptake and insulin sensitivity by 20%, but did not impact the pancreatic insulin secretion [17]. These studies strongly indicate the therapeutic potentials of targeting BAT for glucose metabolism in humans, but leave the question of the relative contribution of activated BAT in whole body glucose metabolism to be answered.

Cold challenge can also enhance BAT lipid metabolism. Studies showed that cold challenge led to significantly enhanced lipid mobilization, increased plasma fatty acid levels, as well as upregulated genes for lipid metabolism in human BAT [12, 58, 156, 160]. It is reported that circulating fatty acids were uptaken by BAT in cold-challenged humans by using the ^{18}FTHA tracing method [88]. In addi-tion, it has been shown that cold-activated BAT significantly contributed to whole body fatty acid utilization in healthy humans [17]. The fatty acid uptake is sig-nificantly lower in overweight human subjects compare to healthy humans [156]. Importantly, BAT activation upon mild cold challenge significantly increased systemic lipid metabolism, whole body lipolysis, triglyceride-fatty acid cycling, and fatty acid oxidation in overweight/obese subjects [14]. These studies suggest that BAT contributes to whole body lipid metabolism and homeostasis in healthy humans as well as in humans with metabolic disorders, suggesting that BAT can serve as a target for lipid abnormality in humans.

The significance of BAT/beige adipose tissue in human whole body metabolism has been studied and remarkable progress has been made in recent years. The acknowledgment that BAT can be activated and subsequently contributes to the human whole body energy expenditure is encouraging. Although the capacity and relative significance of BAT's contribution to whole body energy substrate metabolism has not been elucidated, it should be noticed that majority of the studies in humans were conducted for relative short time period with mild cold challenge conditions. Given, humans usually live under thermoneutrality, we can assume that their BAT functions are repressed under this condition. In addition, the prevalence of BAT in humans varies significantly, which depends on individual's life style, physical activity, and health conditions, which makes it harder to evaluate the contribution of BAT in whole body metabolism [7, 9–12, 21, 22, 24–27, 88, 156, 157].

Hence, a sustainable chronic mild cold challenge strategy aiming to recruit more BAT/beige adipose tissue and enhance their oxidative capacity might provide more significant therapeutic potential in humans over time, especially the human subjects with metabolic disorders. Future studies should also delineate the relative contribution from glucose and fatty acid in human BAT under physiological and pathological conditions; as well as compare the glucose and fatty acid partitioning in different tissues and organs, including skeletal muscle, heart, liver and BAT. These details will guide us to establish better strategies targeting metabolic disorders through BAT activation in the future.

6. Conclusion

The energy metabolism plays critical role in maintaining BAT thermogenic function in mammals. Through the energy metabolism, the chemical energy stored in nutrients (i.e., fatty acids and glucose) can be converted to thermoenergy and dissipated as heat in the BAT. Activated brown adipocytes not only contribute to the intracellular substrate homeostasis, but also contribute significantly to the whole body energy metabolism. BAT with functional thermoactivity is present in adult humans. Thus, BAT activation has remarkable therapeutic implications in human metabolic disorder management.

Acknowledgements

This work was supported by American Heart Association National Scientist Development grant, AHA SDG grant, 12050558 to Yuan Lu; and NIH R35HL135789 to Mukesh K. Jain.

Author details

Yuan Lu[1,2*]

1 Case Western Reserve University School of Medicine, Cleveland, Ohio, USA

2 Harrington Heart and Vascular Institute, University Hospitals Cleveland Medical Center, Cleveland, Ohio, USA

*Address all correspondence to: yuan.lu@case.edu

References

[1] Himms-Hagen J. Nonshivering thermogenesis. Brain Research Bulletin. 1984;**12**(2):151-160

[2] Li Y et al. Taking control over intracellular fatty acid levels is essential for the analysis of thermogenic function in cultured primary brown and brite/beige adipocytes. EMBO Reports. 2014;**15**(10):1069-1076

[3] Oelkrug R, Polymeropoulos ET, Jastroch M. Brown adipose tissue: Physiological function and evolutionary significance. Journal of Comparative Physiology. B. 2015;**185**(6):587-606

[4] Cannon B, Nedergaard J. Brown adipose tissue: Function and physiological significance. Physiological Reviews. 2004;**84**(1):277-359

[5] Kajimura S, Seale P, Spiegelman BM. Transcriptional control of brown fat development. Cell Metabolism. 2010;**11**(4):257-262

[6] Winther S et al. Restricting glycolysis impairs brown adipocyte glucose and oxygen consumption. American Journal of Physiology. Endocrinology and Metabolism. 2018;**314**(3):E214-E223

[7] Cohade C et al. Uptake in supraclavicular area fat ("USA-fat"): Description on 18F-FDG PET/CT. Journal of Nuclear Medicine. 2003;**44**(2):170-176

[8] Virtanen KA et al. Functional brown adipose tissue in healthy adults. The New England Journal of Medicine. 2009;**360**(15):1518-1525

[9] Cypess AM et al. Identification and importance of brown adipose tissue in adult humans. The New England Journal of Medicine. 2009;**360**(15):1509-1517

[10] van Marken Lichtenbelt WD et al. Cold-activated brown adipose tissue in healthy men. The New England Journal of Medicine. 2009;**360**(15):1500-1508

[11] Orava J et al. Different metabolic responses of human brown adipose tissue to activation by cold and insulin. Cell Metabolism. 2011;**14**(2):272-279

[12] Ouellet V et al. Brown adipose tissue oxidative metabolism contributes to energy expenditure during acute cold exposure in humans. The Journal of Clinical Investigation. 2012;**122**(2):545-552

[13] Chondronikola M et al. Brown adipose tissue improves whole-body glucose homeostasis and insulin sensitivity in humans. Diabetes. 2014;**63**(12):4089-4099

[14] Chondronikola M et al. Brown adipose tissue activation is linked to distinct systemic effects on lipid metabolism in humans. Cell Metabolism. 2016;**23**(6):1200-1206

[15] Foster DO. Quantitative contribution of brown adipose tissue thermogenesis to overall metabolism. Canadian Journal of Biochemistry and Cell Biology. 1984;**62**(7):618-622

[16] Lee P et al. A critical appraisal of the prevalence and metabolic significance of brown adipose tissue in adult humans. American Journal of Physiology. Endocrinology and Metabolism. 2010;**299**(4):E601-E606

[17] Iwen KA et al. Cold-induced brown adipose tissue activity alters plasma fatty acids and improves glucose metabolism in men. The Journal of Clinical Endocrinology and Metabolism. 2017;**102**(11):4226-4234

[18] Lowell BB et al. Development of obesity in transgenic mice after genetic ablation of brown adipose tissue. Nature. 1993;**366**(6457):740-742

[19] Feldmann HM et al. UCP1 ablation induces obesity and abolishes diet-induced thermogenesis in mice exempt from thermal stress by living at thermoneutrality. Cell Metabolism. 2009;**9**(2):203-209

[20] Poekes L et al. Activation of brown adipose tissue enhances the efficacy of caloric restriction for treatment of nonalcoholic steatohepatitis. Laboratory Investigation. 2018

[21] Saito M et al. High incidence of metabolically active brown adipose tissue in healthy adult humans: Effects of cold exposure and adiposity. Diabetes. 2009;**58**(7):1526-1531

[22] Yoneshiro T et al. Age-related decrease in cold-activated brown adipose tissue and accumulation of body fat in healthy humans. Obesity (Silver Spring). 2011;**19**(9):1755-1760

[23] Pfannenberg C et al. Impact of age on the relationships of brown adipose tissue with sex and adiposity in humans. Diabetes. 2010;**59**(7):1789-1793

[24] Ouellet V et al. Outdoor temperature, age, sex, body mass index, and diabetic status determine the prevalence, mass, and glucose-uptake activity of 18F-FDG-detected BAT in humans. The Journal of Clinical Endocrinology and Metabolism. 2011;**96**(1):192-199

[25] Hany TF et al. Brown adipose tissue: A factor to consider in symmetrical tracer uptake in the neck and upper chest region. European Journal of Nuclear Medicine and Molecular Imaging. 2002;**29**(10):1393-1398

[26] Enerback S. The origins of brown adipose tissue. The New England Journal of Medicine. 2009;**360**(19):2021-2023

[27] Carpentier AC et al. Brown adipose tissue energy metabolism in humans. Frontiers in Endocrinology (Lausanne). 2018;**9**:447

[28] Lindberg O et al. Studies of the mitochondrial energy-transfer system of brown adipose tissue. The Journal of Cell Biology. 1967;**34**(1):293-310

[29] Kozak UC et al. An upstream enhancer regulating brown-fat-specific expression of the mitochondrial uncoupling protein gene. Molecular and Cellular Biology. 1994;**14**(1):59-67

[30] Ishibashi J, Seale P. Medicine. Beige can be slimming. Science. 2010;**328**(5982):1113-1114

[31] Petrovic N et al. Chronic peroxisome proliferator-activated receptor gamma (PPARgamma) activation of epididymally derived white adipocyte cultures reveals a population of thermogenically competent, UCP1-containing adipocytes molecularly distinct from classic brown adipocytes. The Journal of Biological Chemistry. 2010;**285**(10):7153-7164

[32] Ikeda K et al. UCP1-independent signaling involving SERCA2b-mediated calcium cycling regulates beige fat thermogenesis and systemic glucose homeostasis. Nature Medicine. 2017;**23**(12):1454-1465

[33] Kazak L et al. A creatine-driven substrate cycle enhances energy expenditure and thermogenesis in beige fat. Cell. 2015;**163**(3):643-655

[34] Gottlieb E. p53 guards the metabolic pathway less travelled. Nature Cell Biology. 2011;**13**(3):195-197

[35] Golozoubova V et al. Only UCP1 can mediate adaptive nonshivering thermogenesis in the cold. The FASEB Journal. 2001;**15**(11):2048-2050

[36] Labbe SM et al. In vivo measurement of energy substrate

contribution to cold-induced brown adipose tissue thermogenesis. The FASEB Journal. 2015;**29**(5):2046-2058

[37] Fedorenko A, Lishko PV, Kirichok Y. Mechanism of fatty-acid-dependent UCP1 uncoupling in brown fat mitochondria. Cell. 2012;**151**(2):400-413

[38] Haemmerle G et al. Defective lipolysis and altered energy metabolism in mice lacking adipose triglyceride lipase. Science. 2006;**312**(5774):734-737

[39] Lee J, Ellis JM, Wolfgang MJ. Adipose fatty acid oxidation is required for thermogenesis and potentiates oxidative stress-induced inflammation. Cell Reports. 2015;**10**(2):266-279

[40] Shabalina IG et al. Within brown-fat cells, UCP1-mediated fatty acid-induced uncoupling is independent of fatty acid metabolism. Biochimica et Biophysica Acta. 2008;**1777**(7-8):642-650

[41] Osuga J et al. Targeted disruption of hormone-sensitive lipase results in male sterility and adipocyte hypertrophy, but not in obesity. Proceedings of the National Academy of Sciences of the United States of America. 2000;**97**(2):787-792

[42] Lock EA, Mitchell AM, Elcombe CR. Biochemical mechanisms of induction of hepatic peroxisome proliferation. Annual Review of Pharmacology and Toxicology. 1989;**29**:145-163

[43] Zechner R, Madeo F, Kratky D. Cytosolic lipolysis and lipophagy: Two sides of the same coin. Nature Reviews. Molecular Cell Biology. 2017;**18**(11):671-684

[44] Zimmermann R et al. Fate of fat: The role of adipose triglyceride lipase in lipolysis. Biochimica et Biophysica Acta. 2009;**1791**(6):494-500

[45] Londos C et al. Role of PAT proteins in lipid metabolism. Biochimie. 2005;**87**(1):45-49

[46] Brasaemle DL. Thematic review series: Adipocyte biology. The perilipin family of structural lipid droplet proteins: Stabilization of lipid droplets and control of lipolysis. Journal of Lipid Research. 2007;**48**(12):2547-2559

[47] Ducharme NA, Bickel PE. Lipid droplets in lipogenesis and lipolysis. Endocrinology. 2008;**149**(3):942-949

[48] Steinberg D, Huttunen JK. The role of cyclic AMP in activation of hormone-sensitive lipase of adipose tissue. Advances in Cyclic Nucleotide Research. 1972;**1**:47-62

[49] Sztalryd C et al. Perilipin A is essential for the translocation of hormone-sensitive lipase during lipolytic activation. The Journal of Cell Biology. 2003;**161**(6):1093-1103

[50] Granneman JG et al. Analysis of lipolytic protein trafficking and interactions in adipocytes. The Journal of Biological Chemistry. 2007;**282**(8):5726-5735

[51] Su CL et al. Mutational analysis of the hormone-sensitive lipase translocation reaction in adipocytes. The Journal of Biological Chemistry. 2003;**278**(44):43615-43619

[52] Miyoshi H et al. Perilipin promotes hormone-sensitive lipase-mediated adipocyte lipolysis via phosphorylation-dependent and -independent mechanisms. The Journal of Biological Chemistry. 2006;**281**(23):15837-15844

[53] Miyoshi H et al. Control of adipose triglyceride lipase action by serine 517 of perilipin A globally regulates protein kinase A-stimulated lipolysis in adipocytes. The Journal of Biological Chemistry. 2007;**282**(2):996-1002

[54] Zimmermann R et al. Fat mobilization in adipose tissue is promoted by adipose triglyceride lipase. Science. 2004;**306**(5700):1383-1386

[55] Jenkins CM et al. Identification, cloning, expression, and purification of three novel human calcium-independent phospholipase A2 family members possessing triacylglycerol lipase and acylglycerol transacylase activities. The Journal of Biological Chemistry. 2004;**279**(47):48968-48975

[56] Haemmerle G et al. Hormone-sensitive lipase deficiency in mice causes diglyceride accumulation in adipose tissue, muscle, and testis. The Journal of Biological Chemistry. 2002;**277**(7):4806-4815

[57] Fredrikson G, Tornqvist H, Belfrage P. Hormone-sensitive lipase and monoacylglycerol lipase are both required for complete degradation of adipocyte triacylglycerol. Biochimica et Biophysica Acta. 1986;**876**(2):288-293

[58] Blondin DP et al. Inhibition of intracellular triglyceride lipolysis suppresses cold-induced Brown adipose tissue metabolism and increases shivering in humans. Cell Metabolism. 2017;**25**(2):438-447

[59] Schreiber R et al. Cold-induced thermogenesis depends on ATGL-mediated lipolysis in cardiac muscle, but not Brown adipose tissue. Cell Metabolism. 2017;**26**(5):753 e7-763 e7

[60] Shin H et al. Lipolysis in brown adipocytes is not essential for cold-induced thermogenesis in mice. Cell Metabolism. 2017;**26**(5):764 e5-777 e5

[61] Ahmed K, Tunaru S, Offermanns S. GPR109A, GPR109B and GPR81, a family of hydroxy-carboxylic acid receptors. Trends in Pharmacological Sciences. 2009;**30**(11):557-562

[62] Mulder H et al. Hormone-sensitive lipase null mice exhibit signs of impaired insulin sensitivity whereas insulin secretion is intact. The Journal of Biological Chemistry. 2003;**278**(38):36380-36388

[63] Neely JR, Rovetto MJ, Oram JF. Myocardial utilization of carbohydrate and lipids. Progress in Cardiovascular Diseases. 1972;**15**(3):289-329

[64] Mitchell GA et al. Medical aspects of ketone body metabolism. Clinical and Investigative Medicine. 1995;**18**(3):193-216

[65] Britton CH et al. Human liver mitochondrial carnitine palmitoyltransferase I: Characterization of its cDNA and chromosomal localization and partial analysis of the gene. Proceedings of the National Academy of Sciences of the United States of America. 1995;**92**(6):1984-1988

[66] McGarry JD, Brown NF. The mitochondrial carnitine palmitoyltransferase system. From concept to molecular analysis. European Journal of Biochemistry. 1997;**244**(1):1-14

[67] Furuhashi M, Hotamisligil GS. Fatty acid-binding proteins: Role in metabolic diseases and potential as drug targets. Nature Reviews. Drug Discovery. 2008;**7**(6):489-503

[68] Hunt CR et al. Adipocyte P2 gene: Developmental expression and homology of 5′-flanking sequences among fat cell-specific genes. Proceedings of the National Academy of Sciences of the United States of America. 1986;**83**(11):3786-3790

[69] Haunerland NH, Spener F. Fatty acid-binding proteins–Insights from genetic manipulations. Progress in Lipid Research. 2004;**43**(4):328-349

[70] Makowski L, Hotamisligil GS. The role of fatty acid binding proteins in metabolic syndrome and atherosclerosis. Current Opinion in Lipidology. 2005;**16**(5):543-548

[71] Shen WJ et al. Interaction of rat hormone-sensitive lipase with adipocyte lipid-binding protein. Proceedings of the National Academy of Sciences of the United States of America. 1999;**96**(10):5528-5532

[72] Scheja L et al. Altered insulin secretion associated with reduced lipolytic efficiency in aP2−/− mice. Diabetes. 1999;**48**(10):1987-1994

[73] Coe NR, Simpson MA, Bernlohr DA. Targeted disruption of the adipocyte lipid-binding protein (aP2 protein) gene impairs fat cell lipolysis and increases cellular fatty acid levels. Journal of Lipid Research. 1999;**40**(5):967-972

[74] Syamsunarno MR et al. Fatty acid binding protein 4 and 5 play a crucial role in thermogenesis under the conditions of fasting and cold stress. PLoS One. 2014;**9**(6):e90825

[75] Esser V et al. Expression of a cDNA isolated from rat brown adipose tissue and heart identifies the product as the muscle isoform of carnitine palmitoyltransferase I (M-CPT I). M-CPT I is the predominant CPT I isoform expressed in both white (epididymal) and brown adipocytes. The Journal of Biological Chemistry. 1996;**271**(12):6972-6977

[76] Nyman LR et al. Homozygous carnitine palmitoyltransferase 1a (liver isoform) deficiency is lethal in the mouse. Molecular Genetics and Metabolism. 2005;**86**(1-2):179-187

[77] Yamazaki N et al. High expression of a novel carnitine palmitoyltransferase I like protein in rat brown adipose tissue and heart: Isolation and characterization of its cDNA clone. FEBS Letters. 1995;**363**(1-2):41-45

[78] Ji S et al. Homozygous carnitine palmitoyltransferase 1b (muscle isoform) deficiency is lethal in the mouse. Molecular Genetics and Metabolism. 2008;**93**(3):314-322

[79] Gonzalez-Hurtado E et al. Fatty acid oxidation is required for active and quiescent brown adipose tissue maintenance and thermogenic programing. Molecular Metabolism. 2018;**7**:45-56

[80] Schuler AM et al. Synergistic heterozygosity in mice with inherited enzyme deficiencies of mitochondrial fatty acid beta-oxidation. Molecular Genetics and Metabolism. 2005;**85**(1):7-11

[81] Thorpe C, Kim JJ. Structure and mechanism of action of the acyl-CoA dehydrogenases. The FASEB Journal. 1995;**9**(9):718-725

[82] Gregersen N et al. Mutation analysis in mitochondrial fatty acid oxidation defects: Exemplified by acyl-CoA dehydrogenase deficiencies, with special focus on genotype-phenotype relationship. Human Mutation. 2001;**18**(3):169-189

[83] Bartelt A et al. Brown adipose tissue activity controls triglyceride clearance. Nature Medicine. 2011;**17**(2):200-205

[84] Khedoe PP et al. Brown adipose tissue takes up plasma triglycerides mostly after lipolysis. Journal of Lipid Research. 2015;**56**(1):51-59

[85] Berbee JF et al. Brown fat activation reduces hypercholesterolaemia and protects from atherosclerosis development. Nature Communications. 2015;**6**:6356

[86] Merkel M et al. Apolipoprotein AV accelerates plasma hydrolysis

of triglyceride-rich lipoproteins by interaction with proteoglycan-bound lipoprotein lipase. The Journal of Biological Chemistry. 2005;**280**(22):21553-21560

[87] Pennacchio LA et al. An apolipoprotein influencing triglycerides in humans and mice revealed by comparative sequencing. Science. 2001;**294**(5540):169-173

[88] Blondin DP et al. Dietary fatty acid metabolism of brown adipose tissue in cold-acclimated men. Nature Communications. 2017;**8**:14146

[89] Amri EZ et al. Maturation and secretion of lipoprotein lipase in cultured adipose cells. II. Effects of tunicamycin on activation and secretion of the enzyme. Biochimica et Biophysica Acta. 1986;**875**(2):334-343

[90] Carneheim C, Nedergaard J, Cannon B. Cold-induced beta-adrenergic recruitment of lipoprotein lipase in brown fat is due to increased transcription. The American Journal of Physiology. 1988;**254**(2 Pt 1):E155-E161

[91] Ong JM, Kern PA. The role of glucose and glycosylation in the regulation of lipoprotein lipase synthesis and secretion in rat adipocytes. The Journal of Biological Chemistry. 1989;**264**(6):3177-3182

[92] Stahl A. A current review of fatty acid transport proteins (SLC27). Pflügers Archiv. 2004;**447**(5):722-727

[93] Anderson CM, Stahl A. SLC27 fatty acid transport proteins. Molecular Aspects of Medicine. 2013;**34**(2-3):516-528

[94] Wang H, Eckel RH. Lipoprotein lipase: From gene to obesity. American Journal of Physiology. Endocrinology and Metabolism. 2009;**297**(2):E271-E288

[95] Beigneux AP et al. Glycosylphosphatidylinositol-anchored high-density lipoprotein-binding protein 1 plays a critical role in the lipolytic processing of chylomicrons. Cell Metabolism. 2007;**5**(4):279-291

[96] Davies BS et al. GPIHBP1 is responsible for the entry of lipoprotein lipase into capillaries. Cell Metabolism. 2010;**12**(1):42-52

[97] Beisiegel U, Weber W, Bengtsson-Olivecrona G. Lipoprotein lipase enhances the binding of chylomicrons to low density lipoprotein receptor-related protein. Proceedings of the National Academy of Sciences of the United States of America. 1991;**88**(19):8342-8346

[98] Mead JR, Irvine SA, Ramji DP. Lipoprotein lipase: Structure, function, regulation, and role in disease. Journal of Molecular Medicine (Berlin, Germany). 2002;**80**(12):753-769

[99] Wang CS, Hartsuck J, McConathy WJ. Structure and functional properties of lipoprotein lipase. Biochimica et Biophysica Acta. 1992;**1123**(1):1-17

[100] Wu Q et al. Fatty acid transport protein 1 is required for nonshivering thermogenesis in brown adipose tissue. Diabetes. 2006;**55**(12):3229-3237

[101] Wu Q et al. FATP1 is an insulin-sensitive fatty acid transporter involved in diet-induced obesity. Molecular and Cellular Biology. 2006;**26**(9):3455-3467

[102] Harmon CM, Abumrad NA. Binding of sulfosuccinimidyl fatty acids to adipocyte membrane proteins: Isolation and amino-terminal sequence of an 88-kD protein implicated in transport of long-chain fatty acids. The Journal of Membrane Biology. 1993;**133**(1):43-49

[103] Tao N, Wagner SJ, Lublin DM. CD36 is palmitoylated on both

N- and C-terminal cytoplasmic tails. The Journal of Biological Chemistry. 1996;**271**(37):22315-22320

[104] Anderson CM et al. Dependence of brown adipose tissue function on CD36-mediated coenzyme Q uptake. Cell Reports. 2015;**10**(4):505-515

[105] Som P et al. A fluorinated glucose analog, 2-fluoro-2-deoxy-D-glucose (F-18): Nontoxic tracer for rapid tumor detection. Journal of Nuclear Medicine. 1980;**21**(7):670-675

[106] Kelloff GJ et al. Progress and promise of FDG-PET imaging for cancer patient management and oncologic drug development. Clinical Cancer Research. 2005;**11**(8):2785-2808

[107] Greco-Perotto R et al. Stimulatory effect of cold adaptation on glucose utilization by brown adipose tissue. Relationship with changes in the glucose transporter system. The Journal of Biological Chemistry. 1987;**262**(16):7732-7736

[108] Vallerand AL, Perusse F, Bukowiecki LJ. Cold exposure potentiates the effect of insulin on in vivo glucose uptake. The American Journal of Physiology. 1987;**253**(2 Pt 1):E179-E186

[109] Vallerand AL, Perusse F, Bukowiecki LJ. Stimulatory effects of cold exposure and cold acclimation on glucose uptake in rat peripheral tissues. The American Journal of Physiology. 1990;**259**(5 Pt 2):R1043-R1049

[110] Shimizu Y, Nikami H, Saito M. Sympathetic activation of glucose utilization in brown adipose tissue in rats. Journal of Biochemistry. 1991;**110**(5):688-692

[111] Yu XX et al. Cold elicits the simultaneous induction of fatty acid synthesis and beta-oxidation in murine brown adipose tissue: Prediction from differential gene expression and confirmation in vivo. The FASEB Journal. 2002;**16**(2):155-168

[112] Shimizu Y et al. Increased expression of glucose transporter GLUT-4 in brown adipose tissue of fasted rats after cold exposure. The American Journal of Physiology. 1993;**264**(6 Pt 1):E890-E895

[113] Wang X, Wahl R. Responses of the insulin signaling pathways in the brown adipose tissue of rats following cold exposure. PLoS One. 2014;**9**(6):e99772

[114] Shibata H et al. Cold exposure reverses inhibitory effects of fasting on peripheral glucose uptake in rats. The American Journal of Physiology. 1989;**257**(1 Pt 2):R96-R101

[115] Ebner S et al. Effects of insulin and norepinephrine on glucose transport and metabolism in rat brown adipocytes. Potentiation by insulin of norepinephrine-induced glucose oxidation. European Journal of Biochemistry. 1987;**170**(1-2):469-474

[116] Hutchinson DS et al. Beta-adrenoceptors, but not alpha-adrenoceptors, stimulate AMP-activated protein kinase in brown adipocytes independently of uncoupling protein-1. Diabetologia. 2005;**48**(11):2386-2395

[117] Olsen JM et al. Glucose uptake in brown fat cells is dependent on mTOR complex 2-promoted GLUT1 translocation. The Journal of Cell Biology. 2014;**207**(3):365-374

[118] Bell GI et al. Molecular biology of mammalian glucose transporters. Diabetes Care. 1990;**13**(3):198-208

[119] Gatenby RA, Gillies RJ. Why do cancers have high aerobic glycolysis? Nature Reviews. Cancer. 2004;**4**(11):891-899

[120] Marette A, Bukowiecki LJ. Noradrenaline stimulates glucose

transport in rat brown adipocytes by activating thermogenesis. Evidence that fatty acid activation of mitochondrial respiration enhances glucose transport. The Biochemical Journal. 1991;**277**(Pt 1): 119-124

[121] Lee P et al. Brown adipose tissue exhibits a glucose-responsive thermogenic biorhythm in humans. Cell Metabolism. 2016;**23**(4):602-609

[122] Shimizu Y et al. Noradrenaline increases glucose transport into brown adipocytes in culture by a mechanism different from that of insulin. The Biochemical Journal. 1996;**314**(Pt 2):485-490

[123] Ma SW, Foster DO. Uptake of glucose and release of fatty acids and glycerol by rat brown adipose tissue in vivo. Canadian Journal of Physiology and Pharmacology. 1986;**64**(5):609-614

[124] Jeong JH, Chang JS, Jo YH. Intracellular glycolysis in brown adipose tissue is essential for optogenetically induced nonshivering thermogenesis in mice. Scientific Reports. 2018;**8**(1):6672

[125] Saggerson ED, McAllister TW, Baht HS. Lipogenesis in rat brown adipocytes. Effects of insulin and noradrenaline, contributions from glucose and lactate as precursors and comparisons with white adipocytes. The Biochemical Journal. 1988;**251**(3):701-709

[126] Isler D, Hill HP, Meier MK. Glucose metabolism in isolated brown adipocytes under beta-adrenergic stimulation. Quantitative contribution of glucose to total thermogenesis. The Biochemical Journal. 1987;**245**(3):789-793

[127] Nikami H et al. Cold exposure increases glucose utilization and glucose transporter expression in brown adipose tissue. Biochemical and Biophysical Research Communications. 1992;**185**(3):1078-1082

[128] Smith DM et al. Glucose transporter expression and glucose utilization in skeletal muscle and brown adipose tissue during starvation and re-feeding. The Biochemical Journal. 1992;**282**(Pt 1):231-235

[129] Sugden MC, Holness MJ. Physiological modulation of the uptake and fate of glucose in brown adipose tissue. The Biochemical Journal. 1993;**295**(Pt 1):171-176

[130] Hankir MK et al. Dissociation between brown adipose tissue (18) F-FDG uptake and thermogenesis in uncoupling protein 1-deficient mice. Journal of Nuclear Medicine. 2017;**58**(7):1100-1103

[131] Olsen JM et al. beta3-Adrenergically induced glucose uptake in brown adipose tissue is independent of UCP1 presence or activity: Mediation through the mTOR pathway. Molecular Metabolism. 2017;**6**(6):611-619

[132] Weir G et al. Substantial metabolic activity of human brown adipose tissue during warm conditions and cold-induced lipolysis of local triglycerides. Cell Metabolism. 2018;**27**(6):1348 e4-1355 e4

[133] Chakrabarty K, Chaudhuri B, Jeffay H. Glycerokinase activity in human brown adipose tissue. Journal of Lipid Research. 1983;**24**(4):381-390

[134] Kawashita NH et al. Glycerokinase activity in brown adipose tissue: A sympathetic regulation? American Journal of Physiology. Regulatory, Integrative and Comparative Physiology. 2002;**282**(4):R1185-R1190

[135] Festuccia WT et al. Control of glyceroneogenic activity in rat brown adipose tissue. American Journal of Physiology. Regulatory, Integrative and Comparative Physiology. 2003;**285**(1):R177-R182

[136] Sanchez-Gurmaches J et al. Brown fat AKT2 is a cold-induced kinase that stimulates ChREBP-mediated *De novo* lipogenesis to optimize fuel storage and thermogenesis. Cell Metabolism. 2018;**27**(1):195 e6-209 e6

[137] Song Z, Xiaoli AM, Yang F. Regulation and metabolic significance of *De novo* lipogenesis in adipose tissues. Nutrients. 2018;**10**(10):1383

[138] Guilherme A et al. Adipocyte lipid synthesis coupled to neuronal control of thermogenic programming. Molecular Metabolism. 2017;**6**(8):781-796

[139] Farkas V, Kelenyi G, Sandor A. A dramatic accumulation of glycogen in the brown adipose tissue of rats following recovery from cold exposure. Archives of Biochemistry and Biophysics. 1999;**365**(1):54-61

[140] Hurtado del Pozo C et al. ChREBP expression in the liver, adipose tissue and differentiated preadipocytes in human obesity. Biochimica et Biophysica Acta. 2011;**1811**(12):1194-1200

[141] Witte N et al. The glucose sensor ChREBP links *De novo* lipogenesis to PPARgamma activity and adipocyte differentiation. Endocrinology. 2015;**156**(11):4008-4019

[142] Iizuka K et al. Deficiency of carbohydrate response element-binding protein (ChREBP) reduces lipogenesis as well as glycolysis. Proceedings of the National Academy of Sciences of the United States of America. 2004;**101**(19):7281-7286

[143] Herman MA et al. A novel ChREBP isoform in adipose tissue regulates systemic glucose metabolism. Nature. 2012;**484**(7394):333-338

[144] Vijayakumar A et al. Absence of carbohydrate response element binding protein in adipocytes causes systemic insulin resistance an impairs glucose transport. Cell Reports. 2017;**21**(4):1021-1035

[145] Eberle D et al. SREBP transcription factors: Master regulators of lipid homeostasis. Biochimie. 2004;**86**(11):839-848

[146] Osborne TF. Sterol regulatory element-binding proteins (SREBPs): Key regulators of nutritional homeostasis and insulin action. The Journal of Biological Chemistry. 2000;**275**(42):32379-32382

[147] Kim JB, Spiegelman BM. ADD1/SREBP1 promotes adipocyte differentiation and gene expression linked to fatty acid metabolism. Genes & Development. 1996;**10**(9):1096-1107

[148] Horton JD et al. Overexpression of sterol regulatory element-binding protein-1a in mouse adipose tissue produces adipocyte hypertrophy, increased fatty acid secretion, and fatty liver. The Journal of Biological Chemistry. 2003;**278**(38):36652-36660

[149] Shimano H et al. Elevated levels of SREBP-2 and cholesterol synthesis in livers of mice homozygous for a targeted disruption of the SREBP-1 gene. The Journal of Clinical Investigation. 1997;**100**(8):2115-2124

[150] Yahagi N et al. Absence of sterol regulatory element-binding protein-1 (SREBP-1) ameliorates fatty livers but not obesity or insulin resistance in Lep(ob)/Lep(ob) mice. The Journal of Biological Chemistry. 2002;**277**(22):19353-19357

[151] Guerra C et al. Emergence of brown adipocytes in white fat in mice is under genetic control. Effects on body weight and adiposity. The Journal of Clinical Investigation. 1998;**102**(2):412-420

[152] Cederberg A et al. FOXC2 is a winged helix gene that counteracts obesity, hypertriglyceridemia, and

diet-induced insulin resistance. Cell. 2001;**106**(5):563-573

[153] Kiefer FW et al. Retinaldehyde dehydrogenase 1 regulates a thermogenic program in white adipose tissue. Nature Medicine. 2012;**18**(6):918-925

[154] Schulz TJ et al. Brown-fat paucity due to impaired BMP signalling induces compensatory browning of white fat. Nature. 2013;**495**(7441):379-383

[155] Orava J et al. Blunted metabolic responses to cold and insulin stimulation in brown adipose tissue of obese humans. Obesity (Silver Spring). 2013;**21**(11):2279-2287

[156] Blondin DP et al. Selective impairment of glucose but not fatty acid or oxidative metabolism in brown adipose tissue of subjects with type 2 diabetes. Diabetes. 2015;**64**(7):2388-2397

[157] Au-Yong IT et al. Brown adipose tissue and seasonal variation in humans. Diabetes. 2009;**58**(11):2583-2587

[158] Leitner BP et al. Mapping of human brown adipose tissue in lean and obese young men. Proceedings of the National Academy of Sciences of the United States of America. 2017;**114**(32):8649-8654

[159] Hanssen MJ et al. Short-term cold acclimation improves insulin sensitivity in patients with type 2 diabetes mellitus. Nature Medicine. 2015;**21**(8):863-865

[160] Hoeke G et al. Short-term cooling increases serum triglycerides and small high-density lipoprotein levels in humans. Journal of Clinical Lipidology. 2017;**11**(4):920 e2-928 e2

8

Diabetes Mellitus: A Group of Genetic-Based Metabolic Diseases

Lilian Sanhueza, Pilar Durruty, Cecilia Vargas, Paulina Vignolo and Karina Elgueta

Abstract

Diabetes mellitus (DM) is a disease characterized by defects in action and/or secretion of insulin that results in chronic hyperglycemia and long-term severe vascular complications. The main clinical presentations with the proven genetic base are covered. Type 1 diabetes (DM1) is an autoimmune, heterogeneous, multifactorial, and polygenic-based disease. Selectively destroys 90% of beta cells of the pancreas, mediated by activated T lymphocytes in patients with haplotypes linked to major histocompatibility complex (MHC). Genetic and genomic studies have been carried out in family groups, demonstrating up to 15 affected chromosomal regions. Type 2 diabetes (DM2) presents genes with various polymorphisms which, together with postgenomic and environmental factors, make it more complex to understand the pathogenesis. Monogenic diabetes comprises neonatal diabetes (ND), maturity onset diabetes in young (MODY), an autosomal dominant transmission which is inherited directly in three successive generations, and the very rare mitochondrial diabetes. Latent autoimmune diabetes in adults (LADA) mainly affects patients with normal weight and initially diagnosed as DM2. Its characteristics are low levels of C-peptide in both fasting and postglucagon tests. They present MHC alleles of susceptibility and positive autoantibodies: Anti-decarboxylase glutamic acid.

Keywords: diabetes mellitus type 1, diabetes mellitus type 2, monogenic diabetes, LADA

1. Introduction

Diabetes mellitus (DM) is a group of metabolic diseases of different etiologies characterized by chronic hyperglycemia resulting from a deficit in both the secretion and the action of insulin hormone. Now-a-days, there is a genetic basis of these clinical manifestations. In this chapter we, describe the most important ones such as diabetes mellitus type 1 (DM1), diabetes mellitus type 2 (DM2), monogenic diabetes, and latent autoimmune diabetes in adults (LADA).

2. Diabetes mellitus type 1

DM1 is characterized by autoimmune destruction of the beta cells of the pancreatic islets, which leads to an extreme insulinopenia. The character of autoimmunity confirm the presence of islet cell antibodies (ICA), insulin auto antibodies (IAA), glutamic acid decarboxylase auto antibodies (GAD65), protein tyrosine phosphatase 2 (IA-2), and zinc transporter gene ZnT8 [1]. There is an interaction between genetic and environmental factors in the development of DM1 (**Figure 1**).

2.1 Genetic factors of DM1

DM1 is a polygenic disease, with at least 15 associated chromosomal regions. The leading group of genes that predispose to type 1 diabetes is located on human chro-mosome 6 specifically at 6p21, and this chromosomal region contains a group of genes called major histocompatibility complex (MHC), responsible for the immune response and the antigen presentation of the beta cell to T lymphocytes [2]. The classic histocompatibility genes are extremely polymorphic (amino acid sequence differs among individuals) and include MHC-A, B, and C molecules (class I his-tocompatibility antigens) and the immune-response genes DP, DQ, and DR (class II histocompatibility antigens). Numbers (DR3, DR4; A1, A2; B1, B8) are given to distinguish different alleles of any given gene. The designation w (workshop) with numbers is given for provisionally named alleles (DQw8, DQw7) [3]. DM1 has been associated mainly with allelic variants of MHC-DR (DR3/DR4). The MHC locus is a genetic factor of great importance in DM1, and it was first shown in an association study that revealed that about 95% of all patients with DM1 were heterozygous for MHC-DR3/DR4. The majority of type 1 diabetics have the MHC-DR3, MHC-DR4 haplotype, or both [4]. Susceptibility to DM type 1 is associated with these linked DQ alleles that are often in linkage disequilibrium with DR. The closest associa-tion in DM1 occurs with the haplotypes DQA1*0301, DQB1*0302, DQA1*501, and DQB1*0201. It has been shown that the beta DQ chains of those affected have valine, alanine, or serine at position 57; near the peptide-binding gap, presence of aspartic acid is normal [5] Factors involved in the pathophysiology of DM1 are shown in (**Figures 2** and **3**).

However, there are exceptions to this association, which indicates that amino acids other than Asp57 at position 57 of the beta chain play an essential but not exclusive role in the susceptibility to DM1 [5]. There is an interaction between genetic and environmental factors in the development of DM1.

2.2 Heritage in type 1 DM

The inheritance of DM1 is unknown. Several hypotheses have been suggested, such as that of dominant inheritance, but it is ruled out by the rarity of DM1 in relatives, children, and descendants. The possibility of recessive inheritance was also

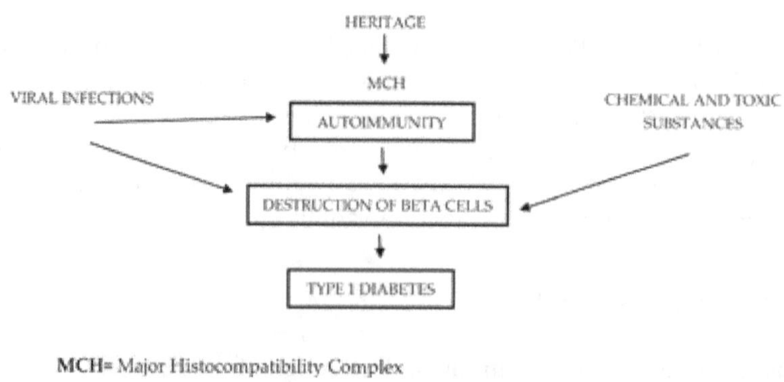

Figure 1.
Pathogenesis of type 1 diabetes.

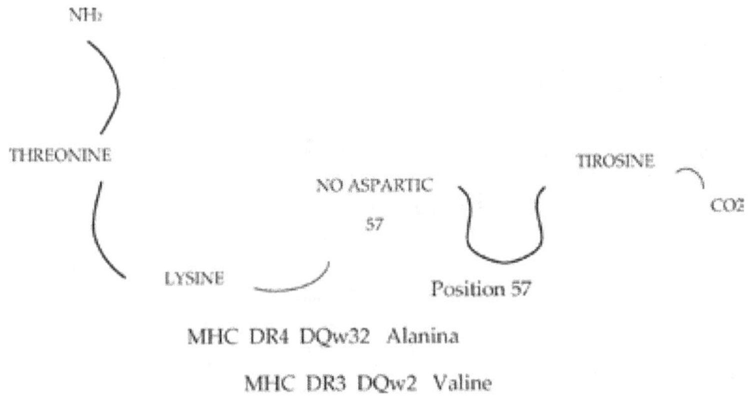

Figure 2.
Etiology of type 1 diabetes. MHC-DQ B antigen amino acid sequence.

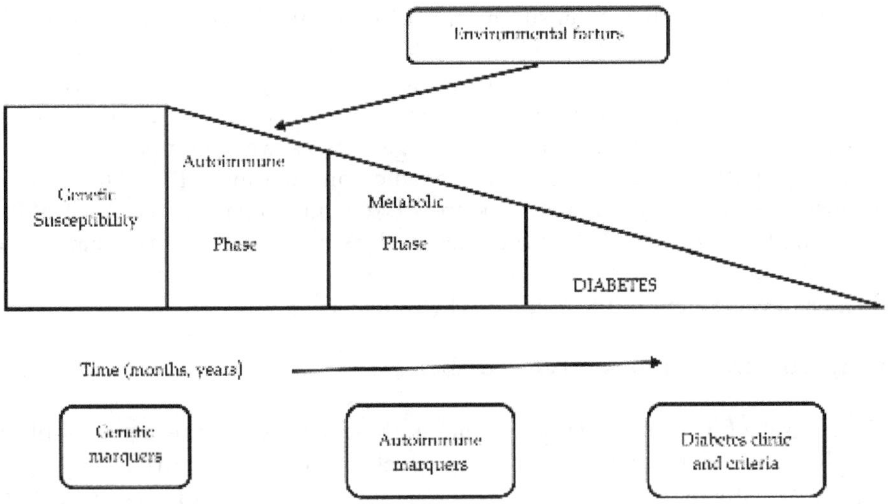

Figure 3.
Pathophysiology of type 1 diabetes.

considered, however, invalidated because in homozygotes for the DR3 or DR4 alleles, the susceptibility to the disease is not increased. The observation that heterozygosis DR3/DR4 increases the risk for diabetes, compared with that presented by homozygotes from other high-risk alleles, suggests a polygenic way of inheritance. The diabetogenic MHC haplotype is necessary for the susceptibility to DM1, but must be positively or negatively influenced by genes not linked to MHC, such as the gene located close to the repeated DNA sequence minisatellite, in the promoter region of the gene of the insulin (chromosome 11p15); one gene on chromosome 11q and another on chromosome 6q [6]. Some genes seem to confer protection against the development of the disease. For example, the DQA1*0102 and DQB1*0602 haplotypes are present in 20% of the United States population but is extremely rare in individuals with DM1 (<1%). This situation indicates that several genes are interacting to determine the DM1 phenotype, so this disease presents genetic heterogeneity (**Figure 4**).

DM1 is uncommon in Chile and usually does not occur in native Chilean families. A study of a family with an affected female child was carried out in a Mapuche community in the Southern (VIII region). This case is a unique and sporadic DM case with Mapuche heritage. Genetic analysis by PCR was done

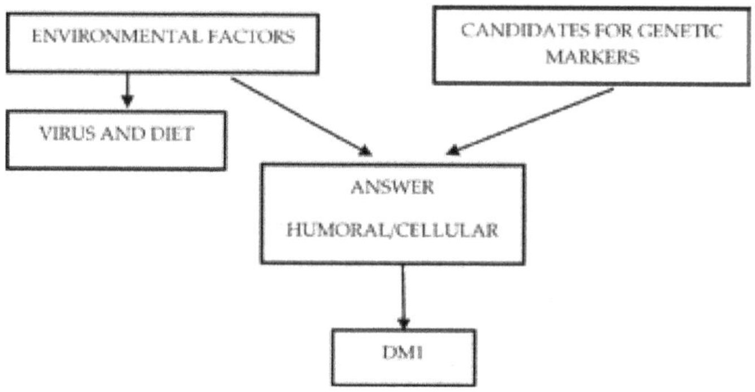

Figure 4.
Interaction of genetic and environmental factors about the immune response in type 1 diabetes.

to detect class I and II HLA genes by reverse dot blot. The proband, her mother, and sister had positive islet cell antibodies (ICA). Her father and brother were negative. All the family was positive for anti-glutamic decarboxylase antibodies (GAD65). All subjects had HLA-DRB1 0407/0407 and HLA-DQB1 0302/0302 alleles. The index case and her father were homozygotes for the HLA-A1: A*68012/A*68012 allele. No evidence of viral infections such as rubella, mumps, or measles was found in this family. All genotypes were compared with the European population, where the diabetogenic combination DR4/DQB1*0302 is the most prevalent [5]. Finally, despite of the high relative risk of DM1 in subjects with certain MHC class II alleles, it does not develop in the majority of people who inherit these alleles, which also suggests that environmental factors influence development of the disease [1] (**Figure 4**).

2.3 Clinical manifestations of type 1 diabetes

The onset of DM1 is usually abrupt, with severe symptoms attributable to hyperglycemia maintained for days or weeks, such as polyuria, polydipsia, polyphagia, asthenia, and progressive weight loss, and manifests as diabetic ketoacidosis. Two pathophysiological situations must be present to establish this condition: extreme insulinopenia and increase of counterregulatory hormones, principally glucagon. DM1 is observed mainly in children, adolescents, and young adults, generally under 30 years, although it can also appear in individuals of more advanced ages.

3. Diabetes mellitus type 2

DM2 has an important genetic component in its pathogenesis. There are multiple genes involved in different metabolic pathways that contribute to the pathogenesis of the disease, in addition to environmental factors such as obesity, an unhealthy diet, and a sedentary lifestyle. These genes have various polymorphisms, which added to post-genomic factors related to their expression and inhibition are respon-sible for this complex disease, as seen in **Figure 5**.

3.1 Heritage in type 2 diabetes

The genetic nature of DM2 has been based on high heritability, estimated at 30–70% and high prevalence in some ethnic groups; 39% of patients with DM2 have at least one relative with the disease. There are 1.5–3 times higher risk of presenting

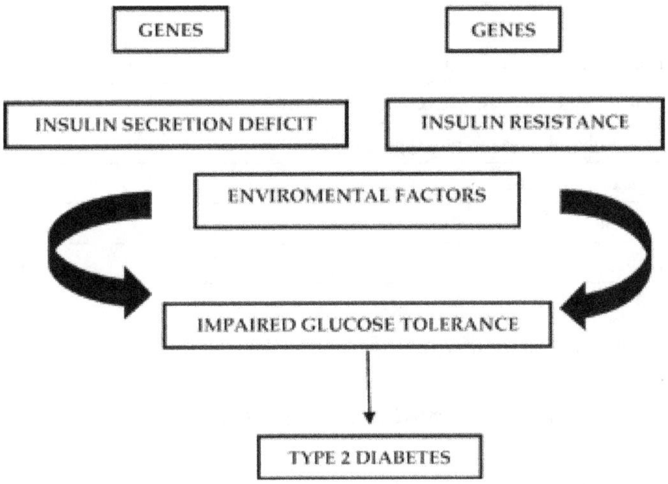

Figure 5.
Pathogenesis of type 2 diabetes.

DM2 if there is a history of the disease in the family; if the affected one is the mother or the brother, the risk is of 2 and 3 times greater, respectively. The heritability described is mainly in middle-aged people (35–60 years) decreasing markedly in larger groups [6]. First-degree relatives of people with DM2 show early defects in insulin withdrawal and action [7]. If other factors such as obesity and impaired fasting glycemia are added to the family history, the risk of presenting the disease is 16 times higher. The presentation of DM2 involves genes that code for proteins or related enzymes in the process of pancreatic formation, synthesis, sectioning, and insulin action. Genome-wide association studies (GWAS) have shown more than 100 locus susceptible to DM2, most related to insulin sequestration, suggesting that this alteration, essential for the presentation of the disease, is more strongly deter-mined by genetic factors.

The genetic susceptibility given by the presence of risk variants is responsible for about 10% of the family aggregation of the disease. There is a considerable percent-age of "missing heritability" that could be explained by: frequent variants of low power, others of low frequency but powerful effect, interaction between gene–gene, gene–environment interaction, epigenetic factors, and others.

3.2 Genetic defects in the action of insulin

The genetic variants associated with the action of insulin and DM2 are related to the transcription of the intracellular signal of insulin. Due directly to protein mutations of the signaling cascade intracellularly or indirectly due to mutations in genes associated with metabolic syndrome, such as those related to obesity and lipid metabolism. Yaghootkar et al. evolved cluster of 17 genetic variables associated with insulin resistance related to the development of DM2 [8]. Some of the genes that are most related to DM2 appear in **Table 1** [9].

3.2.1 Insulin receptor substrate-1 gene (IRS-1)

Located on chromosome 2, it encodes peptides related to the insulin signaling cascade. Arg972Gly mutation, a common variant of IRS-1, is more prevalent in Caucasian DM2 than in non DM2, and in obese adults, it has been associated with increased insulin resistance.

Insulin resistance genes
• Insulin receptor substrate gene-1 (IRS-1)
• Peroxisome proliferator-activating receptor gene (PPARϒ)
• Protein tyrosine phosphatase receptor type D (PTPRD)
• β-3 adrenergic receptor gene
• Adiponectin gene
• Leptin gene

Table 1.
Candidate genes of DM2 for insulin resistance.

3.2.2 Peroxisome proliferator-activated receptor gamma 2 gene (PPARG)

Located on chromosome 3, it codes for the peroxisome proliferator-activated receptor. It has a key role in adipocyte differentiation. The presence of a type of polymorphism is associated with 1.25 odds ratio (OR) for DM2.

3.2.3 Protein tyrosine phosphatase receptor type D gene (PTPRD)

Located on chromosome 9, it is encoded for PTPRD. Its overexpression in the skeletal muscle generates insulin resistance. Diabetes-related polymorphism has been evidenced in Chinese with an OR 1.57.

3.2.4 β-3 adrenergic receptor gene

It regulates lipolysis of visceral fat and is related to thermogenesis. It is associated with risk of obesity and early presentation of DM2.

3.2.5 Adiponectin gene

It is located in chromosome 3q27. Low levels of adiponectin have a role in the pathogenesis of insulin resistance and obesity. Insulinosensitivity is a consistent and independent predictor factor of DM2. Variants in the genes that code for adiponec-tin receptor have proven to be a risk factor for presenting DM2 in some populations.

3.2.6 Leptin gene

Mutations related to this gene are involved with the pathogenesis of obesity and glucose metabolism, thereby decreasing insulin sensitivity and inhibiting the expression of the pre-proinsulin gene in the pancreatic β-cells. Recent evidence suggests that high circulating levels of leptin probably independent of adiposity are associated with an increased risk of type 2 diabetes in men.

3.3 Genetic defects in insulin secretion

There are multiple loci associated with this defect that have been found in GWAS studies. Among them most relevant are those presented in **Table 2** [9].

3.3.1 Calpain 10 gene (CAPN10)

Encodes a family of calpain enzymes, it was one of the first to study in linkage, but it is currently known that the risk of this association is low OR 1.17 [10].

Insulin secretion genes
• Calpain 10 gene (CAPN10)
• Transcription factor 7-like 2 (TCF7L2)
• Potassium voltage gated channel subfamily
• Q member 1 (KCNQ1)
• J member 11 (KCNJ11)

Table 2.
Candidate genes of DM2 for insulin secretion.

3.3.2 Transcription factor 7-like 2 gene (TCF7L2)

It has appeared to be more relevant in the genetic susceptibility to DM2, since a polymorphism of this gene has been found in several ethnic groups of DM2 patients. The increased expression of the gene in the pancreatic beta cell causes secretion alteration due to a decrease in the incretin effect. In liver and adipose tissue, it generates insulin resistance. The risk of DM is consistent with an OR up to 2.5 for homozygous variable [11].

3.3.3 Potassium voltage-gated channel subfamily

Q member 1 (KCNQ1) located on chromosome 11, it codes for the same name channel present in the cell membrane. There are four variants associated with DM2 in various populations. Studies suggest that the effect linked to DM2 is related to epigenetic modifications. J Member 11 (KCNJ11): code for Kir6.2 ATP-sensitive potassium channel. Variant E23K increases the risk of DM2 by 1.2 times associated with decreased insulin sequestration [12].

3.4 Epigenetics in DM2

Epigenetics or genetic modifications not associated to nucleotide mutations that influence the expression of a gene play a key role in the pathogenesis and T2DM complications. There are prenatal factors that induce epigenetic changes that increase the risk of T2DM by altering the secretion and sensitivity of insulin, hepatic glucose production, and the release of hormones involved in glucose metabolism.

The sustained activation of inflammatory-related genes in T2DM patients by epigenetic mechanisms contribute to the progression of vascular complications, arteriosclerosis, and retinopathy.

Types of epigenetic modifications and relation to DM2 are shown in **Figure 6**.

3.4.1 Methylation and histone-modification

These are the epigenetic modifications most associated to vascular complications related to DM2. Both hypomethylations and hypermethylations generate persistent activation of proaterogenic genes such as NF-kB-dependent oxidative and inflam-matory signaling pathway.

3.4.2 Non-coding RNAs (ncRNAs) and chromatin remodeling

Non-coding nRNAs play an essential role in post-transcriptional regulation of gene expression.

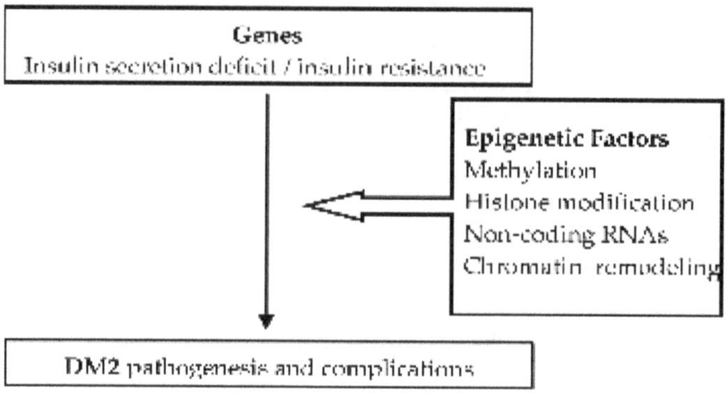

Figure 6.
DM2 epigenetic factors.

The most extensively studied are short nucleotide sequences (18–25) called MicroRNA (miRNA) and represent the principal epigenetic regulators of gene expression.

Deregulation in epithelial cells is correlated to the risk of developing vascular complications in DM2.

There are miRNAs associated with inflammation; for example, miR-155, which is associated to the progression of kidney disease in DM2 patients and miR-126, that when inhibited in pre-diabetes patients is correlated with the increase of activation of the NF-kb pathway in endothelial cells.

Long non-coding RNAs (lncRNAs) have been associated with pancreatic B cell damage, increase of inflammatory processes, alterations in the immune response, and insulin resistance in TSDM. Chromatin remodeling that regulates gene expression, such as p66Shc, has been linked with insulin resistance, increase of vascular risk in DM2, and obesity [13].

DM2 is a polygenic disease, of high heritability, which involves genes related to insulin and action, in addition to those that code for the components of the metabolic syndrome. What has been discovered so far is broad but only accounts for a part of the complex relationship of genetics and its phenotypic expression in DM2. The study of epigenetics in DM2 has opened the possibility to find pathogenetic markers at the onset of the disease and during the development of chronic complications, which will allow early screening and individualized treatment in the near future.

4. Monogenic diabetes

It is caused by one or more defects in a single gene. The disease can be inherited within a family by the genetic transmission of a dominant or recessive nature and not Mendelian. It can also be presented as a spontaneous case due to a de novo mutation [14]. Monogenic diabetes includes neonatal diabetes, maturity onset diabetes young (MODY), and mitochondrial diabetes (**Figure 7**).

4.1 Neonatal diabetes

It is defined as diabetes that appears before 6 months of age and is subdi-vided into transitory (TNDM) and permanent (PNDM). TNDM develops in the first weeks of life and resolves within a few months, but 50% have a relapse in adolescence or adulthood. TNDM is most frequently caused by abnormalities

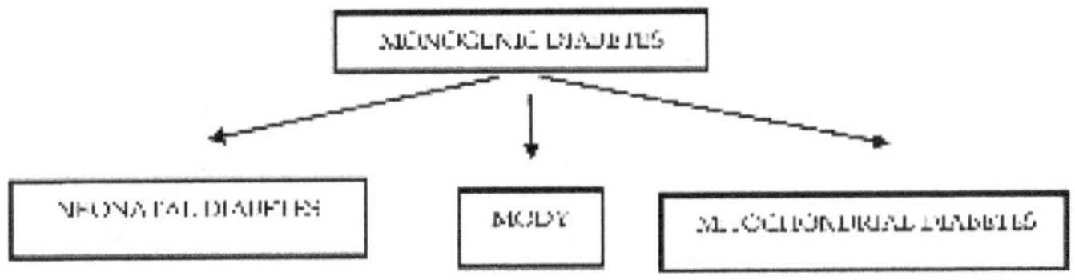

Figure 7.
Subtypes of monogenic diabetes.

in the imprinted region of chromosome 6q24 (spanning two candidate genes PLAGL1 and HYMAI), thereby leading to overexpression of paternally derived genes. Activating mutations in either the KCNJ11 or ABCC8 genes encoding the two subunits (Kir6.2 and SUR1, respectively) of the adenosine triphosphate-sensitive potassium channel on the beta-cell membrane prevent insulin secretion in response to hyperglycemia and can cause both PNDM and TNDM. Diabetes caused by mutations in KCNJ11 and ABCC8 often responds to sulfonylureas. Mutations in other genes critical to beta-cell function and regulation, and in the insulin gene itself, also cause PNDM. Heterozygous coding mutations in the pre-proinsulin gene (INS) are the second common cause of PNDM after K_{ATP} channel mutations [15, 16].

4.2 MODY

Described in 1975 by Tattersall and named in 1976 by Fajans, MODY is recognized as a form of mild-presenting family diabetes that is diagnosed during adolescence or early adulthood. Currently, other types are identified with a less classic presentation [4, 5]. It is estimated that its prevalence is underestimated, and it would correspond about 5% of DM2 and a similar percentage in DM1.

It is presented in 3.6% of the population with diabetes under 30 years [4]. MODY is a heterogeneous group of disorders caused by mutations in genes essential for beta-cell development, function and regulation, glucose sens-ing, and in the insulin gene itself. Although at least 14 genes are associated with MODY, we describe the four most frequent types (**Table 3**). Mutations in HNF-1α, HNF-4α, HNF-1β, and GCK genes account for over 80% of all known MODY cases.

Heterozygous mutations in three of them are responsible for the majority of cases of this type of diabetes: glucokinase gene (GCK); two genes encoding hepato-cyte nuclear factor (HNF) transcription factors HNF-1α and HNF-4α. Most MODY-causative genes, except GCK, encode transcription factors expressed in pancreatic beta-cells (**Table 4**). The majority of patients with MODY exhibit isolated diabetes or stable mild fasting hyperglycemia, but some MODY subtypes have additional features, such as renal abnormalities (MODY 5) and pancreatic exocrine dysfunc-tion (MODY 6) [17–20].

4.2.1 MODY 2 (GCK)

GCK, a glucose sensor expressed in pancreatic beta-cells, is a key enzyme in glucose metabolism that catalyzes the conversion of glucose to glucose-6-phosphate and thus controls glucose-mediated insulin secretion. As such, GCK serves to facilitate insulin release that is both appropriate and proportional to the blood glucose concentration. Heterozygous inactivating mutations in GCK (MODY 2) increase the set point for insulin secretion in response to increased blood sugar, thereby causing stable, mild fasting hyperglycemia. More than 600 mutations have been

	Gene	Frequency %	Physiopathology
MODY 2	GCK	15–20	Glucose sensing defect
MODY 3	HNF-1α	30–50	Insulin secretion deficit
MODY 1	HNF-4α	5	Insulin secretion deficit
MODY 5	HNF-1β	5	Beta cell dysfunction

GCK, glucokinase; HNF, hepatocyte nuclear factor.

Table 3.
Frequent types of MODY and its pathophysiological alteration.

MODY subtypes	Age at the onset of diabetes (years)	Hyperglycemia	Other clinical features	Possible treatment
1	17 (5–18)	High	Neonatal diabetes, neonatal hyperinsulinemic hypoglycemia, low triglycerides	Sensitive to sulphonylureas
2	10 (0–18)	Mild	Mild fasting long-term stability asymptomatic	Diet no medication
3	|4 (4–18)	High	Glycosuria—very low C-reactive protein of <0.5 mg/dl	Sensitive to sulphonylureas
5	<25	Variable	Renal malformations, genital anomalies, pancreatic hypoplasia, low birth weight	Insulin

Table 4.
Clinical characteristics of most common MODY subtypes.

reported. Patients with MODY 2 are usually asymptomatic, and they do not require treatment. In pregnancy, insulin may be required to prevent fetal complications, such as high birth weight and neonatal hypoglycemia. These neonatal complications are dependent on whether the mutation is inherited.

In 2018, a MODY family study was published in Chile [21]. The case is about a 17-year-old woman with DM, fasting blood glucose 130 mg/dl, without ketosis or weight loss, and BMI 18 kg/m^2. No signs of insulin resistance were seen, C-peptide 2.3 ng/ml (normal) and negative DM1 autoantibodies. In a family study, diabetic father and brother with impaired fasting blood glucose (**Figure 8**). The genetic-molecular analysis of the GKC gene, the patient, the father, and the brother presented a mutation at position 1343 of exon 10 corresponding to a heterozygous exchange of guanine for adenine (1343 G > A). The change is not synonymous and determines that at position 448 of the GKC enzyme, the amino acid glycine is substituted by aspartic acid. Diagnosis of MODY 2 was confirmed, and it was established that the mutation was by paternal line.

4.2.2 MODY 3 (HNF-1α)

The transcription factor HNF-1α is expressed in the liver, kidney, intes-tine, and pancreatic beta-cells. Heterozygous HNF-1α mutations result in progressive beta-cell dysfunction that leads to diabetes in early adult life.

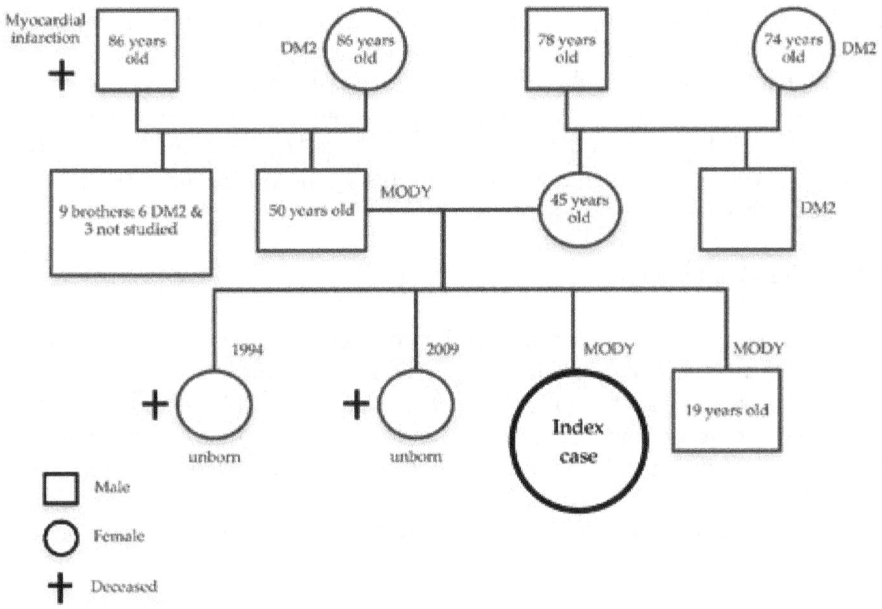

Figure 8.
Chilean family inheritance pattern MODY 2.

According to studies, a total of 414 different HNF-1α mutations were identified in 1200 families, where a mutation (P291fsinsC) in exon 4 was the most common. Hyperglycemia associated with MODY 3 may be severe, and the risk of micro-vascular and macrovascular complications is similar to DM1 and DM2. Because of this, patients require strict glycemic control and close monitoring of possible complications. There is a defect in the renal resorption of glucose, characterized by a decreased glucose threshold for glycosuria and reduced tubular reabsorption of glucose. Patients are sensitive to sulfonylurea therapy, but most of them even-tually progress to insulin treatment. This subtype of MODY is the most frequent in Europe and the US.

4.2.3 MODY 1 (HNF-4α)

This MODY was the first described. The transcription factor HNF-4α is expressed in the liver, kidney, and pancreatic beta-cells. HNF-4α gene encodes a transcription factor important for pancreatic development and beta-cell differentiation and function. Heterozygous HNF-4α mutations cause a similar clinical phenotype observed in MODY 3. Most patients have a progressive insulin deficiency, diabetes onset before age 25 years, and a response to relative low-dose sulfonylurea therapy. Fetal HNF-4α heterozygosity results in macrosomia due to hyperinsulinemia in utero and subsequent neonatal hyperinsulinemic hypogly-cemia, which is responsive to diazoxide. MODY 1 is associated with triglyceride metabolism, and mutation carriers may exhibit reduced levels of apoproteins (apoAII, apoCIII, and apoB).

4.2.4 MODY 5 (HNF-1β)

The transcription factor HNF-1β is involved in the organogenesis of the kidney, genitourinary tract, liver, lungs, gut, and pancreas. Patients with heterozygous mutations in HNF-1β rarely present with isolated diabetes. By contrast, patients usually have renal disorders (especially renal cyst and renal dysplasia). Urogenital tract abnormalities and atrophy of pancreas may also occur. The sensitivity to sulphonylureas is absent, and early insulin therapy is required. At least 50% of HNF-1β MODY cases are due to microdeletion of chromosome 17 (17q12) involving between 15 and 29 genes, including HNF1β. De novo mutations are frequent (up to 50% of cases) and hence family may be absent.

4.3 Mitochondrial diabetes

This disease is a mitochondrial disorder characterized by maternally transmitted diabetes and sensorineural deafness. The most common form is caused by an exchange between an adenine for guanine (3243A/G) in DNA. This mutation also causes a severe neuromuscular disease syndrome called MELAS (mitochondrial myopathy, encephalopathy, lactic acidosis, and stroke). Diabetes onset is usually insidious, but 20% of patients have an acute presentation, even in diabetic ketoacidosis. It usually occurs in the third to fourth decades of life in non-obese individuals [14].

5. Diagnostic algorithm of diabetes in young adult

Figure 9 shows a diagnostic algorithm for diabetes in patients under 30 years of age.

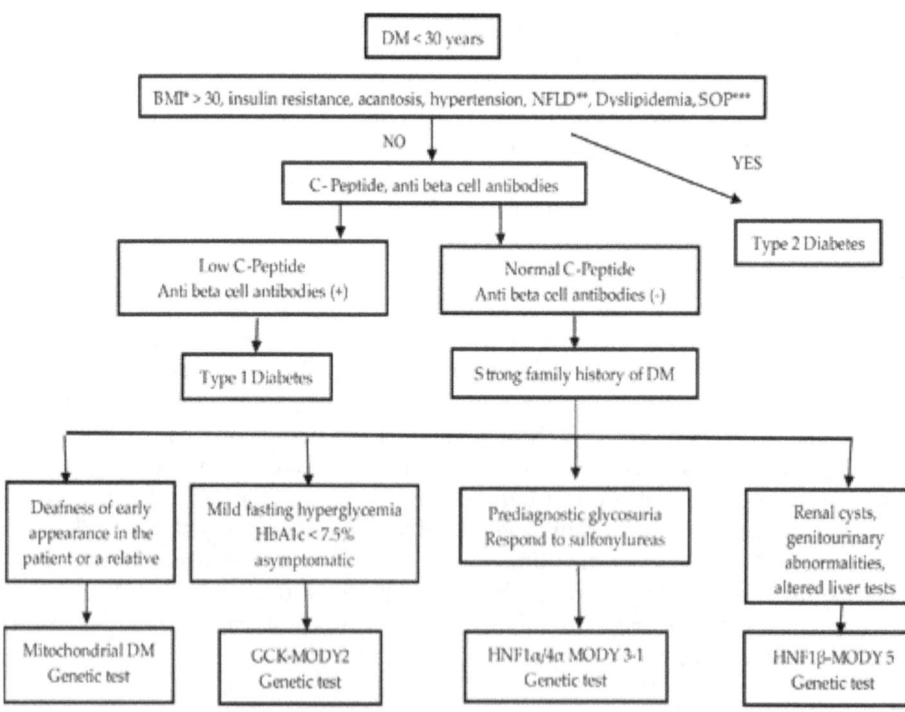

*BMI= Body mass index (kg/m²). **NFLD= Non-alcoholic fatty liver disease, ***POS = polycystic ovary syndrome

Figure 9.
Diagnostic algorithm of diabetes in young adult.

6. LADA: Latent autoimmune diabetes adult

Diabetes is a complex disease, which makes its classification difficult. DM1 is c aused by autoimmune destruction of the beta, cell which leads to absolute insulin deficiency. DM2 is secondary to the progressive loss of insulin secretion by the beta cell, in the context of insulin resistance. Other types of diabetes are gestational diabetes, MODY, post-transplant, and exocrine, among others [22]. DM1 is a heterogeneous disease, the incidence of which is higher in children and adolescents, characterized by the presence of specific immunological markers. However, no less than a percentage of adults experience the disease; so, the term latent adult diabetes (LADA) has been coined. In this case, the disease is even more heterogeneous, since there is a variable proportion of destruction of the beta cell, with the presence of immunological markers, but zero or deficient initial insulin requirements, so they can initially be misclassified as DM2 [23]. These patients probably have pathophysiological processes similar to DM1, but with differences in genetic penetrance and immune factors. The term LADA was introduced in the 90s, to define a subgroup of patients with diabetes initially not requiring insulin, but with immunological markers of DM1 detectable in the serum [24]. In 2015, the Immunology of Diabetes Society proposed three diagnostic criteria for LADA: age of onset>30 years, presence of any DM1 marker antibody, and lack of need for insulin treatment for at least 6 months from diagnosis [25].

Currently, the American Association of Diabetes (ADA) does not recognize this entity, but instead classifies it within the group of patients with DM1, but every day there is more information about its clinical and pathophysiological characteristics, which keep the debate open.

6.1 Epidemiology

The available data show that the prevalence of LADA is higher than previ-ously recognized. About 40% of cases of DM1 occur in adults over 30 years [26]. Scandinavian studies show that 7.5–10% of the population with apparent DM2 have circulating antibodies against the beta cell (ICA or GAD 65) [27]. The Action LADA study, conducted in Europe, which evaluated 6000 adults attended in primary and secondary care centers, reported a frequency of 9.7% of LADA [28]. A Chinese study, LADA China study reported a 5.9% positive antibody in adults previously diagnosed with DM2 [29]. The prevalence of LADA is, therefore generally under-estimated, due to the lack of study with antibodies in adult patients; so, the level of clinical suspicion should be high.

6.2 Pathophysiology

6.2.1 Genetics

In genotype analysis, patients with LADA have been shown to share genetic characteristics with DM1 (HLA, INS VNTR, CTLA4, and PTPN22) and DM2 (TCF7L2) [30], which might suggest that LADA is a spectrum of insulin deficiency between DM1 and DM2. The HLA-DRB1*04-DQB1*0302 and HLA-DRB1*0301DQB1*0201 haplotypes, which confer high susceptibility to DM1, and decrease progressively with increasing age, have been further diminished in elderly DM1 patients and have been described less frequently even in patients with LADA [31].

6.3 Autoimmunity

DM1 is a known autoimmune disease, mediated by cells. The presence of T lymphocytes that are reactive to islet cells, in LADA, gives us some evidence that there is a cell-mediated immune response as well [32]. Adult autoimmune diabetes has a "lower genetic load," characterized by lower circulating antibodies than early-onset DM1 in childhood or adolescence, which correlates with less intense beta cell destruction and lower HLA genetic susceptibility. Many studies have compared circulating antibodies in early DM1 in childhood with LADA, finding ICA, AAI, IA-2, and anti ZnT8 more frequently in children than in adults, while anti GAD and IA-2 were found with similar frequency in both ages [33]. Anti-GAD is the antibody most frequently found in patients with LADA, up to 90% positive, also being the most persistent over time. The beta cell function in early DM1 in childhood and adolescence is severely compromised since diagnosis, a difference in LADA in that the deficit is less severe. It has also been found that there is a correlation between age of diagnosis and fasting C-peptide levels, which is related to the latest age of LADA. To explain the later and less aggressive presentation compared to DM1, several theories are postulated, among others: intermittent crisis of autoimmune aggression (**Figure 10**) or greater capacity to regenerate beta cells and protection against the apoptotic process.

6.4 Clinical characteristics

Patients with LADA are a heterogeneous group, with antibody titers and body mass index (BMI). In general, the appearance of the condition is 35 years later, with cases described since the age of 25. Patients with LADA tend to have a BMI higher than DM1, but less than DM2. The existence of other autoimmune comorbidity or their family history is common, mainly thyroid disease. A higher frequency of anti-peroxidase antibodies (TPO) has been seen, in up to 27% of patients, compared to those with anti-GAD negative, which makes it necessary to monitor thyroid function and perform screening for other autoimmune diseases. The initial response to oral therapy is satisfactory, progressing in varying degrees to insulin requirements, from 6 months to several years, depending mainly on antibody titers (**Table 5**).

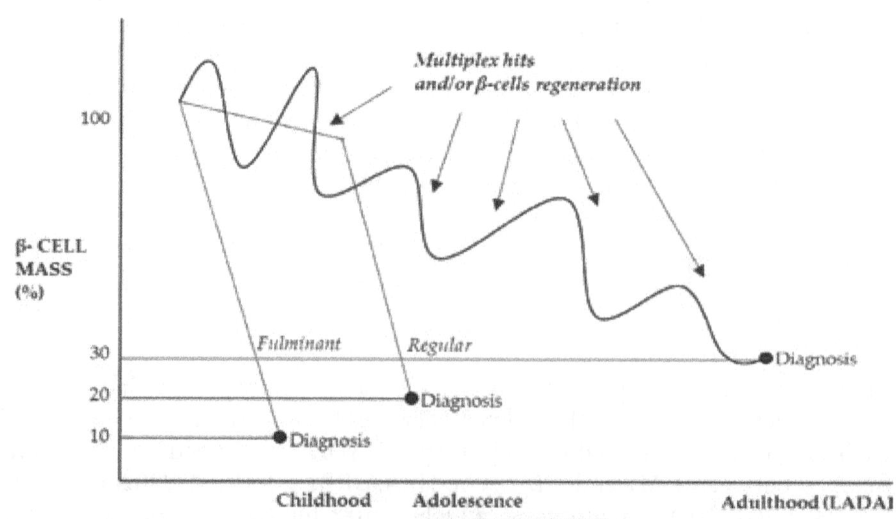

Figure 10.
The destruction of beta cell and the appearance of DM1 according to the age of onset and the putative pathogenetic mechanisms.

Clinical features	DM1	LADA	DM2
Age at diagnosis	Childhood and adolescence rare adulthood	>30 years	Adulthood, rare childhood, and adolescence
Start	Acute	Rare acute	Slow
Autoimmunity	Greatly increased	Increased	No modifications
Ketosis	Frequently	Rare	Rare
Insulin resistance	No modifications	Increased or without changes	Severely increased
Beta cell function	Very reduced	Reduced	Increased or unchanged
Insulin requirements	From the diagnostic	>6 months from the diagnostic	Years from the diagnostic
Body mass index	Low or normal	Normal or overweight	Overweight or obesity
MHC susceptibility	Severely increased	Increased	No changes

Adapted to [35].

Table 5.
LADA, DM1, and DM2 clinical features.

High titers of anti-GAD compared to low ones, have lower BMI, less endogenous insulin secretion, and faster progression to insulin-dependence. The presence of anti-GAD antibodies (or ICA) may be useful to identify patients with a previous diagnosis of DM2, who respond partially to treatment with oral antidiabetics and who quickly require insulin therapy. Regarding the metabolic profile, patients with LADA have advantages regarding DM2 with a better profile, that is, lower triglyceride levels, higher levels of HDL, lower BMI, and lower waist circumference. There are no specific guidelines for the treatment of patients with LADA. However, the metabolic goals are the same as for DM1 and DM2 patients, so you should try to achieve HbA1c <7%. The diet and exercise recommendations do not show differences with the classic presentations. Despite the extensive use of oral antidiabetics in DM2, especially metformin, there are no studies of this drug in patients with LADA. Glibenclamide and insulin were compared in LADA patients, finding that the group that used GBC had worse metabolic control and faster deterioration of CP secretion at a follow-up of 30 months. Therefore, the use of sulfonylureas as a first-line drug in this type of diabetes is not recommended. TZD combined with insulin show preservation of beta cell function in a small group of Chinese patients. The use of other agents such as insulin sensitizers could be used in combination with insulin in patients who share characteristics with DM2, that is, BMI >30 kg/m^2 and signs of insulin resistance. The role of the iDPP4 is not established. Patients with LADA treated with insulin glargine, the effect of adding sitagliptin or placebo was compared, the group with sitagliptin had a minimal decrease in C-peptide at one-year follow-up, compared to placebo. However, more studies support this evidence. We should keep in mind, like DM1 and DM2, patients.

Today, evidence indicates that early insulin therapy, along with changes in lifestyle, is the therapy of choice in patients with LADA when metabolic control is impaired primarily in young patients with elevated antibody titers since this treatment slows the deterioration of beta cell function, controls hyperglycemia, and prevents glucotoxicity [34]. TZD combined with insulin shows preservation of beta cell function in a small group of Chinese patients. The use of other agents such as insulin sensitizers could be used in combination with insulin in patients who share characteristics with DM2, that is, BMI >30 kg/m^2 and signs of insulin resistance. The role of the iDPP4 is not established. Patients with LADA treated with insulin glargine, the effect of adding sitagliptin or placebo was compared, the group with sitagliptin had a minimal decrease in C-peptide at one-year follow-up, compared to placebo. However, more studies support this evidence. We should keep in mind, like DM1 and DM2, patients with LADA require a multidisciplinary approach to proper treatment.

6.5 Clinical features of DM1, LADA, and DM2

This Table summarizes major clinical features of DM1, LADA, and DM2.

Acknowledgements

Special thanks to Miss Macarena Darby for her contribution in the translation of the chapter.

Author details

Lilian Sanhueza[1,2*], Pilar Durruty[3,4], Cecilia Vargas[4], Paulina Vignolo[5,6] and Karina Elgueta[1,7]

1 Diabetes Unit, Internal Medicine Service, San Juan de Dios Hospital, Santiago, Chile

2 Department of Medicine, Faculty of Medical Sciences, University of Santiago of Chile, Santiago, Chile

3 Diabetes Unit, Department of Medicine, Faculty of Medicine, San Juan de Dios Hospital, University of Chile, Santiago, Chile

4 Endocrinonology and Diabetes Section, Department of Medicine, Faculty of Medicine, Clinical Hospital University of Chile, Santiago, Chile

5 Diabetes Unit, Department of Internal Medicine, Padre Hurtado Hospital, Santiago, Chile

6 Faculty of Medicine, Development University, Santiago, Chile

7 Adult Diabetes Unit, Department of Internal Medicine Las Condes Clinic, Santiago, Chile

*Address all correspondence to: lilianllay@yahoo.es

References

[1] Durruty P, Pérez-Bravo F. Patogénesis de la diabetes mellitus. García de los Ríos M y Durruty P. Diabetes Mellitus. Santiago, Chile. Ed Mediterráneo 2014: 25-39

[2] Salas F, Santos J, Pérez-Bravo F. Genética de la diabetes mellitus tipo 1. Revista Chilena de Endocrinología y Diabetes. 2013;**6**(1):15-22

[3] Paschou S, Papadopoulou-Marketou N, Chrousos G et al. On type 1 diabetes mellitus pathogenesis. Endocrine Connections. 2018;**7**:R38-R-46

[4] Thompson M, Mclnnes R, Willard H. Genética en Medicina, Thompson & Thompson. 4th Edición ed. Barcelona, España: Editorial Masson; 1996. pp. 345-349

[5] Asenjo S, Gleisner A, Pérez-Bravo F. Marcadores genéticos (HLA) y perfil de auto-anticuerpos en una familia mapuche con un caso de diabetes tipo 1. Revista Médica de Chile. 2004;**132**:47-50

[6] Lyssenko V, Jonsson A, Almgren P, et al. Clinical risk factors, DNA variants, and the development of type 2 diabetes. The New England Journal of Medicine. 2008;**359**(21):2220-2232

[7] Lyssenko V, Almgren P, Anevski D, et al. Predictors of and longitudinal changes in insulin sensitivity and secretion preceding onset of type 2 diabetes. Diabetes. 2005;**54**(1):166-174

[8] Yaghootkar H, Scott RA, White CC, et al. Genetic evidence for a normal-weight metabolically obese phenotype linking insulin resistance, hypertension, coronary artery disease, and type 2 diabetes. Diabetes. 2014;**63**(12):4369-4377

[9] Durruty P, Pérez-Bravo F. Etiopatogenia de la diabetes mellitus. In: García de los Ríos M, Durruty P, editors. Diabetes Mellitus. Santiago, Chile: Ed Mediterráneo; 2003. pp. 42-55

[10] Weedon MN, Schwarz PE, Horikawa Y, et al. Meta-analysis and a large association study confirm a role for calpain-10 variation in type 2 diabetes susceptibility. American Journal of Human Genetics. 2003;**73**:1208-1212

[11] Sale MM, Smith SG, Mychaleckyj JC, et al. Variants of the transcription factor 7-like 2 (TCF7L2) gene are associated with type 2 diabetes in an African–American population enriched for nephropathy. Diabetes. 2007;**56**(10):2638-2642

[12] Gloyn AL, Weedon MN, Owen KR, et al. Large-scale association studies of variants in genes encoding the pancreatic beta-cell KATP channel subunits Kir62 (KCNJ11) and SUR1 (ABCC8) confirm that the KCNJ11 E23 K variant is associated with type 2 diabetes. Diabetes. 2003;**52**(2):568-572

[13] Coco C, Sgarra L, et al. Can epigenetics of endothelial dysfunction represent the key to precision medicine in type 2 diabetes mellitus? International Journal of Molecular Sciences. 2019;**20**:2949

[14] Hattersley AT, Greeley SAW, Polak M, et al. ISPAD clinical practice consensus guidelines 2018: The diagnosis and management of monogenic diabetes in children and adolescents. Pediatric Diabetes. 2018;**19**(Suppl. 27):47-63

[15] Timsit J, Saint-Martin C, Dubois-Laforgue D, Bellanné- Chantelot C.Searching for maturity-onset diabetes of the young (MODY): When and what for? Canadian Journal of Diabetes. 2016;**40**:455-461

[16] Kim SH. Maturity-onset diabetes of the young: What do clinicians need

to know? Diabetes and Metabolism Journal. 2015;**39**(6):468-477

[17] Shields BM, Shepherd M, Hudson M, et al. Population-based assessment of a biomarker-based screening pathway to aid diagnosis of monogenic diabetes in young-onset patients. Diabetes Care. 2017;**40**(8):1017-1025

[18] Thomas CC, Philipson LH. Update on diabetes classification. The Medical Clinics of North America. 2015;**99**(1):1-16

[19] Urakami T. Maturity-onset diabetes of the young (MODY): Current perspectives on diagnosis and treatment. Diabetes, Metabolic Syndrome and Obesity: Targets and Therapy. 2019;**12**:1047-1056

[20] McDonald TJ, Ellard S. Maturity onset diabetes of the young: Identification and diagnosis. Annals of Clinical Biochemistry. 2013;**50**:403-415

[21] Estica M, Selenfreund D, Durruty P, Briones G. Hallazgo de una nueva mutación en una familia chilena con diabetes monogénica. Caso Clínico. Revista Médica de Chile. 2018;**146**:929-932

[22] Classification and diagnosis of diabetes. Standards of medical care in diabetes 2019. Diabetes Care. 2019;**42**(suppl 1):S13-S28

[23] Hernandez M, Mollo A, Marsal JP, et al. Insulin secretion in patients with latent autoimmune diabetes (LADA): Half way between type 1 and type 2 diabetes: Action LADA 9. BMC Endocrine Disorders. 2015;**15**:1-6

[24] Tuomi T, Groop LC, Zimmet P, et al. Antibodies to glutamic acid decarboxylase reveal latent autoimmune diabetes mellitus in adults with a non- insulin dependent onset of disease. Diabetes. 1993;**42**:359-362

[25] Fourlanos S, Dotta F, Greenbaum CJ, et al. Latent autoimmune diabetes in adults (LADA) should be less latent. Diabetologia. 2005;**48**:2206-2212

[26] Mølbak AG, Christau B, Marner B, Borch-Johnsen K, Nerup J. Incidence of insulin- dependent diabetes mellitus in age groups over 30 years in Denmark. Diabetic Medicine. 1994;**11**:650-655

[27] Landin-Olsson M, Nilsson KO, Lernmark A, Sundkvist G. Islet cell antibodies and fasting C-peptide predict insulin requirement at diagnosis of diabetes mellitus. Diabetologia. 1990;**33**(9):561-568

[28] Hawa MI, Kolb H, Schloot N, et al. Adult onset autoinmune diabetes in Europe is prevalent with broad clinical phenotype: Action LADA 7. Diabetes Care. 2013;**36**:908-913

[29] Zhou Z, Xiang Y, Ji L, et al. Frequency, immunogenetics, and clinical characteristics of latent autoimmune diabetes in China (LADA China study): A nationwide, multicenter, clinic-based cross-sectional study. Diabetes. 2013;**62**:543-550

[30] Cervin C, Lyssenko V, Bakhtadze E, Lindholm E, Nilsson P, Tuomi T, et al. Genetic similarities between latent autoimmune diabetes in adults, type 1 diabetes, and type 2 diabetes. Diabetes. 2008;**57**:1433-1437

[31] Leslie RD, Delli Castelli M. Age-dependent influences on the origins of autoimmune diabetes: Evidence and implications. Diabetes. 2004;**53**:3033-3040

[32] Brooks-Worrell BM, Juneja R, Minokadeh A, Greenbaum CJ, Palmer JP. Cellular immune responses to human islet proteins in antiboby-positive type 2 diabetic patients. Diabetes. 1999;**48**:983-988

[33] Lampasona V, Petrone A, Tiberti C, Capizzi M, Spoletini M, Di PS, et al. Zinc transporter 8 antibodies complement GAD and IA–2 antibodies in the identification and characterization of adult-onset autoimmune diabetes: Non insulin requiring autoimmune diabetes (NIRAD). Diabetes Care. 2010;**33**:104-108

[34] Zhou Z, Li X, Huang G, et al. Rosiglitazone combines with insulin preserves islet beta cell function in adult- onset latent autoimmune diabetes. (LADA). Diabetes/Metabolism Research and Reviews. 2005;**21**:203-208

[35] de Buzzeti R et al. Nature Reviews. 2017;**13**:674-686

9

Metabolic Syndrome

Armindo Miguel de Jesus Sousa de Araújo Ribeiro

Abstract

Metabolic syndrome (MS) is recognized by a set of cardiovascular risk factors (CVRF) that usually coincide with insulin resistance and hyperglycemia. These risk factors include hyperglycemia (fasting glucose > 100 mg/dl), high blood pressure (SAD ≥ 130 mmHg and TAD ≥ 85 mmHg), triglyceride increase (≥ 150 mg/dl), decreased HDL levels <40 mg/dl, and central obesity (waist circumference ≥102 cm in men and ≥88 cm in women). The purpose of this chapter is to review the natural history of metabolic syndrome and epidemiology and to review risk factors for the appearance of metabolic syndrome, pathophysiology, and biochemistry among the various cardiovascular risk factors and their importance within the metabolic syndrome.

Keywords: diabetes, hypertension, dyslipidemia

1. Introduction

The purpose of this chapter is to provide a review of the pathophysiology of metabolic syndrome (MS) and the relationship between its different components. Because this is a very broad subject, not all aspects will be discussed. Throughout the chapter, a bibliographic reference is left for each topic so that this can be looked into more deeply. This is a non-systematic review of the literature through research on PubMed and Google.

At an epidemiologic study conducted by Armindo Sousa Ribeiro et al, in a Portuguese population, was found that 23.8% of the population has MS more prevalent in 24 males, no smoking, no significant alcohol consumption, sedentary life style, with a high body mass index (BMI) and its prevalence increases with age [1]. Each CVRF has different importance in the metabolic syndrome [2].

In Western societies, cardiovascular diseases (CVD) are the main cause of mortality and morbidity for both sexes. This fact entails very high social and economic costs [3].

CVDs are the biggest cause of mortality worldwide. In the 2010 Global Burden of Disease Study, CVD is estimated to cause 15.6 million deaths per year worldwide, corresponding to 29.6% of all deaths [4]. Although the tendency is to reduce the number of deaths in Europe due to CVD, this remains the main cause of death, cor-responding to more than 4 million deaths per year, that is to say 45% of all deaths recorded in Europe [4]. Of the CVD, coronary heart disease remains the most important cause of death in both genders, which corresponds to 19% in the male sex and 20% in the female sex.

2. Metabolic syndrome and cardiovascular risk factors (CVRF)

2.1 Definition

There are several definitions of MS. MS groups a constellation of pathophysiological factors that increase the risk of developing diabetes mellitus (DM) type 2 and CVD [5–10]. At present, attempts have been made to unify criteria to have a consensus on its diagnosis [11].

The World Health Organization (WHO), International Diabetes Federation (IDF), National Cholesterol Education Program Adult Treatment Panel III

	ATP III	WHO	AACE	IDF
Triglycerides ≥ 150 mg / dl	X	X	X	X
HDL <40 mg / dl in men and <50 mg / dl in women	X	X	X	X
Blood pressure ≥ 130/85 mmHg	X	X	X	X
Insulin Resistance (IR)		X		
Fasting glucose ›100 mg / dl	X		X	X
Glucose 2h: 140 mg / dl			X	
Abdominal Obesity	X			X
High body mass index		X	X	
Microalbuminuria		X		
Risk factors and or diagnosis	3 plus IR	More than 2	Clinical Criteria	Abdominal Obesity

Figure 1.
Components of the metabolic syndrome considering its definition: according to the National Cholesterol Education Program Adult Treatment Panel III (ATP III), World Health Organization (WHO), American Association of Clinical Endocrinologists (AACE), and International Diabetes Federation (IDF).

(NCEP-ATP III), and the American Association of Clinical Endocrinologists (AACE) have proposed their diagnostic criteria or components of MS (**Figure 1**) [11].

In 2009, representatives of the IDF and the American Heart Association/National Heart, Lung, and Blood Institute (AHA/NHLBI)-ATP III Guidelines discussed resolving differences between definitions of MS, unifying criteria [6, 11]. The definition of MS according to the unification of criteria (harmonizing the metabolic syndrome) is used for many international works and publications and requires the presence of three of the following five criteria:

1. Elevation of fasting blood glucose (≥100 mg/dl) or receiving antidiabetic treatment with insulin or oral antidiabetics

2. Elevation of systolic blood pressure (SBP) ≥ 130 mmHg or diastolic blood pressure (PAD) ≥ 85 mmHg or receiving antihypertensive drug treatment

3. High-density lipoprotein (HDL) cholesterol values <40 mg/dl (men) or <50 mg/dl (women)

4. Triglycerides ≥150 mg/dl or under treatment with specific lipid lowering agents

5. Abdominal perimeter ≥102 cm (men) or ≥88 cm (women) [6, 12]

The concept of MS was carried out in a progressive manner. Its clinical importance is enormous due to the fact that it can identify patients with high risk of suffering some CVD and/or DM, allowing a preventive intervention [6, 13, 14].

The notion that CVRF has a tendency to aggregation allows an earlier diagnosis of MS, thus also leading to the early introduction of therapeutic, pharmacological, or non-pharmacological measures.

2.2 Historical evolution

The interest in the MS is not something recent. Its nomenclature has evolved over the years.

About 260 years ago, before the description of MS, the Italian anatomist and doctor, Morgagni, identified the association between visceral obesity, arterial hypertension (AHT), atherosclerosis, hyperuricemia, and frequent episodes of sleep apnea. In the mid-twentieth century, French physicist Vague first identified the relationship between android obesity and the increased prevalence of DM and CVD. In the 1960s Avogaro and Crepaldi attributed the name "plurimetabolic syndrome" to the simultaneous presence of obesity, AHT, and DM. They verified that all these abnormalities were frequent among the population and contributed to an increase in cardiovascular risk [15]. A decade later, Haller related this set of risk factors with atherosclerosis [15].

In 1980, Vague suggested that adipose tissue alone had little effect on the onset of DM in obese individuals. It is currently known that central or android obesity is a predisposing factor for DM and atherosclerosis and affects insulin secretion unlike gynoid obesity [15].

In 1988, Reaven described this phenomenon as syndrome X, the set of hypergly-cemia and insulin resistance, obesity, dyslipidemia, and hypertension. Reaven also suggested that insulin resistance and the consequent increase in blood insulin levels were the central pathophysiological features in syndrome X which in itself was a risk factor for CVD [15, 16].

In 1991, Ferrannini and his colleagues referred to this as insulin resistance syndrome [15, 17].

In 1993, Van Gaal attributed for the first time the name MS to all comorbidities associated with visceral obesity and currently this nomenclature is being used [15].

The concept of MS has evolved significantly in the last decade, which resulted in the presentation of multiple clinical definitions by different scientific societies. In general terms as it was widely treated, MS is defined as an aggregation of several CVRF in the same individual. The nuclear elements for the classification and diagnosis of MS are basically obesity, abnormal glycemic metabolism, iatrogenic dyslipidemia, and hypertension.

The metabolic syndrome was defined by the World Health Organization (WHO) in 1998 [13, 15]. In 2001, there was a new revision of the definition through the National Cholesterol Education Program-Adult Treatment Panel III (NCEP-ATP III), where glycemia is not considered an essential factor; in this way it only becomes one of the diagnostic components of MS [18].

With the verification of evidence of the relationship between central obesity and cardiovascular risk, there was a tendency to value this diagnostic component for MS more. Thus, in 2004, the International Diabetes Federation (IDF) created a new definition of MS, where central obesity, defined by the value of the abdominal circumference, would be essential for diagnosis [7, 19]. With the adoption of this definition, an increase in the prevalence of MS was observed in a large part of the populations studied, particularly in the elderly [20].

In 2005, a new review of the criteria by the American Heart Association/National Heart, Lung, and Blood Institute (AHA/NHLBI) maintained the criteria of the NCEP-ATP III [21]. The justification was the fact that in that criterion a single etiology for MS does not stand out and is of simpler application. They modified only the cutoff point of fasting blood glucose from 110 to 100 mg/dl, an adjustment promoted by the American Diabetes Association (ADA) in the diagnosis of DM. However, the first Brazilian Guideline for the Diagnosis of Treatment of MS, of 2005, uses the criteria of the NCEP-ATP III, of 2001, for diagnosis [22, 23]. In the criteria of the IDF, the component of the abdominal circumference becomes essential, using a smaller waist circumference, and categorized more people as having the MS than the ATP III definition. However, higher values of abdominal circumference in older people have been related to lower values of body mass index (BMI), in relation to younger adults [24–28].

2.3 Prevalence of metabolic syndrome

The prevalence of MS varies according to geographic area, with age, race, sex, and classification used for diagnosis [29].

The DARIOS study performs a pooled analysis of 11 studies in 24,670 Spanish individuals aged 35–74, using the criteria of the IDF/NHLBI/AHA-2009, and shows a prevalence of 32% in men and 29% in women [12, 30].

The West of Scotland Coronary Prevention Study (WOSCOPS), one of the largest in Europe, reports a general prevalence of 26.2% [31].

The Third National Health and Nutrition Examination Survey (NHANES III), in the USA showed that 7% of individuals between the ages of 20 and 29, 42% of individuals with ages between 60 and 69 years, and 44% of individuals with ages over 70 years presented MS. The overall prevalence was 23.7% in the 8814 adults who entered the study. This study showed that the incidence increased with age, mainly after 40 years, with significant variations consistent with ethnicity, being the most frequent MS in Hispanics (31.9%) compared with Caucasians (23.8%) and African-Americans (21.6%). A similar prevalence was recorded in both sexes, although it was more prevalent in males in Caucasians and in females in African-Americans and Hispanics [32].

Based on the Diabetes Epidemiology Collaborative Analysis of Diagnostic Criteria in Europe (DECODE) study of Europe in 2004, the prevalence of MS was 15.7% in male and 14.2% in the female [33]. Epidemiological studies on MS have suggested that the prevalence of MS in Western societies is high and is increasing due to the increase in obesity, especially in younger individuals [34].

In Portugal there are few studies on the prevalence of MS. In 2004, an epidemio-logical study was carried out in Porto where the prevalence of MS was 23.9%, being more prevalent in females (27%) than in males (19.1%) [35].

In 2008, the first study was conducted on the prevalence of MS and its cardiovascular complications in the Portuguese adult population at the primary care level in continental Portugal, Azores, and Madeira—VALSIM study (epidemiological study of the prevalence of metabolic syndrome in the Portuguese population). In this study, 719 family doctors participated. The study was carried out between April 2006 and November 2007. A total of 16,856 individuals were evaluated between the ages of 18 and 96, with averages of 58.1 ± 15.1 years. This study showed a high prevalence of MS in adults affecting 27.5% of the population analyzed, being more prevalent in females (28.7%) compared with males (26%). This study revealed an increase in the prevalence of MS related to age, body mass index, and abdominal perimeter. This study also demonstrated the association between metabolic risk fac-tors, including MS and the occurrence of cardiac events, stroke, and DM [34]. The prevalence of MS exhibited a regional variation, being more prevalent in Alentejo (30.99%), Madeira (29.38%), Central Region (28.79%), and North Region (28.17%) and less prevalent in Algarve (24.42%), Lisbon and Tagus Valley (25.71%), and Azores (26.05%) [34].

3. Insulin resistance and metabolic syndrome

The pathophysiology of MS is not yet fully understood, but the most accepted hypothesis is insulin resistance, so MS is also known as insulin resistance syndrome. Insulin resistance is defined as a defect in its action at the peripheral level that results in hyperinsulinemia, required to maintain normal blood glucose [36, 37]. Predisposing factors for insulin resistance are central obesity and the release of large concentrations of free fatty acids from adipose tissue. Free fatty acids in the liver will determine the increase in glycogenesis and gluconeogenesis, increased triglyceride production, and the secretion of very low-density lipoproteins (VLDL).

Insulin is a hormone with anti-lipolytic action that is produced in pancreatic beta cells and is released into the bloodstream starting its metabolic effects after binding to its receptor. The insulin receptor is a transmembrane glycoprotein composed of four subunits linked by disulfide bridges, that is, two extracellular alpha subunits that contain the insulin binding domain and two β subunits, which contain an extracellular domain, a transmembrane domain, and a intracellular domain insulin binding to the α subunit and β subunit, and tyrosine kinase activation in the β subunit is the first stage of insulin action on glucose metabolism [38, 39]. The activation of tyrosine kinase, and phosphorylation of the β subunit of the insulin receptor, catalyzes the phosphorylation reaction of several tyrosine residues in a family of proteins called insulin receptor substrate (IRS), which include IRS-1, IRS-2, IRS-3, and IRS-4, which are the most specific in the insulin signaling cas-cade. Other isoforms are growth factor receptor-bound protein 2 (GRB2)-associated binding protein 1 (Gab-1), Shc, and p62 [40, 41].

Structurally, the four IRS proteins have many similarities, in particular, the phos-photyrosine-binding domain (PTB), whose function is the recognition sequence of asparagine-proline glutamic phosphotyrosine acid (NPEpY) located in the juxta membrane region of the β receptor of insulin. It is involved in the interaction of IRS proteins with the insulin receptor and with the COOH-terminal domain of the proteins, capable of binding to the SH2 domains [39]. In addition to their similari-ties, the four IRS proteins differ in tissue distribution [39]. The SH2 domain consists of 100 amino acids, and it is the phosphotyrosine binding site [41]. The activation of proteins that contain the SH2 domain is important for the transmission of the insulin signal and thus carry out its functions [39].

The IRS-1 gene located on chromosome 2q36–37 was the first substrate that was identified. IRS-1 is involved in many of the actions of insulin, in the activation of nitric oxide (NO) synthetase, in the activation of the activity of the iN + pump, and in vascular relaxation through activated protein kinases by mitogens (MAP kinase) [39, 42].

The IRS-2 gene is found on chromosome 13q34 [43]. The IRS-2 protein and the IRS-1 are very similar in structure and functions, and the main difference lies in the region between amino acids 591 and 789 of the IRS-2. IRS-2, unlike IRS-1, is more in the cellular cytosol [44].

IRS-3 does not appear to be expressed in human cells [39].

IRS-4 is a protein that consists of 1257 amino acids located in the cell membrane of various organs in which it is expressed. It was discovered initially in embryonic kidney cells. The IRS-4 gene is found on the X chromosome [39].

Insulin causes vascular relaxation through stimulation of NO production in the endothelial cells of blood vessels and by reducing the concentration of intracellular calcium in muscle and smooth muscle cells by decreasing sensitization of the light chains of calcium (Ca^{2+})-myosin. These effects are mediated by the activation of the phosphatidylinositol 3-kinase/protein kinase B (PI3K/Akt) signaling pathway, as well as the stimulation of glucose transport in skeletal muscle and vascular adipose tissue [45]. In the case of insulin resistance, the increase in lipolysis produces greater amounts of fatty acids, promoting greater inhibition of the lipolytic effect of insulin, causing an increase in lipolysis [46, 47].

Insulin resistance is defined as the inability of glucose uptake by skeletal muscle and adipose tissue and the inability of the liver to inhibit gluconeogenesis in response to an increase in insulin [38, 43].

The evidence of the causal relationships between insulin resistance and hypertension is increasing [48]. There are still many uncertainties about the mechanisms that link the two conditions, but in addition to the common genetic predisposition between insulin resistance and AHT, there are several other possible mechanisms that can explain it [48, 49].

The renin-angiotensin-aldosterone system (RAAS) is a therapeutic target for AHT. Changes in the RAAS are also important in insulin resistance, and the common molecular mechanisms of insulin resistance and AHT are not well understood [50, 51].

The RAAS thus plays an important role in MS. Angiotensinogen is a 58 kDa protein produced primarily in the liver under physiological conditions, but it can be produced in smaller amounts in adipocytes, mesangial cells, and epithelial cells of the proximal contoured tubule and in the brain (neurons and glial cells). Angiotensinogen by the action of the renin enzyme produced in the glomerular juxta cells in the kidney is subjected to cleavage in the amino-terminal acid of 10 NH2 to form angiotensin I. Angiotensin I is an inactive peptide that is hydroxylated by the enzyme angiotensin converter, forming an octapeptide designated angiotensin II [50].

The RAAS has autocrine/paracrine activity in various tissues, especially in the skeletal muscle, smooth muscle of blood vessels, kidneys, heart, and pancreas [52]. The angiotensin (AT) II receptor is a transmembrane protein that belongs to the family of receptors coupled to the G protein which is divided into type I (AT1) and type II (AT2) [40, 42, 53]. The genes encoding the AT1 receptor were cloned in 1992, and the genes encoding the AT2 receptor were cloned in 1994 [54]. The binding of angiotensin II to the AT1 receptor leads to the phosphorylation of various proteins such as IRS-1 and IRS-2, making the P110/PI3K subunit the common activation mechanism for the insulin signaling pathway. Resistance to insulin is caused by the inhibition of insulin through the inhibition of the mechanisms involved in glucose transport and vasodilation [42, 45].

Skeletal muscle is important in the development of insulin resistance, which is responsible for 75–95% glucose metabolism [55, 56]. Skeletal muscle constitutes the largest insulin-sensitive tissue in the body and is the primary site for insulin-stimulated glucose utilization [55]. Skeletal muscle resistance to insulin is fundamental to the metabolic dysregulation associated with obesity and physical inactivity and contributes to the development of the MS [55]. Potential mechanisms contributing to reduced insulin signaling and action in skeletal muscle include adipose tissue expansion and increased inflammatory adipokines, increased RAAS activity, decreases in muscle mitochondrial oxidative capacity, increased intramuscular lipid accumulation, and increased reactive oxygen species (ROS) [55].

Angiotensin II has prooxidant and pro-inflammatory effects that regulate apoptosis, inflammation, cell growth, fibrosis, and insulin sensitivity by inducing the formation of oxygen free radicals through the enzyme nicotinamide adenine dinucleotide phosphate (NADPH) oxidase [56]. ROS are formed by the electron transport chain in the mitochondria, and it increases in situations that require the oxidation of substrates such as glucose [56]. Multiple ROS activate transcription factors such as factor nuclear kappa B (NF-KB) protein-1 and hypoxia-inducible factor-1 HIF-1, which are involved in the mechanism of insulin resistance in skeletal muscle [57].

Hyperinsulinemia can alter some of the components of the RAAS such as pro-fibrotic stimulants and pro-inflammatory actions mediated by angiotensin II and can cause cardiovascular disease development [50].

3.1 Relationship between inflammation and insulin resistance

Obesity is a worldwide epidemic that has generated many scientific publications in recent years. Thus, new paradigms were emerging, while the old challenges are still to be unraveled [58]. Obesity is a chronic disease and has become a major problem in most industrialized countries due to its increased prevalence and association with various diseases and due to its great economic impact [59]. To preserve the energy reserves of body fat during periods of negative energy balance, the body has devel-oped various regulatory mechanisms [60]. This control is achieved by balancing the intestinal absorption of glucose, the production of glucose by the liver, and the absorption and metabolism of glucose by peripheral tissues [61]. Insulin regulates blood glucose due to its action in stimulating glucose uptake in the muscle and liver and inhibiting hepatic gluconeogenesis [61]. Despite the change in tolerance to the action of insulin initially, blood glucose is normal because pancreatic β cells have the ability to increase the ability of insulin secretion to overcome existing insulin resistance [9]. Excess nutrients induce hypertrophy of adipocytes that promote the development of various chronic morbidities such as type 2 DM and insulin resistance, glucose intolerance, and dyslipidemia with the consequent development of cardiovas-cular diseases [60]. There is also strong evidence to suggest that cell hypoxia may be an important factor in the pathophysiology of adipocytes and may be one of the causes of this dysfunction contributing to the metabolic alteration associated with obesity [62].

Obesity is defined according to the WHO, as an abnormal or excessive accumulation of fat that can cause imbalances in the health of an individual [63]. Males are more likely to accumulate intra-abdominal adipose tissue compared with females [32, 64]. On average, men have twice the amount of visceral fat than women and have a higher prevalence of metabolic diseases associated with obesity and MS [65].

Obesity is associated with a higher incidence of type 2 DM, insulin resistance, AHT, dyslipidemia, some types of cancer, and CVD [66]. Fat tissue is deposited in two main compartments, subcutaneous and central or visceral [63]. Based on this, obesity can be divided into central and peripheral obesity. Central obesity is charac-terized by hyperplasia and hypertrophy of adipocytes around the intra-abdominal organs and is associated with greater development of MS [65].

Adipocyte hypertrophy is the main consequence of excess nutrient intake, promoting the development of insulin resistance and glucose intolerance [63]. Triglycerides present in fatty tissue are the body's largest energy reserve, so adipose tissue is a very effective energy storage mechanism that allows survival of living beings in times of famine [59].

Recent studies have shown that hypoxia in adipose tissue can play an important role in the cellular mechanisms of chronic inflammation, macrophage infiltra-tion, adiponectin reduction, leptin elevation, adipocyte apoptosis, and reticulum endoplasmic and mitochondrial dysfunction in white adipose tissue in obese [67].

Hypoxia inhibits pre-adipocyte differentiation and stimulates the secretion of leptin and vascular epithelial growth factor (VEGF) of mature adipocytes [68].

There are several genes that regulate the action of hypoxia, such as HIF-1α (hypoxia-inducible factor 1α), vascular endothelial growth factor VEGF, glucose transporter-1 (GLUT-1), heme oxygenase-1, and pyruvate dehydrogenase kinase 1 [67]. It was found that, under conditions of hypoxia in adipose tissue, all genes increased their expression except for VEGF messenger ribonucleic acid (mRNA) [67].

Adipose tissue is thus divided into white adipose tissue and brown adipose tissue, which have different functions. Brown adipose tissue has the function of producing heat. The understanding and vision we currently have in white adipose tissue have changed significantly in the last 15 years [69]. Currently, white adipose tissue is considered, not only as a tissue in which energy is stored in the form of triglycerides but also as the main secretory and endocrine organ of the organism [70]. White adipose tissue produces and secretes various proteins. Initially, these proteins were designated adipocytokines [71]. Currently, it is universally adopted the name of adipokines to the set of proteins synthesized and secreted by adipose tissue, since not all proteins are cytokines [71].

In addition to adipokines and fatty acids, adipose tissue produces other lipid substances such as steroid hormones, prostaglandins and prostanoids, cholesterol, and retinol (cholesterol and retinol are not synthesized by adipocytes but are stored and released from them) [71].

In 1994, Friedman and his colleagues discovered leptin, a hormone produced by white adipose tissue. Leptin, also called OB protein, is a hormone composed of 16 kDa and is produced from a protein with molecular weight of 18 kDa. It is cleaved leading to the production of leptin protein. Leptin is expressed in many tissues including the white and brown adipose tissue, stomach, placenta, mammary gland, ovarian follicles, and others [72]. It is expressed primarily in adipose tissue, but it is also expressed in smaller amounts in the brain and in the cerebrospinal fluid [70].

Leptin acts both in the central nervous system (hypothalamus) and in the peripheral organs. This discovery allowed us to realize that adipose tissue is not only a tissue in which energy storage occurs but also an endocrine organ [73].

White adipose tissue produces various adipokines, many of which are inflammatory mediators such as tumor necrosis factor alpha (TNF-α), interleukin (IL)-β, and IL-6. The production of inflammatory interleukins increases with obesity and is involved in the development of insulin resistance and in the metabolic syndrome [62, 74]. Adipokines have various functions, in particular in the energy balance (e.g., leptin) on insulin sensitivity and glucose metabolism (e.g., adiponectin), inflammation (e.g., TNF-α), immunity (e.g., adipsin), lipid metabolism (e.g., cholesterol ester transfer protein), blood pressure control (e.g., angiotensinogen) hemostasis (e.g., plasminogen activation inhibitor-1), and angiogenesis (e.g., vascular epidermal growth factor) [62, 71].

3.2 Relationship between hypoxia and insulin resistance

Inflammation is an important process in obesity and in type 2 DM, as it promotes insulin resistance by inhibiting the function of IRS-1 and IRS-2 in the pathway insulin signaling [61, 75].

Obesity is characterized by a state of chronic inflammation with increased inflammatory parameters, for example, haptoglobin, IL-6, C-reactive protein (CRP), plasminogen activator inhibitor-1 (PAI-1), and TNF-α [69]. Large concentrations of cytokines produced by adipose tissue, such as TNF-α, IL-6, and IL-1β, and low concentrations of certain adiponectins cause chronic hyperinsulinemia and insulin resistance [63].

3.3 Relationship between endothelial dysfunction and insulin resistance

During periods of positive energy balance, there is an increase in white adipose tissue in order to be able to store excess triglycerides. Consequently, adipose tissue becomes hypoxic due to the decrease in vascularization of the same [62]. The role of hypoxia in adipose tissue in obesity and insulin resistance is still unclear [76]. Regulatory responses mediated by HIF-1 depend on its degree and duration [76].

HIF-1 is a heterodimeric transcription factor induced by hypoxia and composed of two subunits, the α subunit consisting of 120 kDa and the β subunit, consisting of 91 to 94 kDa [77]. The α subunit (HIF-1α) determines the transcriptional activity of HIF-1, and its increase occurs in response to hypoxia. The β (HIF-1β) subunit is constitutively expressed and may also be referred to as an aryl hydrocarbon receptor nuclear translocator (tRNA) [78]. Under normoxia conditions, HIF-1α is hydroxylated in proline and asparaginyl residues, catalyzed by the enzyme prolylhydroxylase (PHD), which promotes binding to an ubiquitin ligase complex. This process leads to proteosomal degradation mediated by HIF-1 [76, 78]. Under conditions of hypoxia, hydroxylation and activation of HIF-1 target genes are inhibited [76].

HIF-1α activates fibrosis of adipose tissue, causing an increase in macrophage infiltration into adipose tissue that mediates a greater increase in inflammation and concomitant sensitivity decrease in insulin [79, 80].

VEGF is a target gene of HIF-1. VEGF promotes angiogenesis, a necessary process for the differentiation of adipocytes and the growth of adipose tissue [81].

4. Obesity and metabolic syndrome

The endothelium is a layer of cells that lines the inner surface of the blood vessel [82]. Endothelial dysfunction is defined as the interruption of the release of vasodilator factors such as NO and prostacyclin (PGI2) and vasoconstrictor factors such as endothelin-1 (ET-1) and angiotensin II [74]. Alterations in vascular endothelial function at the level of adipose tissue can cause insulin resistance [81]. In insulin resistance, the release rate of free fatty acids from the adipose tissue, which contribute to diabetic dyslipidemia, increased. HDL-cholesterol decreased, increasing low-density lipoprotein (LDL) cholesterol and free fatty acids. This stimulates the inflammatory response, causing the adhesion of monocytes and lymphocytes to the endothelial cells, and increases the flow of glucose and free fatty acids to the cells, causing excessive formation of oxygen free radicals, with an increase in metabolic and hemodynamic degradation. There is an increase in free radicals and nitrites by intramitochondrial enzymatic oxy-reduction which promotes apoptosis and impaired vascular endothelial function [82]. This alteration causes insulin metabolic dysfunction, promoting greater resistance to insulin [74, 82].

4.1 Relationship between insulin resistance and hypertension in obesity

Obesity alone is the cause of AHT in 78% of men and 65% of women [83]. Evidence of the causal relationship between insulin resistance and AHT is increasing [40]. Under physiological circumstances, insulin and insulin growth factor-1 (IGF-1) increase vasodilation by stimulating the production of NO and reducing the concentration of intravascular calcium in vascular smooth muscle cells. It also increases the sensitization of myosin-Ca^{2+} light chains. These actions are mediated through the activation of the PI3K enzyme and the Akt protein. Vascular relaxation in response to activation of PI3K/Akt pathway is mediated by endothelial cells producing NO, which involves phosphorylation of endothelial NO synthetase [41].

Patients with AHT have an increase in fasting and postprandial insulin levels in relation to non-hypertensive patients with the same body mass index [41]. It does not occur with secondary AHT but with essential AHT. This is related in part to the action of insulin that changes at the level of muscle tissue, which is the predominant site of glucose use stimulated by insulin [40].

There is an association between blood pressure and the proportion of type II fibers in skeletal muscle, which are less sensitive to insulin than type I fibers [40]. Under normal physiological conditions, there is a relationship between glucose-mediated insulin availability and increased blood flow in response to insulin. This response decreases in obese people with insulin resistance, which suggests resistance to insulin action in reference to vascular NO production [41]. Due to the difficulty of assimilating the rapid growth of adipose tissue in a hypoxia at an early stage, conditions are created for an increase in the expression of HIF-1α [84–87].

There is evidence that insulin resistance and hyperinsulinemia predispose patients to the development of AHT due to cellular abnormalities in insulin signaling pathways and are associated with metabolic and hemodynamic alterations [84]. Metabolic abnormalities are linked to hypertension caused by pathophysiological processes that involve the sympathetic-adrenergic system, the imbalance of cell cations, the increase in inflammation, the oxidative stress, and the RAAS [40].

RAAS plays an important role in insulin resistance by being more active in visceral adipose tissue compared to peripheral adipose tissue [83]. Insulin resistance in obese individuals is also associated, in part, with an antagonism to the action of angiotensin II (AT II) [83]. This produces a decrease in NO production with an increase in vasoconstriction and a decrease in GLUT-4 in skeletal muscle. It also reduces glucose uptake and decreases vasodilation in tissues [83]. Under normal conditions, the RAAS system is a mechanism for regulating blood pressure [88]. With the increase in the activity of the RAAS, there is a greater production of ATII, which leads to stimulation of the sympathetic nervous system, insulin resistance, sodium retention, and increased intravascular volume. It also contributes to kidney disease, hypertension, insulin resistance, and left ventricular hypertrophy [83].

There is a common genetic predisposition to insulin resistance and hypertension. Genetic defects have been described in people who had insulin resistance and AHT and mutations found on chromosome 7q, where other genes important for glycemic control, blood pressure, and obesity are also found [49].

Hyperinsulinemia can directly stimulate the reabsorption of sodium and water in the kidney, which increases the volume in the extracellular space, causing AHT. Another mechanism by which insulin can cause hypertension involves stimulation of the sympathetic nervous system (SNS) by increasing the concentration of norepinephrine. This increase is directly related to the increase in pulse and blood pressure by direct effect on the reabsorption of sodium in the kidney, peripheral vasoconstriction, and increased cardiac output [9].

5. Conclusion

The correlation of CVRF with the metabolic syndrome differs from one another. The notion that CVRF has a tendency to aggregation allows an earlier diagnosis of MS, thus also leading to the early introduction of therapeutic, pharmacological, or non-pharmacological measures. The prevalence of MS increases with age and is present in people of working age, increasing the risk of cardiovascular diseases, work-related absences, and socioeconomic costs. The concept of MS was carried out in a progressive manner. Its clinical importance is enormous due to the fact that it can identify patients with high risk of suffering some CVD and/or DM, allowing a preventive intervention. This may explain the importance of understanding the pathophysiology of the MS, and the author hopes that at the end of the chapter, the reader can understand it. This chapter is the beginning of an explanation of meta-bolic syndrome that should be complemented by reading other articles.

Author details

Armindo Miguel de Jesus Sousa de Araújo Ribeiro[1,2]

1 Unidade Local de Saúde do Litoral Alentejano, Hospital Litoral Alentejano, Santiago do Cacém, Portugal

2 Universidad de Extremadura, Badajoz, Spain

*Address all correspondence to: amsr29@gmail.com

References

[1] Ribeiro AS, Seixas R, Gálvez JM, Climent V. Cardiovascular risk factors: Is the metabolic syndrome related to aging? Epidemiology in a Portuguese population. Diabetes and Metabolic Syndrome. 2018;**12**(6):885-891

[2] Ribeiro AS, Gálvez JM, Vicente C. Equality or disparity. Cardiovascular risk factors and their correlation within the metabolic syndrome. The secret of Gods. Diabetes and Metabolic Syndrome. 2019;**13**(2):1393-1398

[3] Villar Álvarez F, Maiques Galán A, Brotons Cuixart C, Torcal Laguna T, Sánchez-Pinilla O, Vilaseca Canals J, et al. Grupos de expertos del PAPPS. Actividades preventivas cardiovasculares en atención primaria. Atencion Primaria. 2003; **32**(2):27-41

[4] Townsend N, Nichols M, Scarborough P, Rayner M. Cardiovascular disease in Europe-epidemiological update 2015. European Heart Journal. 2015;**36**:2696-2705

[5] Cruz WS, Siguenza OA, Guamancela SF, Piedra C, Andrade GC, Valdez MT. Prevalencia del síndrome metabólico en individuos adultos de las parroquias urbanas de la ciúdad de Cuenca, Ecuador. Síndrome Cardiometabólico. 2013;**3**:113-125

[6] Alberti KGMM, Eckel RH, Grundy SM, Zimmet PZ, Cleeman JI, Donato KA, et al. Harmonizing the metabolic syndrome. A Joint interim Statement of the International diabetes Federation Task Force on Epidemiology and Prevention; National Heart, Lung and Blood Institute; American Heart Association; World Heart Federation; International Atherosclerosis Society; and International Association for the Study of Obesity. Circulation. 2009;**120**:1640-1645

[7] Reaven GM. Banting lecture. Role of insulin resistance in human disease. Diabetes. 1988;**37**(12):1595-1607

[8] Isomaa B, Almgren P, Tuomi T, Forsen B, Lahti K, Nissen M, et al. Cardiovascular morbidity and mortality associated with the metabolic syndrome. Diabetes Care. 2001;**24**(4):683-689

[9] DeFronzo RA, Ferrannini E. Insulin resistance. A multifaceted syndrome responsible for NIDDM, obesity, hypertension, dyslipidemia, and atherosclerotic cardiovascular disease. Diabetes Care. 1991;**14**(3):173-194

[10] Garber AJ, Handelsman Y, Einhorn D, Bergman DA, Bloomgarden ZT, Fonseca V, et al. Diagnosis and management of prediabetes in the continuum of hyperglycemia: When do the risks of diabetes begin? A consensus statement from the American College of Endocrinology and the American Association of Clinical Endocrinologists. Endocrine Practice. 2008;**14**(7):933-946

[11] Robles JCL. Síndrome metabólico: Concepto y aplicación práctica. Anales de la Facultad de Medicina. 2013;**74**(4):315-320

[12] Fernandéz-Bergés D, Cabrera de León C, Sanz H, Elosuad R, Guembef MJ, Alzamorah M, et al. Síndrome metabólico en España: Prevalencia y riesgo coronario asociado a la definición armonizada y a la propuesta por la OMS. Estudio DARIOS. Revista Española de Cardiología. 2012;**65**:241-248

[13] Alberti KG, Zimmet PZ. Definition, diagnosis and classification of diabetes mellitus and its complications. Part 1: Diagnosis and classification of diabetes mellitus. Provisional report of a WHO consultation. Diabetic Medicine. 1998;**15**(7):539-553

[14] Mykkanen L, Zaccaro DJ, Wagenknecht LE, Robbins DC, Gabriel M, Haffner SM. Microalbuminuria is associated with insulin resistance in nondiabetic subjects: The insulin resistance atherosclerosis study. Diabetes. 1998;**47**(5):793-800

[15] Crepaldi G, Maggi S. The metabolic syndrome: A historical context. Diabetes Voice. 2006;**51**:8-10

[16] Kraemer FB, Ginsberg HN, Gerald M, Reaven MD. Demonstration of the central role of insulin resistance in type 2 diabetes and cardiovascular disease. Diabetes Care. 2014;**37**:1178-1181

[17] Haffner S, Taegtmeyer H. Epidemic obesity and the metabolic syndrome. Circulation. 2003;**108**:1541-1545

[18] Himsworth HP. Insulin deficiency and insulin inefficiency. British Medical Journal. 1940;**1**:719-721

[19] Fronzo R, Tobin J, Andres R. Glucose clamp technique: A method for quantifying insulin secretion and resistance. The American Journal of Physiology. 1979;**237**:E214-E223

[20] Alberti KG, Zimmet P, Shaw J. The metabolic syndrome-a new worldwide definition. Lancet. 2005;**366**:1059-1062

[21] Grundy SM, Cleeman JI, Daniels SR, et al. Diagnosis and management of the metabolic syndrome: An American Heart Association/National Heart, Lung, and Blood Institute scientific statement. Circulation. 2005;**112**(17):2735-2752

[22] Sociedade Brasileira de Cardiologia. I Diretriz brasileira de diagnóstico e tratamento da síndrome metabólica. Arquivos Brasileiros de Cardiologia. 2005;**84**(1):1-28

[23] Expert Panel on Detection, Evaluation, and Treatment of High Blood Cholesterol in Adults. Executive summary of the third report of the National Cholesterol Education Program (NCEP) expert panel on detection, evaluation, and treatment of high blood. Cholesterol in adults (adult treatment panel III). JAMA. 2001;**285**:2486-2497

[24] The IDF worldwide definition of the metabolic syndrome. 2006. Available from: http://www.idf.org/home/index. cfm?node=1429 [Accessed: 30 April 2007]

[25] Woo J, Ho SC, Yu AL, Sham A. Is waist circumference a useful measure in predicting health outcomes in the elderly? International Journal of Obesity and Related Metabolic Disorders. 2002;**26**(10):1349-1355

[26] Adams RJ, Appleton S, Wilson DH, Taylor AW, Dal Grande E, Chitleborough C, et al. Population comparison of two clinical approaches to the metabolic syndrome: Implications of the new International Diabetes Federation consensus definition. Diabetes Care. Nov 2005;**28**(11):2777-2779

[27] Qiao Q, Laatikainen T, Zethelius B, et al. Comparison of definition of metabolic syndrome in relation to the risk of developing stroke and coronary heart disease in Finnish and Swedish cohorts. Stroke. 2009;**40**:337-343

[28] The DECODE Study Group. Comparison of different definitions of the metabolic syndrome in relation to cardiovascular mortality in Europe men and women. Diabetologia. 2006;**49**:2837-2846

[29] Villalobos SC, Mosquera CP, Tovar CH. Prevalencia del Síndrome metabólica en consulta de medicina interna Hospital de San José de Bogotá DC. Repertorio de Medicina y Cirugía. 2011;**20**:93-102

[30] Grau M, Elosua R, León AC, Guembe MJ, Baena-Díez JM, Alonso TV.

Factores de riesgo cardiovascular en España en la primera década del siglo XXI: Análises agrupado con datos individuales de 11 estudios de base poblacional, estudio DARIOS. Revista Española de Cardiología. 2011;**64**(4):295-304

[31] Elias CC, Arnold DL, Trimiño FY, Armas RA. Epidemiologia y prevención del sindrome metabólico. Revista Cubana de Higiene y Epidemiología. 2012;**50**:250-256

[32] Ford ES, Giles WH, Dietz WH. Increasing prevalence of metabolic among US adults. Findings from the third national health and nutrition examination survey. JAMA. 2002;**287**(3):356-359

[33] Hu G, Qiao Q, Tuomiehto J, et al. Prevalence of metabolic syndrome and its relation to all-cause and cardiovascular mortality in nondiabetic European men and women. Archives of Internal Medicine. 2004;**164**(10):1066-1076

[34] Fiuza M, Cortez-Dias N, Martins S, et al. Síndrome metabólico em Portugal: Prevalência e implicações no risco cardiovascular-resultados do estudo VALSIM. Revista Portuguesa de Cardiologia. 2008;**27**(12):1495-1529

[35] Santos AC, Lopes C, Barros H. Prevalence of metabolic syndrome in the city of Porto. Revista Portuguesa de Cardiologia. 2004;**23**(1):45-52

[36] Haffner SM, Valdez RA, Hazuda HP, et al. Prospective analysis of the insulin-resistance syndrome (syndrome X). Diabetes. 1992;**41**(6):715-722

[37] Petersen KF, Dufour S, Savage DB, et al. The role of skeletal muscle insulin resistance in the pathogenesis of the metabolic syndrome. Proceedings of the National Academy of Sciences of the United States of America. 2007;**104**:12587-12594

[38] Khan AH, Pessin JE. Insulin regulation of glucose uptake: A complex interplay of intracellular signalling pathway. Diabetologia. 2002;**45**:1475-1483

[39] Sesti G, Federici M, Hribal ML, Lauro D, Sbraccia P, Lauro R. Defects of the insulin receptor substrate (IRS) system in human metabolic disorders. The FASEB Journal. 2001;**15**:2099-2111

[40] DeFronzo RA. Pathogenesis of type 2 diabetes mellitus. Medical Clinics of North America. 2004;**88**:787-835

[41] Virkamaki A, Ueki K, Kahn CR. Protein-protein interaction in insulin signaling and the molecular mechanisms of insulin resistance. Journal of Clinical Investigation. 1999;**103**(7):931-943

[42] Sowers JR, Epstein M, Frohlich ED. Diabetes, hypertension, and cardiovascular disease: An update. Hypertension. 2001;**37**:1053-1059

[43] Ribeiro AS, Bicho MP. Insulin resistance and essential hypertension. Which mechanisms? Revista Portuguesa de Hipertensão e Risco Cardiovascular. 2012;**30**:12-15

[44] Inoue G, Cheatham B, Emkey R, Kahn CR. Dynamic of insulin signaling in 3T3-L1 adipocytes. Differential compartmentalization and trafficking of insulin receptor substrate (IRS)-1 and IRS-2. Journal of Biological Chemistry. 1998;**27**(3):11548-11555

[45] Sowers J. Insulin resistance and hypertension. American Journal of Physiology-Heart and Circulatory Physiology. 2004;**286**:H1597-H1602

[46] Boden G, Jadali F, White J, et al. Effects of fat on insulin-stimulated carbohydrate metabolism in normal men. Journal of Clinical Investigation. 1991;**88**:960-966

[47] Lann D, Laroith D. Insulin resistance as the underlying cause of

metabolic syndrome. Medical Clinics of North America. 2007;**91**:1063-1077

[48] Stump CS, Hamilton MT, Sowers JR. Effect of antihypertensive agents on the development of type 2 diabetes mellitus. Mayo Clinic Proceedings. 2006:796-806

[49] Cheng LSC, Davis RC, Raffel LJ, Xiang AH, Wang N, Quinones M. Coincident linkage of fasting plasma insulin and blood pressure to chromosome 7q in hypertensive Hispanic families. Circulation. 2001;**104**:1255-1260

[50] Putman K, Shoemaker R, Yiannikouris F, Cassis LA. The renin-angiotensin system: A target of and contributor to dyslipidemias, altered glucose homeostasis, and hypertension of the metabolic syndrome. American Journal of Physiology. Heart and Circulatory Physiology. 2012;**302**:H1219-H1230

[51] Giani JF, Mayer MA, Munoz MC, Silberman EA, Hocht C, Taira CA, et al. Chronic infusion of angiotensin-(1-7) improves insulin resistance and hypertension induced by a high-fructose diet in rats. American Journal of Physiology. Endocrinology and Metabolism. 2009;**296**:E262-E271

[52] Habibi J, Whaley-Connel A, Hayden MR, DeMarco VG, Scneider R, Sowers SD, et al. Renin inhibition attenuates insulin resistance, oxidative stress, and pancreatic remodeling in the transgenic Ren2 rat. Endocrinology. 2008;**149**:5643-5653

[53] Marinho C, Ferreira J, Levy P, Coelho C, Gomes A, Laires MJ. Contribution of genetic polymorphisms involved in the control of blood pressure and cell growth/oxidative stress for the modulation of angiotensin converting enzyme activity. Revista Portuguesa de Cardiologia. 2011;**30**(1):18

[54] Haywood Guy A, Gullestad L, Katsuya T, Hutchinson HG, Pratt RE, Horiuchi M, et al. AT1 and AT2 angiotensin receptor gene expression in human heart failure. Circulation. 1997;**95**:1201-1206

[55] Stump CS, Henriksen EJ, Wei Y, Sowers JR. The metabolic syndrome: Role of skeletal muscle metabolism. Annals of Medicine. 2006;**38**:389-402

[56] Ndisang JF, Lane N, Syed N, Jadhav A. Up-regulating the heme oxygenase system with hemin improves insulin sensitivity and glucose metabolism in adult spontaneously hypertensive rats. Endocrynology. 2010;**151**(2):549-560

[57] Wei Y, Sowers JR, Clark SE, Li W, Ferrario CM, Stump C. Angiotensin II-induced skeletal muscle insulin resistance mediated by NF-kB activation. American Journal of Physiology-Endocrinology and Metabolism. Feb 2008;**294**(2):E345-E351

[58] Ribeiro AS, Bicho MP. Obesidade: Novos paradigmas e velhos desafios. Revista Portuguesa de Hipertensão e Fatores de Risco Cardiovasculares. 2013;**34**:20-25

[59] American Gastroenterological Association Clinical Practice Committee. AGA technical review on obesity. Gastroenterology. 2002;**123**:882-932

[60] Kopecky J, Rossmeisl M, Flachs P, Brauner P, Sponarová J, Matejková O, et al. Energy metabolism of adipose tissue-physiological aspects and target in obesity treatment. Physiological Research. 2004;**53**:S225-S232

[61] Saltiel AR, Kahn CR. Insulin signalling and the regulation of glucose and lipid metabolism. Nature. 2001;**414**:799-806

[62] Wood IS, Heredia FP, Wang B, Trayhum P. Cellular hypoxia and adipose tissue dysfunction in obesity. Proceedings of the Nutrition Society. 2009;**68**(4):370-377

[63] Donohoe CL, Doyle SL, Reynolds JV. Visceral adiposity, insulin resistance and cancer risk. Diabetology and Metabolic Syndrome. 2011;**3**:1-13

[64] Nedungadi TP, Clegg DJ. Sexual dimorphism in body fat distribution and risk for cardiovascular diseases. Journal of Cardiovascular Translational Research. 2009;**2**:321-327

[65] Ford eS, Li C, Zhao G, Pearson WS, Mokdad AH. Prevalence of the metabolic syndrome among U.S. adolescents using the definition from the International Diabetes Federation. Diabetes Care. 2008;**31**:587-589

[66] Krishnan J, Danzer C, Simka T, Ukropec J, Walter KM, Kumpf S, et al. Dietary obesity-associated HIF1x activation in adipocytes restricts fatty acid oxidation and energy expenditure via suppression of the Sirt2-NAD+ system. Genes and Development. 2011;**26**:259-270

[67] Ye J. Emerging role of adipose tissue hypoxia in obesity and insulin resistance. International Journal of Obesity. 2009;**33**(1):54-66

[68] Ye J, Gao Z, Yin J, He H. Hypoxia is a potential risk factor for chronic inflammation and adiponectin reduction in adipose tissue of ob/ob and dietary obese mice. American Journal of Physiology. Endocrinology and Metabolism. 2007;**293**:E1118-E1128

[69] Trayhurn P, Wang B, Wood SI. Hypoxia in adipose tissue: A basis for the dysregulation of tissue function in obesity? British Journal of Nutrition. 2008;**100**:227-235

[70] Rajala MW, Scheerer PE. Minireview: The adipocyte-at the crossroads of energy homeostasis, inflammation, and atherosclerosis. Endocrinology. 2004;**144**(9):3765-2773

[71] Trayhurn P, Wood S. Adipokines: Inflammation and the pleiotropic role of white adipose tissue. British Journal of Nutrition. 2004;**92**(3):347-355

[72] Trayhurn P, Beattie JH. Physiological role of adipose tissue: White adipose tissue as an endocrine and secretory organ. Proceedings of the Nutrition Society. 2001;**60**:329-339

[73] Zhang YY, Proenca R, Maffei M, Barone M, Leopold L, Friedman JM. Positional cloning of the mouse obese gene and its human homolog. Nature. 1994;**372**:425-432

[74] Jansson PA. ENdothelial dysfunction in insulin resistance and type 2 diabetes. Journal of Internal Medicine. 2007;**262**:173-183

[75] Fantin VR, Sparling JD, Slot JW, Keller SR, Lienhard GE, Lavan B. Characterization of insulin receptor substrate 4 in human embryologic kidney 293 cells. The Journal of Biological Chemistry. 1998;**273**:10726-10732

[76] Jiang C, Qu A, Matsubara T, Chanturiya T, Jou W, Gavrilova O, et al. Disruption of hypoxia-inducible factor 1 in adipocytes improves insulin sensitivity and decreases adiposity in high-fat diet-fed mice. Diabetes. 2011;**60**:2484-2495

[77] Huang LE, Arany Z, Livingston DM, Bunn HF. Activation of hypoxia-inducible transcription factor depends primarily upon redox-sensitive stabilization of its alpha subunit. The Journal of Biological Chemistry. 1996;**271**(50):32253-32259

[78] He Q , Gao Z, Yin J, Zhang J, Yun Z, Ye J. Regulation of HIF-1α activity in adipose tissue by obesity-associated factors: Adipogenesis, insulin, and hypoxia. American Journal of Physiology. Endocrinology and Metabolism. 2001;**300**:E877-E885

[79] Halberg N, Khan T, Trujillo ME, Wernstedt-Asterholm I, Attie AD, Sherwani S, et al. Hypoxia-inducible factor 1α induces fibrosis and insulin resistance in white adipose tissue. Molecular and Cellular Biology. 2009;**29**(16):4467-4483

[80] Michailidou Z, Turban S, Miller E, Zou X, Schrader J, Ratcliffe P, et al. Increased angiogenesis protects against adipose hypoxia and fibrosis in metabolic disease-resistant 11β-hydroxysteroid dehydrogenase type 1 (HSD-1)-deficient mice. Journal of Biological Chemistry. 3 Feb 2012;**287**(6):4188-4197

[81] Cao Y. Adipose tissue angiogenesis as a therapeutic target for obesity and metabolic diseases. Nature Reviews. Drug Discovery. 2010;**9**:107-115

[82] Cersosimo E, DeFronzo RA. Insulin resistance and endothelial dysfunction: The road map to cardiovascular diseases. Diabetes/Metabolism Reviews. 2006;**22**:423-436

[83] El-Atat FA, Stas SN, Mcfarlane SI, Sowers JR. The relationship between hyperinsulinemia, hypertension and progressive renal disease. Journal of the American Society of Nephrology. 2004;**15**:2816-2827

[84] Kalidas K, Wasson J, Glaser B, Meyer JM, Duprat LJ, White MF, et al. Mapping of the human insulin receptor substrate-2 gene, identification of a linked polymorphic marker and linkage analysis in families with type II diabetes: No evidence for a major susceptibility role. Diabetologia. 1998;**41**:1389-1391

[85] Cantley J, Grey ST, Maxwell PH, Withers DJ. The hypoxia response pathway and B-cell function. Diabetes, Obesity and Metabolism. 2010;**12**(2):159-167

[86] Huang LE, Gu J, Schau M, Bunn HF. Regulation of hypoxia-inducible factor 1alpha is mediated by an O_2-dependent degradation domain via the ubiquitin-proteasome pathway. Proceedings of the National Academy of Sciences of the United States of America. 1998;**95**:7987-7992

[87] Ribeiro AS, Bicho M, Mascarenhas MR. Hipertensão arterial e densitometria óssea: Que relação? Revista Portuguesa de Hipertensão. 2014;**43**:6-8

[88] Kotsis V, Stabouli S, Papakatsika S, Rizos Z, Parati G. Mechanisms of obesity-induced hypertension. Hypertension Research. 2010;**33**:386-393

10

Pathogenesis of Insulin Resistance

Gaffar S. Zaman

Abstract

Insulin resistance is interpreted as being a normal or raised insulin level giving rise to a biological reaction which is attenuated in effect; classically this cites to the weakened sensitivity to the disposal of insulin arbitrate glucose. *Compensatory hyperinsulinemia* eventuates when the secretion of the β cells of pancreas gets elevated to sustain the level of blood glucose in normal levels. The term *insulin resistance syndrome* is used to refer to a group of abnormalities and interconnected physical consequences that eventuate in long-standing insulin-resistant persons. Under standard situations of insulin reactivity, the response of insulin triggers the intake of glucose into the body cells, for utilization as energy, and impedes the utilization of fat for energy, as a result of which, the concentration of glucose circulating in the blood decreases. There are a number of risk factors for insulin resistance. Four major metabolic abnormalities characterize type 2 diabetes mellitus (T2DM): impaired insulin action, obesity, increased endogenous glucose output, and insulin secretory dysfunction. The evolution (and subsequent progression) of type 2 diabetes mellitus is delineated by a gradual deterioration of glucose tolerance over several years. Glucose tolerance testing, hyperinsulinemic euglycemic clamp, modified insulin suppression test, homeostatic model assessment (HOMA), and quantitative insulin sensitivity check index (QUICKI) method for insulin assessment are some of the methods by which insulin resistance can be measured. Moreover, longer-term effective researches as well are essential to preferably ascertain the significance of the glycemic index in the blood glucose regulation and to prevent the complications of diabetes, particularly in relations to insulin resistance risk factors. The possible role of insulin resistance in the glycemic index in depleting oxidative stress postprandially and related pro-inflammatory situations also merits further appraisal.

Keywords: insulin resistance, compensatory hyperinsulinemia, type 2 diabetes mellitus, obesity

1. Introduction

One of the most renowned hormones of our body is insulin which enables glucose to go inside the cells which additionally decreases blood glucose. The hor-mone insulin is secreted by the pancreas in response to glucose entering the blood-stream after a meal. Insulin resistance (IR) is contemplated as a pathological situation in which our body cells decline to react normally to insulin hormone [1]. To avert hyperglycemia and apparent damage to our body organs in the future [2], insulin production by the body starts when glucose enters into the bloodstream, predominantly from the dietary carbohydrate digestion and absorption. Under standard situations of insulin reactivity, the response of insulin triggers the intake of glucose into the body cells, for utilization as energy, and impedes the utilization of fat for energy, as a result of which, the concentration of glucose circulating in the blood decreases, which results in glucose remaining within the normal range in case of consumption of a substantial amount of carbohydrates. Carbohydrates contains sugars, i.e., from only one glucose containing monosaccharides, such as fructose and glucose; two glucose containing disaccharides, like cane sugar; and many glucose containing polysaccharides (e.g., starches) and glycoprotein, glycolipids, etc. Fructose, ultimately metabolized into triglycerides inside the liver, stimulates the production of insulin and is seen to have a more impressive sequel than other carbohydrates. A customarily increased intake of carbohydrates, and specifically fructose, imparts to insulin resistance and has been connected to gain of weight and obesity [3–5]. If surplus blood glucose is not adequately transported into the cells even in the insulin's presence, the augmented level of blood glucose can elicit in the classic hyperglycemic triad of polydipsia (increased thirst), polyphagia (increased appetite), and polyuria (increased urination). Circumventing carbohydrates, a zero-carbohydrate diet or conditions of fasting can counteract insulin resistance [6, 7]. The first narration of insulin resistance is found historically in the 1960s, soon after the invention of radioimmunoassays helped in making serum insulin quantification possible and subsequently revealed that people having late-onset diabetes mellitus had high levels of insulin [8, 9]. Drs. Yalow and Berson [8, 9] defined insulin resistance as "a state in which a greater than normal amount of insulin is required to elicit a quantitatively normal response." The next milestone discovery in this context was the detection and ascertaining of the insulin receptor and also the discovery that insulin resistance led to hyperinsulinemia and the fact that this was interconnected very much with atypical/unconventional binding of insulin hormone with its receptor in various rodent models [10, 11]. The succeeding milestone discovery in the history of insulin resistance came with the pioneering recognition of the receptor for insulin and the observation that high insulin in blood, secondary to the insulin resistance, was interconnected with atypical binding of insulin hormone to its receptor [10, 11]. It was not before the year 1976 when the evidence came that defects in receptor for insulin could be interlinked with resis-tance to the insulin hormone in humans, provided genetic translational verification for the significance of insulin resistance [10]. Researchers like Kahn et al. [10] delineated two syndromes that were distinguished by virilization, acanthosis nigricans, hirsutism, anovulation, acne, and flawed binding of insulin on the insulin receptor of lymphocytes circulating in the blood. This initial syndrome was desig-nated as type A insulin resistance when it took place without the existence of anti-insulin antibodies and the corresponding accountability of the insulin receptor was the main factor, and in contrast it was designated as the type B when it occurred with the clinical characteristics of various autoimmune ailments and it always occurred when neutralizing anti-insulin antibodies were present [10]. After this specification of type A and type B syndrome, all rare and extreme levels of insulin resistance like acanthosis nigricans syndrome, hyperandrogenism, the Rabson-Mendenhall syndrome, lipodystrophy, and leprechaunism were discovered [12, 13]. To defend the idea by Himsworth that some cases of diabetes were not secondary to absolute insulin deficiency, it was imperative to prove the truth that insulin circu-lating in the blood was present in the insensitive form in those patients as classified by Himsworth. This was put to rest with the publication by Yalow and Berson in 1960 of an immunoassay research of endogenous blood insulin in humans. During this novel innovative work, Berson and Yalow delineated an excellent immunological technique for measuring the amount of insulin that integrated the degree of specificity with sensitivity required to take the measurement of even the smallest concentrations of insulin contained in the body circulation. Utilizing this novel tech-nique to categorize immunoreactive insulin amounts present in plasma in normal people to those particular patients having maturity-onset type diabetes, it was con-ceived that the amounts of insulin were on the average elevated in the patients with diabetes. Putting stress on the rationale of these particular outcomes, both of them summarized that the tissues of a person with maturity-onset diabetes do not have a good response to the level of insulin; on the contrary, the tissues of a nondiabetic person respond very well to his level of insulin. In the words or terminology of Himsworth, patients with this character in diabetes can be considered as "insulin insensitive." In spite of the fact these results of the novel publication by Yalow and Berson were afterwards authenticated by various research groups, it became evident that the interconnection between insulin concentrations and plasma glucose in patients having type 2 diabetes was not such a simple one. To be precise, in a person with comparatively slight increments of fasting plasma glucose concentration, the responses of plasma insulin to oral glucose were equivalent or prominently higher than the normal but with elevated proportions of glucose intolerance and with the appearance of noteworthy hyperglycemia during fasting [14–17]. It was obligatory to evolve an exploratory perspective that would quantify in an unequivocal way the capability of an individual person to get rid of fixed glucose load under the influence of identical insulin stimuli during steady-state conditions [18].

2. Risk factors for insulin resistance

A number of risk factors are found for insulin resistance, together with being overweight or obese or pursuing a sedentary lifestyle [19]. Numerous genetic constituents can elevate the chances for the same, and there are some particular medical circumstances correlated with insulin resistance [19].

The National Institute of Diabetes and Digestive and Kidney Diseases has specified several hazard factors:

1. Age of 45 or more.
2. Native Alaskan, Asian American, American Indian, African American, Native Hawaiian, or Latino/Hispanic ethnicity.
3. Having abnormal health states such as increased systolic/diastolic pressure and increased levels of cholesterol.
4. Having gestational diabetes history.
5. Having a history stroke or heart disease [19].
6. In addition, some medications and other health conditions can raise the risk [19].

2.1 Types of people more likely to develop insulin resistance

Individuals who have hereditary factors or lifestyle-related factors are bound to have in their later life insulin resistance or prediabetes [20]. Hazard factors incorporate:

- Overweight or obesity.
- Age 45 or more.

- Having a parent, sibling, or sister with diabetes.
- African American, Alaskan Native, American Indian, Asian American, Hispanic/Latino, Native Hawaiian, or Pacific Islander American ethnicity.
- Physical idleness.
- Health conditions, for example, hypertension and high cholesterol levels.
- A history of gestational diabetes.
- A history of coronary illness or stroke.
- Polycystic ovary disorder, also known as PCOS.
- Individuals who have metabolic disorder—hypertension, irregular cholesterol levels, and enormous waist size—are bound to have prediabetes.
- Hormonal imbalances, for example, Cushing's disorder and acromegaly.
- Sleep issues, particularly rest apnea.

In spite of the fact that you cannot change hazard factors, for example, family ancestry, age, or ethnicity, you can change lifestyle factors such as eating, physical activity, and weight. These ways of life changes can bring down your odds of creating insulin resistance or prediabetes [21].

There are a number of other hazard factors that are firmly connected to insulin resistance; however, these factors are yet to give clear answers about how much these variables might be a reason for the same.

The variables include:

- Abundance of fat
- Having hypertension or hypercholesterolemia
- Having a nearby relative with type 2 diabetes
- History of gestational diabetes
- Having potential to develop type 2 diabetes

2.2 Diet

The food that we take are regularly seen as a conspicuous reason for diabetes and frequently supposed as a reason.

Many studies have shown that our diet can have an effect in type 2 diabetes; however, it is one factor among numerous others, and speculations ought not be drawn without the thought of other contributing variables.

2.3 Contributions of genetics

Research has revealed various factors, which are related with an elevated danger of diabetes. There are various elements which can impact our plasma glucose, as well as taking into consideration where we disperse fat in our body and how efficiently our muscles allow glucose to enter from the blood.

It is a well-known fact that genes help control and regulate every metabolic activity in the body, and mutations, in genes, which have influence in the digestion and absorption process can cause problem with controlling blood glucose level. To date scientists have distinguished more than 60 genes related with type 2 diabetes mellitus (T2DM).

2.4 Medication

Corticosteroids treat mainly inflammatory disorders, as rheumatoid joint inflammation, lupus, and hypersensitivities. Normal steroids incorporate hydrocortisone and prednisone. Be that as it may, steroid creams (for a rash) or inhalers (for asthma) aren't an issue.

Medications that treat nervousness, attention deficit hyperactivity disorder (ADHD), sorrow, and other emotional well-being issues can incorporate clozapine, olanzapine, risperidone, and quetiapine:

- Contraception pills
- Medications that treat hypertension, for example, beta-blockers and thiazide diuretics
- Statins to bring down cholesterol
- Adrenaline for serious hypersensitive responses
- High doses of asthma medications or medications that you infuse for asthma treatment
- Isotretinoin for skin breakout
- Tacrolimus, which you get after an organ transplant
- A few meds that treat *human immunodeficiency virus* (HIV) and hepatitis
- Pseudoephedrine, a decongestant in some cold and influenza prescriptions
- Niacin or vitamin B3

Alongside these hazard factors, different things that may add to insulin resistance incorporate:

- Antihypertensive agents such as β-blockers, diuretics, oral contraceptives, corticosteroids, nicotinic acid, and antipsychotic agents are said to increase insulin resistance [22, 23]; in addition, many anti-retroviral protease inhibitors utilized to treat human immunodeficiency virus infection also cause insulin resistance. The mechanisms of actions vary: β-blockers impede the secretion of insulin from the pancreas by blocking the β-adrenoceptors, depletion of the blood levels of potassium is the main action of thiazide diuretics, counter-regulatory hormonal activity is the main action of oral contraceptives and corticosteroids, and loss of peripheral subcutaneous fat with partial lipodystrophy (with resultant accumulation of truncal adipose tissue) is the main action of HIV-1 protease inhibitors. All these ultimately lead to insulin resistance [24].

2.5 Lifestyle factors

Many hormones can instigate insulin resistance, prominent among them being human placental lactogen, growth hormone, and cortisol [25]. Counteraction of insulin is done by cortisol and can cause elevated hepatic gluconeogenesis, decreasing peripheral use of glucose and elevating insulin resistance [26]. Cortisol does this by diminishing the translocation of glucose transporters (especially GLUT4) to the respective cell membrane [27, 28]. This is based on the noteworthy augmentation in the sensitivity of insulin in humans after doing bariatric surgery and surgical removal of the duodenum in rats [29]; it has been speculated that some substance is manufactured in the mucosa of duodenum, which gives a signal to the body cells to become insulin resistant. With removal of the producing tissue, cessation of the signal occurs, and reverting back of the body cells to normal insulin sensitivity is seen. No such particular substance has been discovered as yet, and the certainty of such a substance remains speculative. Leptin is a hormone derived from the adipocytes and ob gene [30] whose role is the regulation of hunger by forewarning the body when it is full [31]. Researches have depicted that dearth of leptin leads to severe obesity and is intensely associated with insulin resistance [32].

3. Pathogenesis

Four major metabolic abnormalities characterize type 2 diabetes mellitus: impaired insulin action, obesity, increased endogenous glucose output, and insulin secretory dysfunction [33–35]. In spite of the fact that there is considerable authentication that three of these idiosyncrasies exist in most people before the commencement of diabetes, the concatenation with which they evolve and their corresponding contributions to the advancement from normal glucose tolerance (NGT) to impaired glucose tolerance (IGT), and ultimately to type 2 diabetes [36–38], cannot be confirmed for sure even though some detailed longitudinal studies have provided some information [39–44]. Contemporaneous comprehension of the pathogenesis of type 2 diabetes is established on a wide-reaching number of cross-sectional [45–57] and prospective [58–72] studies. The evolution (and subsequent progress) of type 2 diabetes mellitus is delineated by a gradual degeneration of glucose tolerance over several years [33–38]. Prospective and cross-sectional data demonstrate that defects in insulin secretion, body weight gain, insulin action, and an elevation in endogenous glucose output are cardinal in this decline [45–56]. The pathogenetic history of diabetes—the corporeal chain of events with which these metabolic aberrations evolve in relation to one another throughout the various stages of the illness—remains unrevealed. Many authors have suggested that a flaw in insulin activity is the principal aberration in the premature stages of the evolution of type 2 diabetes and that secretory dysfunction of insulin takes place only at a later stage.

3.1 Molecular mechanism of insulin resistance in the muscle

Cellular contents of lipids inside myocytes (of muscles) are referred to as intramyocellular lipid (IMCL) and basically reflect intramyocellular triglyceride content. Although IMCL emphatically relates with muscle insulin resistance in inactive people, triglycerides themselves have been disassociated from insulin resistance, recommending that other lipids (e.g., diacylglycerols, ceramides, and so on) intervene insulin resistance [73]. Various investigations have depicted the interrelationship insulin resistance in muscles and between diacylglycerol (DAG) content. Insulin-animated tyrosine phosphorylation of insulin receptor substrate 1 (IRS-1) and IRS-1-related phospho-inositide 3-kinase (PI3K) actuation were intensely debilitated in skeletal muscle of lipid-injected people [74] and rodents [75, 76]. In rodents, lipids and high-fat intake bring about transient increments in muscle DAG content [75], bringing about continued appearance of protein kinase C-theta (PKCθ) that constrained phosphorylation of IRS-1 by insulin receptor substrate 1 (IRTK). Lipid mixtures in normal human volunteers correspondingly elevated skeletal muscle DAG [77, 78] and caused muscle insulin resistance. The improvement of muscle insulin resistance can prompt metabolic ailment. This has been seen in hereditary mouse models of particular muscle insulin resistance [79], which are inclined to hepatic steatosis [75] and increased adiposity [77]. In young, lean, and people with skeletal muscle insulin resistance, ingested glucose is not taken up by muscle and gets occupied to the liver, where it becomes substrate for liver once more by lipogenesis, increasing liver triglyceride; furthermore, plasma triglyceride increase results in decreasing plasma high-density lipoprotein (HDL) levels [80]. Nonalcoholic fatty liver disease (NAFLD) is unequivocally connected with hepatic insulin resistance. In patients with lipodystrophy, ectopic lipid accumulation in the liver and skeletal muscle was related with extreme hepatic and muscle insulin resistance [81]. Leptin treatment diminished the consumption of calorie, settled hepatic steatosis, and improved insulin activity [82]. Lipodystrophic mice have a comparable phenotype, and fat transplantation protected these mice by permitting redistribution of lipids from ectopic destinations to transplanted fat tissue and standardization of insulin activity [83]. Mice overexpressing lipoprotein lipase in the liver cause hepatic steatosis and liver-explicit insulin resistance [84]. In rodents and mice, high-fat eating regimens lead to hepatic steatosis and hepatic insulin resistance. In this way, these considerations show that ectopic lipid in the liver is explicitly related with hepatic insulin resistance. Magkos et al. showed that hepatic DAG content (not hepatic ceramide content) was the best indicator of hepatic insulin resistance in obese people [85]. There are a few conditions wherein hepatic steatosis shows up disassociated from hepatic insulin resistance. A typical single-nucleotide polymorphism (rs738409, I48M) in the lipid bead protein-like phospholipase domain-containing protein 3 [patatin-like phospholipase domain-containing A3 gene (PNPLA3), likewise called adiponutrin (ADPN)] has been related with expanded hepatic steatosis, but not insulin resis-tance [86–88]. Along these lines, occasions of obvious disassociation of hepatic steatosis and hepatic insulin opposition might be clarified by a superior compre-hension of the subcellular conveyance of DAG [89]. Ceramides are additionally bioactive lipid particles that are embroiled in the advancement of insulin resistance. Increments in hepatic and muscle ceramide content have been related with insulin resistance in rodents, and inhibitors of ceramide blend can forestall insulin resis-tance [90, 91]. In any case, a disassociation between ceramide substance and tissue insulin resistance has been revealed in numerous investigations [85, 92, 93], and the fundamental component connecting ceramides to insulin resistance has not been completely finalized. As of late, some investigations analyzed how a particular ceramide species, C16:0, resists mitochondrial oxidation, permitting triglyceride to accumulate and cause insulin resistance [94, 95]. Despite the fact it has not been concluded, it is conceivable that the equal relationship between C16:0 ceramide and mitochondrial oxidation could additionally cause increments DAG and impede insulin action.

3.2 Insulin resistance in the adipose tissue

Some of the actions of insulin on adipose tissue are (1) stimulation of uptake of glucose and biosynthesis of triglyceride and (2) suppression of triglyceride hydrolysis-cum release of free fatty acids (FFA) and glycerol into the blood [96, 97]. It has been seen that adipose tissue insulin resistance (Adipo-IR), which means diminished suppression of lipolysis when high insulin levels are present, is interlinked with glucose intolerance, and increased plasma FFA amounts have also shown to diminish insulin signaling in muscles, endorse gluconeogenesis in the liver, and diminish glucose-activated insulin response [97–103]. In spite of the fact that the natural history and role of β-cell abnormality (or impairment) and insulin resistance in muscle are firmly established in the evolution of T2DM, the influence of Adipo-IR in the progression from normal glucose tolerance (NGT) to type 2 diabetes mellitus (T2DM) is still not clear. It is possible to quantitate palmitate turnover by the utiliza-tion of tracers [104–106] which can also provide the release rate of glycerol [107, 108] to furnish a lipolysis index. Tracer turnover (i.e., labeled palmitate or glycerol) or FFA suppression during insulin infusion (euglycemic-hyperinsulinemic clamp) or oral glucose tolerance test (OGTT) has led to the development of a number of indices of Adipo-IR [109]. Gastaldelli et al. [104] *confirmed that* fasting Adipo-IR index can be considered as a reliable index of insulin resistance in the fat cell when considered over the entire spectrum from NGT to T2DM. There has been consistent demonstration of weakened suppression of plasma FFA and also glycerol and 14C-palmitate turnover with the stepped hyperinsulinemic clamp [106, 110, 111] (i.e., adipocyte insulin resistance) in persons having T2DM. Therefore a decline in insulin secretion/insulin resistance (disposition) index has been seen with progression from lean NGT to obese NGT to IGT [112]. A decline in the secretion from β-cell of insulin is also interlinked with an elevation in the fasting Adipo-IR. Therefore, it can be said that the fasting adipocyte insulin resistance index (fasting FFA × fasting insulin) increases in a forward-looking, innovative manner over the stretch of glucose tolerance, extending from NGT to T2DM, and furnishes a reliable index of fat cell sensitivity to the actions of insulin [113–115]. In contradiction, there is increment of adipocyte insulin resistance index during OGTT (from NGT to IGT) and decrease with advancement of IGT to T2DM; this is due to gradual deficiency of the secretion of insulin in this group having diabetes. In conclusion, the gradual decrease in β-cell function that progresses from NGT is interlinked with a gradual elevation in fasting Adipo-IR and FFA [104].

Fat insulin resistance is the failure of insulin to activate fat glucose transport, advance lipid take-up, and diminish lipolysis. While diminished fat glucose take-up is exhibited in both in vivo and in vitro models, metabolic effect of hindered insulin-intervened glucose take-up in fat tissue is not well explained. For instance, the loss of fat GLUT4 in mice does not modify adiposity or, on the other hand, how weight gain prompts insulin resistance in skeletal muscle and liver [116]. Glucose transport into fat cells initiates starch reaction component restricting protein (ChREBP), which may affect fat lipid digestion [117]. Adipocytes discharge explicit unsaturated fats that are related with increased insulin affectability, as palmitoleate [118, 119] or monomethyl chain unsaturated fats [120].

4. Methods for diagnosis

4.1 Fasting insulin levels

A fasting serum insulin amount of more than 25 mU/L or in other sense 174 pmol/L designates insulin resistance. The same amounts pertain to 3 hours after taking the last meal [121].

4.2 Glucose tolerance testing

While performing a glucose tolerance test (GTT), a patient who is fasting is given a 75 g of glucose orally. Then plasma glucose levels are continuously monitored (along with urine glucose) for a period of 2 hours.

The elucidation of the test is established on the guidelines of the World Health Organization (WHO). After a period of 2 hours, a plasma glucose amount of less than 7.8 mmol/L (140 mg/dL) is regarded as normal, and a plasma glucose amount of between 7.8 and 11.0 mmol/L (140 to 197 mg/dL) is regarded as impaired glucose tolerance (IGT), and a plasma glucose amount of greater than or equal to 11.1 mmol/L (200 mg/dL) is regarded as diabetes mellitus. Extension of the testing (for several more hours) may reveal a hypoglycemic "dip" that is a result of an overshoot in insulin production after the failure of the physiologic postprandial insulin response.

4.3 Using the hyperinsulinemic euglycemic clamp

"Hyperinsulinemic euglycemic clamp" is also known as the gold standard for investigating and quantifying insulin resistance. It is so-called because it calculates the level of glucose obligatory to reimburse for an elevated insulin level without giving rise to hypoglycemia [122]. It is a kind of glucose clamping technique. The test is seldom carried out in clinical settings but is utilized in medical research [123]. The process takes around 2 hours. Insulin is infused through a peripheral vein. The rate is of infusion is 10–120 mU per m^2/minute. With the intention to recompense for the infusion of insulin, 20% glucose is infused to sustain blood glucose levels between 5 and 5.5 mmol/L. The blood sugar levels every 5 to 10 minutes, to determine the rate of infusion of glucose [123]. The determination of insulin sensitivity is made by the rate of glucose infusion in the last 30 minutes of the test. If greater levels (7.5 mg/min or greater) are required, the patient is considered as insulin sensitive. Low levels such as 4.0 mg/min or lower than that designates that the body is resistant to actions of insulin. Levels between 4.0 and 7.5 mg/min are not conclusive and indicate "impaired glucose tolerance," which is a premature gesticulation of insulin resistance [123, 127]. This basic method may be modified significantly by the utilization of glucose tracers.

4.4 Modified insulin suppression test

Gerald Reaven developed the modified insulin suppression test at Stanford University. The test corresponds well with the euglycemic clamp, with minute operator-dependent error. Particularly, this test has been utilized in research correlating to the metabolic syndrome [123]. A 25 µg of octreotide (Sandostatin) is given to the patient in 5 mL of normal saline over a period of 3–5 minutes through intravenous infusion (IV) as a primary bolus. Subsequently, the patient is continuously infused with an IV infusion of somatostatin (0.27 µg/m^2/min) to repress internal glucose and insulin secretion. Next, 20% glucose and insulin are infused at varying rates of 32 and 267 mg/m^2/min. Plasma glucose is monitored at 0, 30, 60, 90 minutes, and lastly at 120 minutes and subsequently after each 10 minutes for the final 30 minutes of the study. The averages of these final four values are utilized to ascertain the steady-state plasma glucose level (SSPG). People having an SSPG greater than 150 mg/dL are contemplated to have insulin resistance [45].

4.5 Homeostatic model assessment (HOMA)

By this method it is possible to quantify insulin resistance. Also, pancreatic beta-cell function can be possibly elucidated:

$$\text{HOMA-IR} = \frac{\text{Glucose} \times \text{Insulin}}{22.5} \tag{1}$$

and

$$\text{HOMA-}\beta = \frac{20 \times \text{Insulin}}{\text{Glucose} - 3.5\%} \tag{2}$$

where glucose is in mmol/L.
Also,

$$\text{HOMA-IR} = \frac{\text{Glucose} \times \text{Insulin}\,\%}{405} \tag{3}$$

and

$$\text{HOMA-}\beta = \frac{360 \times \text{Insulin}\,\%}{\text{Glucose} - 63} \tag{4}$$

where glucose is in mg/dL.

Note: insulin is taken in μU/mL; both glucose and insulin are taken during fasting; IR means insulin resistance; HOMA-β is the percentage of beta cell function [124–128].

4.6 Quantitative insulin sensitivity check index (QUICKI) method for insulin assessment

QUICKI is obtained utilizing the inverse of the addition of the logarithms of the fasting insulin and fasting glucose:

$$1/[\mathit{log}\,(\mathit{fasting\ insulin\ \mu U/mL}) + \mathit{log}\,(\mathit{fasting\ glucose\ mg/dL})] \tag{5}$$

The QUICKI method corresponds well with glucose clamp researches (r = 0.78) and is very good for the measurement of insulin sensitivity (IS), which is derived by utilizing the reciprocal quantity of insulin resistance (IR).

5. Conclusions

From the time insulin resistance was discovered, the cellular and molecular mechanisms were the considerations for which drugs were tried for diabetes mellitus. Considering the cellular mechanisms of insulin resistance which are mostly concerned with plasma cell membrane glycoprotein-1 (PC-1), also termed as ectonucleotide pyrophosphatase/phosphodiesterase 1 (ENPP1), the mechanism is really complex. The full understanding of the cellular mechanisms will permit the development of novel targets for various treatment modalities. From the therapeutic point of view, we need to have a clear knowledge about the cellular mechanism of insulin resistance in order to treat and also to prevent the occurrence of diabetes from prediabetic stage. From recent studies, it is evident that insulin resistance can be stopped or reversed if the pathophysiology is clear. It is the necessity to implement a huge global strategic plan for identifying and preventing/treatment of insulin resistance in the prediabetic stage.

Author details

Gaffar S. Zaman
Department of Clinical Laboratory Sciences, College of Applied Medical Science, King Khalid University, Abha, Kingdom of Saudi Arabia

*Address all correspondence to: drgszaman@gmail.com

References

[1] Wang G. Raison d'etre of insulin resistance: The adjustable threshold hypothesis. Journal of the Royal Society. 2014;**11**(101):20140892. Available from: https://www.ncbi.nlm.nih.gov/pmc/articles/PMC4223910

[2] Diagnosis and Classification of Diabetes Mellitus. American Diabetes Association Diabetes Care. Jan 2014;**37** (Suppl 1):S81-S90. DOI: 10.2337/dc14-s081

[3] Basciano H, Federico L, Adeli K. Fructose, insulin resistance, and metabolic dyslipidemia. Nutrition and Metabolism. 2005;**2**(1):5. DOI: 10.1186/1743-7075-2-5

[4] Hu F. Obesity Epidemiology. New York: Oxford University Press, Inc.; 2008. pp. 283-285. ISBN: 978-0-19-971847-4. Available from: https://books.google.com/books?id=rv42OY9QLkMC&pg=PA283 [Accessed: 20 April 2013]

[5] Smith TA, Biing-Hwan L, Jonq-Ying L. Taxing Caloric Sweetened Beverages: Potential Effects on Beverage Consumption, Calorie Intake, and Obesity, ERR-100. U.S. Department of Agriculture, Economic Research Service. July 2010

[6] Gallagher J. Fasting Diet 'Regenerates Diabetic Pancreas' (Health and Science Reporter, BBC News Website). BBC. 2017. Available from: https://www.bbc.com/news/health-39070183 [Accessed: 06 August 2018]

[7] Sakzewski E. Fasting Diet Could Regenerate Pancreas and Reverse Diabetes, Researchers Say. ABC. 2017. Available from: http://www.abc.net.au/news/2017-02-26/fasting-diet-could-reverse-diabetes-and-regenerate-the-pancreas/8304426 [Accessed: 06 August 2018]

[8] Yalow RS, Berson SA. Immunoassay of endogenous plasma insulin in man. The Journal of Clinical Investigation. 1960;**39**:1157-1175

[9] Yalow RS, Berson SA. Plasma insulin concentrations in nondiabetic and early diabetic subjects. Determinations by a new sensitive immuno-assay technic. Diabetes. 1960;**9**:254-260

[10] Kahn CR, Neville DM Jr, Roth J. Insulin-receptor interaction in the obese-hyperglycemic mouse. A model of insulin resistance. The Journal of Biological Chemistry. 1973;**248**:244-250

[11] Goldfine ID, Kahn CR, Neville DM Jr, Roth J, Garrison MM, Bates RW. Decreased binding of insulin to its receptors in rats with hormone induced insulin resistance. Biochemical and Biophysical Research Communications. 1973;**53**:852-857

[12] Mantzoros CS, Flier JS. Insulin resistance: The clinical spectrum. Advances in Endocrinology and Metabolism. 1995;**6**:193-232

[13] Fiorenza CG, Chou SH, Mantzoros CS. Lipodystrophy: Pathophysiology and advances in treatment. Nature Reviews. Endocrinology. 2011;**7**:137-150

[14] Buchanan KD, McKiddie MT. Factors determining the plasmainsulin response to oral glucose in diabetes mellitus. Diabetes. 1967;**16**:466-471

[15] Chiles R, Tzagournis M. Excessive serum insulin response to oral glucose in obesity and mild diabetes. Diabetes. 1970;**19**:458-464

[16] Hales CN, Randle PJ. Effects of low-carbohydrate diet and diabetes mellitus on plasma concentrations of glucose, non-esterified fatty acids, and insulin during oral glucose tolerance. Lancet. 1963;**1**:790-794

[17] Reaven GM, Miller R. Study of the relationship between glucose and insulin responses to an oral glucose load in man. Diabetes. 1968;**17**:560-569

[18] Shen S-W, Reaven GM, Farquhar JW. Comparison of impedance to insulin-mediated glucose uptake in normal subjects and diabetic subjects and in subjects with latent diabetes. The Journal of Clinical Investigation. 1970; **49**:2151-2160

[19] Insulin Resistance & Prediabetes. National Institute of Diabetes and Digestive and Kidney Diseases; 2018. Available from: https://www.niddk.nih.gov/health-information/diabetes/overview/what-is-diabetes/prediabetes-insulin-resistance

[20] Available from: https://www.niddk.nih.gov/health-information/diabetes/overview/what-is-diabetes/prediabetes-insulin-resistance

[21] Available from: https://www.diabetes.co.uk/causes-of-type2-diabetes.html

[22] Bressler P, De Fronzo RA. In: Alberti KGMM, Zimmet P, Defronzo RA, Keen H, editors. International Textbook of Diabetes Mellitus (2nd ed). New York: John Wiley & Sons; 1997. p. 213-254

[23] Ananth J, Parameswaran S, Gunatilake S. Side effects of atypical antipsychotic drugs. Current Pharmaceutical Design. 2004;**10**: 2219-2229

[24] Chen D, Misra A, Garg A. Lipodystrophy in human immunodeficiency virus-infected patients. The Journal of Clinical Endocrinology and Metabolism. 2002; **87**:4845-4856

[25] Newbern D, Freemark M. Placental hormones and the control of maternal metabolism and fetal growth. Current Opinion in Endocrinology, Diabetes, and Obesity. 2011;**18**(6):409-416. DOI: 10.1097/MED.0b013e32834c800d

[26] Brown TA, Bracken SJ. USMLE Step 1 Secrets in Color 4th Edition. Saunders; 3 edition (September 6, 2012). ISBN-10: 0323085148

[27] King MW. Lange Q&A USMLE Step 1. 6th ed. New York: McGraw-Hill Medical; 2005. p. 82. ISBN: 978-0-07-144578-8

[28] Piroli GG, Grillo CA, Reznikov LR, Adams S, McEwen BS, Charron MJ, et al. Corticosterone impairs insulin-stimulated translocation of GLUT4 in the rat hippocampus. Neuroendocrinology. 2007;**85**(2):71-80. DOI: 10.1159/000101694

[29] Garrido-Sanchez L, Murri M, Rivas-Becerra J, Ocaña-Wilhelmi L, Cohen RV, Garcia-Fuentes E, et al. Bypass of the duodenum improves insulin resistance much more rapidly than sleeve gastrectomy. Surgery for Obesity and Related Diseases. 2012;**8**(2): 145-150. DOI: 10.1016/j.soard.2011.03.010

[30] Friedman JM. Obesity in the new millennium. Nature. 2000;**404**(6778): 632-634. DOI: 10.1038/35007504

[31] Flier JS. Clinical review 94: What's in a name? In search of leptin's physiologic role. The Journal of Clinical Endocrinology and Metabolism. 1998; **83**(5):1407-1413. DOI: 10.1210/jcem.83.5.4779

[32] Elmquist JK, Maratos-Flier E, Saper CB, Flier JS. Unraveling the central nervous system pathways underlying responses to leptin. Nature Neuroscience. 1998;**1**(6):445-450. DOI: 10.1038/2164

[33] Weir GC, Leahy JL. Pathogenesis of non-insulin-dependent (type II) diabetes mellitus. In: Kahn CR,

Weir GE, editors. Joslin's Diabetes Mellitus. 13th ed. Philadelphia, PA: Lea & Febiger; 1994. pp. 240-264

[34] Bogardus C. Metabolic abnormalities in the development of noninsulin-dependent diabetes mellitus. In: LeRoith D, Taylor SI, Olefski JM, editors. Diabetes Mellitus. Vol. 459. Philadelphia, PA: Lippincot-Raven Publishers; 1996

[35] DeFronzo RA. Lilly Lecture 1987. The triumvirate: B-cell, muscle, liver. A collision responsible for NIDDM. Diabetes. 1988;**37**:667-687

[36] National Diabetes Data Group. Classification and diagnosis of diabetes mellitus and other categories of glucose intolerance. Diabetes. 1979;**28**:1039-1057

[37] World Health Organization. Diabetes mellitus: Report of a WHO Study Group. In: World Health Organization Technical Report Series. Vol. 17. 1985. p. 45. Available from: https://apps.who.int/iris/handle/10665/39592

[38] American Diabetes Association. Report of an expert committee on the diagnosis and classification of diabetes mellitus. Diabetes Care. 1998;**21**(Suppl. 1):S5-S19

[39] Reaven GM. Role of insulin resistance in human disease. Diabetes. 1988;**37**:1595-1607

[40] DeFronzo RA. Pathogenesis of type 2 diabetes: Metabolic and molecular implications for identifying diabetes genes. Diabetes Reviews. 1997;**5**:177-269

[41] DeFronzo RA, Bonadonna RC, Ferrannini E. The pathogenesis of NIDDM. A balanced overview. Diabetes Care. 1992;**15**:318-368

[42] Groop LC, Widen E, Ferrannini E. Insulin resistance and insulin deficiency in the pathogenesis of type 2 (non-insulin-dependent) diabetes mellitus: Errors of metabolism or of methods? Diabetologia. 1993;**36**: 1326-1331

[43] Taylor SI, Accili D, Imai Y. Insulin resistance or insulin deficiency. Which is the primary cause of NIDDM? Diabetes. 1994;**43**:735-740

[44] Ferrannini E. Insulin resistance vs. insulin deficiency in noninsulin-dependent diabetes mellitus: Problems and prospects. Endocrine Reviews. 1998;**19**:477-490

[45] Reaven GM, Olefsky JM. Relationship between heterogeneity of insulin responses and insulin resistance in normal subjects and patients with chemical diabetes. Diabetologia. 1977;**13**:201-206

[46] Reaven GM, Hollenbeck CB, Chen YDI. Relationship between glucose tolerance, insulin secretion, and insulin action in nonobese individuals with varying degrees of glucose tolerance. Diabetologia. 1989;**32**:52-55

[47] Bogardus C, Lillioja S, Howard BV, Reaven G, Mott DM. Relationships between insulin secretion, insulin action and fasting plasma glucose concentration in nondiabetic and noninsulin-dependent diabetic subjects. The Journal of Clinical Investigation. 1984;**74**:1238-1246

[48] Cerasi E, Luft R, Efendic S. Decreased sensitivity of the pancreatic beta cells to glucose in prediabetic and diabetic subjects. A glucose dose-response study. Diabetes. 1972;**21**: 224-234

[49] Eriksson J et al. Early metabolic defects in persons at increased risk for non-insulin dependent diabetes mellitus. The New England Journal of Medicine. 1989;**321**:337-343

[50] Pimenta W et al. Insulin secretion and insulin deficiency in people with

impaired glucose tolerance. Diabetic Medicine. 1996;**13**:S33-S36

[51] Yoneda H et al. Analysis of early-phase insulin responses in nonobese subjects with mild glucose intolerance. Diabetes Care. 1992;**15**:1517-1521

[52] O'Rahilly SP et al. Beta-cell dysfunction, rather than insulin insensitivity, is the primary defect in familial type 2 diabetes. Lancet. 1986;**2**: 360-364

[53] Kosaka K, Kuzuya T, Hagura R, Yoshinaga H. Insulin response to oral glucose load is consistently decreased in established non-insulin dependent diabetes mellitus: The usefulness of decreased early insulin response as a predictor of diabetes mellitus. Diabetic Medicine. 1996;**13**:S109-S119

[54] Pimenta W et al. Pancreatic beta-cell dysfunction as the primary genetic lesion in NIDDM. Evidence from studies in normal glucose-tolerant individuals with a first-degree NIDDM relative. Journal of the American Medical Association. 1995;**273**:1855-1861

[55] Perseghin G, Ghosh S, Gerow K, Shulman GI. Metabolic defects in lean nondiabetic offspring of NIDDM parents. A cross-sectional study. Diabetes. 1997;**46**:1001-1009

[56] Gulli G, Ferrannini E, Stern M, Haffner S, DeFronzo RA. The metabolic profile of NIDDM is fully established in glucose-tolerant offspring of two Mexican-American NIDDM parents. Diabetes. 1992;**41**:1575-1586

[57] Zimmet P, Whitehouse S, Alford F, Chisholm D. The relationship of insulin response to a glucose stimulus over a wide range of glucose tolerance. Diabetologia. 1978;**15**:23-27

[58] Haffner SM, Miettinen H, Gaskill SP, Stern MP. Decreased insulin action and insulin secretion predict the development of impaired glucose tolerance. Diabetologia. 1996;**39**:1201-1207

[59] Wilson PW, McGee DL, Kannel WB. Obesity, very low density lipoproteins, and glucose intolerance over 14 years. The Framingham Study. American Journal of Epidemiology. 1981;**114**:697-704

[60] Knowler WC, Pettitt DJ, Saad MF, Bennett PH. Diabetes mellitus in the Pima Indians: Incidence, risk factors and pathogenesis. Diabetes/Metabolism Reviews. 1990;**6**:1-27

[61] Ohlson LD et al. The influence of body fat distribution on the incidence of diabetes mellitus. 13.5 years of follow-up of the participants in the study of men born in 1913. Diabetes. 1985;**34**: 1055-1058

[62] Lillioja S et al. Insulin resistance and insulin secretory dysfunction as precursors of non-insulin dependent diabetes mellitus. Prospective studies of Pima Indians. The New England Journal of Medicine. 1993;**329**:1988-1992

[63] Martin BC et al. Role of glucose and insulin resistance in development of type 2 diabetes. Results of a 25-year follow-up study. Lancet. 1992;**340**: 925-929

[64] Warram JH, Martin BC, Krolewski AS, Soeldner JS, Kahn CR. Slow glucose removal rate and hyperinsulinemia precede the development of type 2 diabetes in the offspring of diabetic parents. Annals of Internal Medicine. 1990;**113**:909-915

[65] Saad MF et al. A two-step model for development of non-insulin dependent diabetes mellitus. The American Journal of Medicine. 1991;**90**:229-235

[66] Charles MA et al. Risk factors for NIDDM in white population: Paris prospective study. Diabetes. 1991;**40**: 796-799

[67] Chen KW et al. Earlier appearance of impaired insulin secretion than of visceral adiposity in the pathogenesis of NIDDM. Diabetes Care. 1995;**18**:747-753

[68] Nagi DK et al. Early and late insulin response as predictors of NIDDM in Pima Indians with impaired glucose tolerance. Diabetologia. 1995;**38**: 187-192

[69] Warram JH, Sigal RJ, Martin BC, Krolewski AS, Soeldner JS. Natural history of impaired glucose tolerance: Follow-up at Joslin Clinic. Diabetic Medicine. 1996;**13**:S40-S45

[70] Efendic S, Luft R, Wajngot A. Aspects of the pathogenesis of type 2 diabetes. Endocrine Reviews. 1984;**5**: 395-410

[71] Lundgren H, Bengtsson C, Blohme G, Lapidus L, Waldenstroem J. Fasting serum insulin concentration and early insulin response as risk determinants for developing diabetes. Diabetic Medicine. 1990;**7**:407-413

[72] Kadowaki T et al. Risk factors for worsening to diabetes in subjects with impaired glucose tolerance. Diabetologia. 1984;**26**:44-49

[73] Yu C et al. Mechanism by which fatty acids inhibit insulin activation of insulin receptor substrate-1 (IRS-1)-associated phosphatidylinositol 3-kinase activity muscle. The Journal of Biological Chemistry. 2002;**277**(52):50230-50236

[74] Dresner A et al. Effects of free fatty acids on glucose transport and IRS-1-associated phosphatidylinositol 3-kinase activity. The Journal of Clinical Investigation. 1999;**103**(2):253-259

[75] Yu C et al. Mechanism by which fatty acids inhibit insulin activation of insulin receptor substrate-1 (IRS-1)-associated phosphatidylinositol 3-kinase activity in muscle. The Journal of Biological Chemistry. 2002;**277**(52):50230-50236

[76] Griffin ME et al. Free fatty acid-induced insulin resistance is associated with activation of protein kinase C theta and alterations in the insulin signalling cascade. Diabetes. 1999;**48**(6): 1270-1274

[77] Itani SI, Ruderman NB, Schmieder F, Boden G. Lipid-induced insulin resistance in human muscle is associated with changes in diacylglycerol, protein kinase C, and IκB-α. Diabetes. 2002;**51**(7):2005-2011

[78] Szendroedi J et al. Role of diacylglycerol activation of PKCtheta in lipid-induced muscle insulin resistance in humans. Proceedings of the National Academy of Sciences of the United States of America. 2014;**111**(26): 9597-9602

[79] Kim JK et al. Glucose toxicity and the development of diabetes in mice with muscle-specific inactivation of GLUT4. The Journal of Clinical Investigation. 2001;**108**(1):153-160

[80] Petersen KF et al. The role of skeletal muscle insulin resistance in the pathogenesis of the metabolic syndrome. Proceedings of the National Academy of Sciences of the United States of America. 2007;**104**(31): 12587-12594

[81] Petersen KF et al. Leptin reverses insulin resistance and hepatic steatosis in patients with severe lipodystrophy. The Journal of Clinical Investigation. 2002;**109**(10):1345-1350

[82] Petersen KF et al. Leptin reverses insulin resistance and hepatic steatosis in patient with severe lipodystrophy. The Journal of Clinical Investigation. 2002;**109**(10):1465-1474

[83] Kim JK, Gavrilova O, Chen Y, Reitman ML, Shulman GI. Mechanism of insulin resistance in A-ZIP/F-1 fatless mice. The Journal of Biological Chemistry. 2000;**275**(12):8456-8460

[84] Kim JK et al. Tissue-specific overexpression of lipoprotein lipase causes tissue-specific insulin resistance. Proceedings of the National Academy of Sciences of the United States of America. 2001;**98**(13):7522-7527

[85] Magkos F et al. Intrahepatic diacylglycerol content is associated with hepatic insulin resistance in obese subjects. Gastroenterology. 2012;**142**(7): 1444-1446

[86] Romeo S et al. Genetic variation in PNPLA3 confers susceptibility to nonalcoholic fatty liver disease. Nature Genetics. 2008;**40**(12):1461-1465

[87] Smagris E et al. Pnpla3I148M knockin mice accumulate PNPLA3 on lipid droplets and develop hepatic steatosis. Hepatology. 2015;**61**(1): 108-118

[88] Peter A et al. PNPLA3 variant I148M is associated with altered hepatic lipid composition in humans. Diabetologia. 2014;**57**(10): 2103-2107

[89] Coate KC et al. Hepatic glucose uptake and disposition during short-term high-fat vs. high fructose feeding. American Journal of Physiology. Endocrinology and Metabolism. 2014; **307**(2):E151-E160

[90] Holland WL et al. Inhibition of ceramide synthesis ameliorates glucocorticoid-, saturated fat-, and obesity-induced insulin resistance. Cell Metabolism. 2007;**5**(3):167-179

[91] Ussher JR et al. Inhibition of de novo ceramide synthesis reverses diet-induced insulin resistance and enhances whole-body oxygen consumption. Diabetes. 2010;**59**(10):2453-2464

[92] Kumashiro N et al. Cellular mechanism of insulin resistance in nonalcoholic fatty liver disease. Proceedings of the National Academy of Sciences of the United States of America. 2011;**108**(39):16381-16385

[93] Perry RJ et al. Reversal of hypertriglyceridemia, fatty liver disease, and insulin resistance by a liver-targeted mitochondrial uncoupler. Cell Metabolism. 2013;**18**(5):740-748

[94] Raichur S et al. CerS2 haploinsufficiency inhibits β-oxidation and confers susceptibility to diet-induced steatohepatitis and insulin resistance. Cell Metabolism. 2014;**20**(4):687-695

[95] Turpin SM et al. Obesity-induced CerS6-dependent C16:0 ceramide production promotes weight gain and glucose intolerance. Cell Metabolism. 2014;**20**(4):678-686

[96] Saponaro C, Gaggini M, Carli F, Gastaldelli A. The subtle balance between lipolysis and lipogenesis: A critical point in metabolic homeostasis. Nutrients. 2015;7:9453-9474

[97] Boden G. Obesity and free fatty acids. Endocrinology and Metabolism Clinics of North America. 2008;**37**: 635-646

[98] Allister CA, Liu LF, Lamendola CA, et al. In vivo $2H_2O$ administration reveals impaired triglyceride storage in adipose tissue of insulin-resistant humans. Journal of Lipid Research. 2015;**56**:435-439

[99] Roden M, Price TB, Perseghin G, et al. Mechanism of free fatty acid-induced insulin resistance in humans. The Journal of Clinical Investigation. 1996;**97**:2859-2865

[100] Belfort R, Mandarino L, Kashyap S, et al. Dose-response effect of elevated plasma free fatty acid on insulin signaling. Diabetes. 2005;**54**:1640-1648

[101] Kashyap S, Belfort R, Gastaldelli A, et al. A sustained increase in plasma free fatty acids impairs insulin secretion in

nondiabetic subjects genetically predisposed to develop type 2 diabetes. Diabetes. 2003;**52**:2461-2474

[102] Ferrannini E, Barrett EJ, Bevilacqua S, DeFronzo RA. Effect of fatty acids on glucose production and utilization in man. The Journal of Clinical Investigation. 1983;**72**:1737-1747

[103] Roden M, Stingl H, Chandramouli V, et al. Effects of free fatty acid elevation on postabsorptive endogenous glucose production and gluconeogenesis in humans. Diabetes. 2000;**49**:701-707

[104] Gastaldelli A, Gaggini M, DeFronzo RA. Role of adipose tissue insulin resistance in the natural history of type 2 diabetes: Results from the San Antonio Metabolism Study. Diabetes. Apr 2017;**66**(4):815-822. DOI: 10.2337/db16-1167

[105] Fabbrini E, Magkos F, Conte C, et al. Validation of a novel index to assess insulin resistance of adipose tissue lipolytic activity in obese subjects. Journal of Lipid Research. 2012;**53**: 321-324

[106] Groop LC, Bonadonna RC, DelPrato S, et al. Glucose and free fatty acid metabolism in non-insulin-dependent diabetes mellitus. Evidence for multiple sites of insulin resistance. The Journal of Clinical Investigation. 1989;**84**:205-213

[107] Robinson C, Tamborlane WV, Maggs DG, et al. Effect of insulin on glycerol production in obese adolescents. The American Journal of Physiology. 1998;**274**:E737-E743

[108] Gastaldelli A, Casolaro A, Ciociaro D, et al. Decreased whole body lipolysis as a mechanism of the lipid-lowering effect of pioglitazone in type 2 diabetic patients. American Journal of Physiology. Endocrinology and Metabolism. 2009;**297**:E225-E230

[109] Sondergaard E, Jensen MD. Quantification of adipose tissue insulin sensitivity. Journal of Investigative Medicine. 2016;**64**:989-991

[110] Karpe F, Dickmann JR, Frayn KN. Fatty acids, obesity, and insulin resistance: Time for a re-evaluation. Diabetes. 2011;**60**:2441-2449

[111] Jensen MD, Nielsen S. Insulin dose response analysis of free fatty acid kinetics. Metabolism. 2007;**56**:68-76

[112] Gastaldelli A, Ferrannini E, Miyazaki Y, Matsuda M, DeFronzo RA, San Antonio Metabolism Study. Beta-cell dysfunction and glucose intolerance: Results from the San Antonio Metabolism (SAM) study. Diabetologia. 2004;**47**:31-39

[113] Defronzo RA. Banting lecture. From the triumvirate to the ominous octet: A new paradigm for the treatment of type 2 diabetes mellitus. Diabetes. 2009;**58**:773-795

[114] Ferrannini E, Natali A, Muscelli E, et al. Natural history and physiological determinants of changes in glucose tolerance in a non-diabetic population: The RISC Study. Diabetologia. 2011;**54**: 1507-1516

[115] Kahn SE, Cooper ME, Del Prato S. Pathophysiology and treatment of type 2 diabetes: Perspectives on the past, present, and future. Lancet. 2014;**383**: 1068-1083

[116] Abel ED et al. Adipose-selective targeting of the GLUT4 gene impairs insulin action in muscle and liver. Nature. 2001;**409**(6821):729-733

[117] Herman MA et al. A novel ChREBP isoform in adipose tissue regulates systemic glucose metabolism. Nature. 2012;**484**(7394):333-338

[118] Stefan N et al. Circulating palmitoleate strongly and independently

predicts insulin sensitivity in humans. Diabetes Care. 2010;**33**(2):405-407

[119] Cao H, Gerhold K, Mayers JR, Wiest MM, Watkins SM, Hotamisligil GS. Identification of a lipokine, a lipid hormone linking adipose tissue to systemic metabolism. Cell. 2008;**134**(6):933-944

[120] Su X et al. Adipose tissue monomethyl branched chain fatty acids and insulin sensitivity: Effects of obesity and weight loss. Obesity (Silver Spring). 2015;**23**(2):329-334

[121] DeFronzo RA, Tobin JD, Andres R. Glucose clamp technique: A method for quantifying insulin secretion and resistance. The American Journal of Physiology. 1979;**237**(3):E214-E223. DOI: 10.1152/ajpendo.1979.237.3.e214

[122] Muniyappa R, Lee S, Chen H, Quon MJ. Current approaches for assessing insulin sensitivity and resistance in vivo: Advantages, limitations, and appropriate usage. American Journal of Physiology. Endocrinology and Metabolism. 2008; **294**(1):E15-E26. DOI: 10.1152/ajpendo.00645.2007

[123] Matthews DR, Hosker JP, Rudenski AS, Naylor BA, Treacher DF, Turner RC. Homeostasis model assessment: Insulin resistance and beta-cell function from fasting plasma glucose and insulin concentrations in man. Diabetologia. 1985;**28**(7):412-419. DOI: 10.1007/BF00280883

[124] Wallace TM, Levy JC, Matthews DR. Use and abuse of HOMA modeling. Diabetes Care. 2004;**27**(6): 1487-1495. DOI: 10.2337/diacare.27.6.1487

[125] Singh B, Saxena A. Surrogate markers of insulin resistance: A review. World Journal of Diabetes. 2010;**1**(2): 36-47. DOI: 10.4239/wjd.v1.i2.36. PMC 3083884

[126] Hermans MP, Levy JC, Morris RJ, Turner RC. Comparison of tests of beta-cell function across a range of glucose tolerance from normal to diabetes. Diabetes. 1999;**48**(9):1779-1786. DOI: 10.2337/diabetes.48.9.1779

[127] Hermans MP, Levy JC, Morris RJ, Turner RC. Comparison of insulin sensitivity tests across a range of glucose tolerance from normal to diabetes. Diabetologia. 1999;**42**(6):678-687. DOI: 10.1007/s001250051215

[128] Katz A, Nambi SS, Mather K, Baron AD, Follmann DA, Sullivan G, et al. Quantitative insulin sensitivity check index: A simple, accurate method for assessing insulin sensitivity in humans. The Journal of Clinical Endocrinology and Metabolism. 2000; **85**(7):2402-2410

11

Emerging Knowledge from Noninvasive Imaging Studies: Is Ammonia Control Enough?

Andrea L. Gropman

Abstract

Multiple lines of research suggest that ammonia is harmful to the brain if the levels remain elevated for extended periods of time. Several decades ago, there was no testing or standard of care to monitor the effect of hyperammonemia (HA) on neurological function in urea cycle disorders (UCD), and the timing of HA encephalopathy is still not clear. Magnetic resonance imaging (MRI) was not done routinely, if at all, so it was not known what changes were occurring in the brain, during and after recovery from HA. Decades ago, a diagnosis of a UCD meant severe disability and early death. Earlier diagnosis, improved management, and nitrogen scavenger therapy have improved the lives and life span of patients with UCD. However, many patients suffer from learning difficulties under the umbrella "executive function" which comprises neurologically based skills involving mental control and self-regulation. The general agreement of the core elements of executive functions includes inhibition, working memory, and cognitive flexibility and is necessary in development of skills in reasoning, fluid intelligence, problem-solving, and planning. Our research focuses on the use of noninvasive neuroimaging coupled with neuropsychological testing to understand the complex relationship between ammonia, glutamine, cognitive function, seizures, and specifically impact on development of working memory.

Keywords: ammonia, EEG, glutamine, MRI, neuroimaging, urea cycle disorder

1. Introduction

The urea cycle disorders (UCD) represent one of the most common groups of inborn errors of metabolism, with an overall incidence of 1 in 30,000 [1, 2], and involve deficiency of one of six urea cycle enzymes or of a related cofactor or transporter [3, 4]. The most common of these, ornithine transcarbamylase deficiency (OTCD), is the only disorder of ureagenesis inherited in an X-linked manner, with an estimated incidence of 1 in 70,000 [5]. Over 240 missense mutations have been identified in the OTCD gene but overall 400 including nonsense, frameshift, in-frame indels, splice site errors, and one in a regulatory domain [6, 7]. About 60% of hemizygous males harbor a mutation around the enzyme active site and present with hyperammonemic (HA) coma in the newborn period [8, 9]. The remaining 40% of patients demonstrate more peripheral mutations in other parts of the gene, associated with less severe phenotypes and later onset presentation [10].

A majority of children with OTCD have cognitive and motor deficits due to hyperammonemic episodes [8, 11–13]. Neonatal onset disease mortality rate is high. Prior to advances in recognition and treatment, it was not uncommon for survivors of neonatal onset disease to have intellectual disability, cerebral palsy, and seizures [14]. Neonatal survivors have a decreased IQ which may be as low as 43.In males with partial deficiencies, disease onset is later, and outcome is better, although still associated with high mortality and morbidity with many individuals manifesting cognitive, motor, and psychiatric sequelae [15–17], in particular impaired working memory and other measures of executive function which are essential for performing well in school, vocations, and relationships. Treatment of OTCD involves a combination of protein restriction (which is also a restriction in nitrogen, leading to ammonia accumulation) and medications that invoke an alternative pathway of waste nitrogen excretion [18, 19]. Females heterozygous for OTCD have a variable phenotype and display a broad range of symptoms from apparently asymptomatic to fully affected, owing to both allelic heterogeneity and differential X-inactivation patterns.

A common presumption for years has been that approximately 85% of heterozygous females are asymptomatic based on history, whereas the remainder show symptoms ranging from behavioral and learning disabilities and protein intolerance to cyclical vomiting, stroke-like episodes, and hyperammonemic coma [20–23]. Symptomatic women who harbor mutations seen in the neonatal onset disorder in hemizygous males [10] may develop HA due to skewed X-inactivation. There is therefore a range of residual enzyme capacities and urea synthetic capacities that result in this variation [24]. However, advances in neuroimaging and the work of the Urea Cycle Disorders Consortium (UCDC) have demonstrated that many of these previously presumed asymptomatic females have similar brain structural, biochemical, and cognitive biomarkers seen in those who are clinically impacted, yet they may be mild under conditions of low demand. More obvious symptoms were uncovered when cognitive demand increases or there is superimposed illness or stressor [25].

2. Pathophysiology of the UCDs

Ammonia is a product of the metabolism of proteins and other compounds, and it is required for the synthesis of essential cellular compounds. However, a five- to tenfold increase in ammonia in the blood induces toxic effects in most animal species, with alterations in the function of the central nervous system. Ammonia is a normal constituent of all body fluids. At physiologic pH, it exists mainly as ammonium ion. Reference serum levels are less than 35 mmol/L (outside the newborn period, where higher levels are seen). Excess ammonia is excreted as urea, which is synthesized in the liver through the urea cycle. Sources of ammonia include bacterial hydrolysis of urea and other nitrogenous compounds in the intestine, the purine-nucleotide cycle and amino acid transamination in skeletal muscle, and other metabolic processes in the kidneys and liver. Increased entry of ammonia to the brain is a primary cause of neurological disorders associated with HA, such as congenital deficiencies of urea cycle enzymes, hepatic encephalopa-thies, Reye syndrome, several other metabolic disorders, and some toxic encepha-lopathies [26–28].

On the basis of studies in animal models and other preclinical model systems, several mechanisms of ammonia neurotoxicity at the molecular level have been proposed.

While the exact pathophysiology remains unclear, current theories include (1) glutamine accumulation, with associated impaired cerebral osmoregulation, and (2) glutamate/*N*-methyl D-aspartate (NMDA) receptor activation, with resultant excitotoxic injury and energy deficit [26–30]. The WM is preferentially affected in proximal UCD, and the extent of injury has been shown to depend upon the duration of HA coma and the interval between coma and death [28].

Acute ammonia intoxication in an animal model leads to increased extracellular concentration of glutamate in the brain and results in activation of the NMDA receptor. Activation of this receptor mediates ATP depletion and ammonia toxicity; blocking the NMDA receptor with dizocilpine (MK-801) prevents both phenomena. The ATP depletion is due to activation of Na^+/K^+-ATPase, which, in turn, is a consequence of decreased phosphorylation by protein kinase C. Activation of the NMDA receptor may account for seizures seen in some patients during acute HA [29].

High levels of ammonia in the brain also induce other metabolic changes that are not mediated by activation of the NMDA receptor and thus are not involved directly in ammonia-induced ATP depletion or neurotoxicity. These include increases in brain levels of lactate, pyruvate, glutamine, and glucose, with concomitant decreases in brain glycogen, ketone bodies, and glutamate [30].

Chronic HA is associated with an increase in inhibitory neurotransmission as a consequence of two factors. The first involves the downregulation of glutamate receptors secondary to excessive extrasynaptic accumulation of glutamate. The second mechanism implies increased GABAergic tone resulting from benzodiazepine receptor overstimulation by endogenous benzodiazepines and neurosteroids. These changes likely play a role in central nervous system features of intellectual function, decreased consciousness, and coma [26, 30].

In the brain, glutamine represents a storage depot for nitrogen, binding excess ammonia and offering a shortterm buffering of excess ammonia in patients with HA, likely as a protective mechanism. It is these high levels of glutamine in the brain that are also hypothesized to be neurotoxic and one of the factors leading to injury (UCDC unpublished). Brain astrocytes are key players in the interactions of glutamine and ammonia via the Gln/Glu cycle (**Figure 1**).

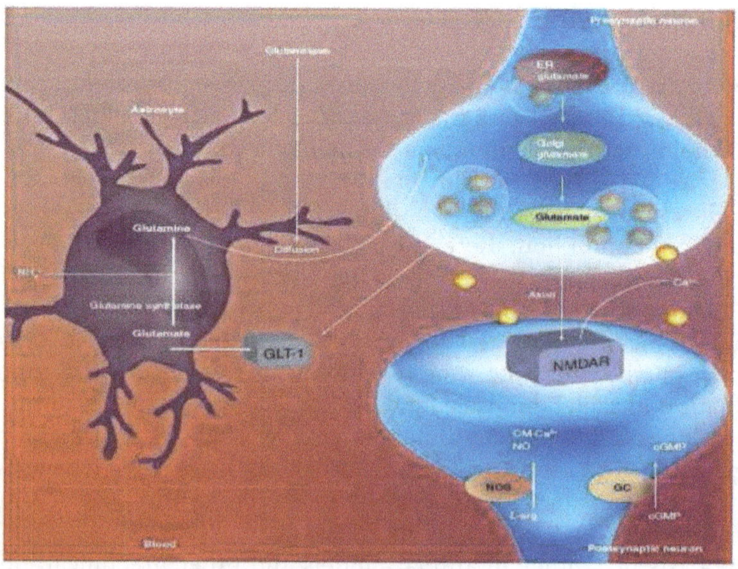

Figure 1.
The glutamine-glutamate cycle. Reprinted with permission from Pediatric Health, 2008, 2(6):701–713.

When ammonia is not adequately detoxified by the hepatic urea cycle, there is an increase in scavenger amino acids, including glutamine. Ammonia entering the brain is rapidly incorporated into the formation of glutamine by glutamine synthetase, present in the astrocyte. Glutamine concentrations increase in hyperammonemic states. While not measured directly, indirect measures with ^{1}H magnetic resonance spectroscopy (MRS) studies of patients with urea cycle disorders have demonstrated elevations of the glutamine/glutamate complex [30]. Glutamine has been implicated in hyperammonemic encephalopathy. It has been shown that a rise in plasma glutamine levels precedes HA. There is a sustained positive correlation between plasma glutamine and ammonia levels.

Inhibition of glutamine synthetase in hyperammonemic rats by treatment with enzyme inhibitors prevents the rise in cortical glutamine levels and cortical water content [31]. Clearance of synaptic glutamate by glial cells is required for the normal function of excitatory synapses and to prevent neurotoxicity. This process occurs in the atrocity, which takes up synaptic glutamate and returns glutamate to the neurons in the form of glutamine, a non-neuroactive amino acid that the neurons subsequently reconvert to glutamate via the action of mitochondrial phosphate-dependent glutaminase.

2.1 Short-term clinical effects of HA

Clinical signs of HA may occur at concentrations >60 micromol/L and are very individual as some patients may tolerate higher levels before symptoms are noticed. The shortterm changes may include initially anorexia, irritability, lethargy, somnolence, disorientation, vomiting, and asterixis (flapping tremor). As symptoms progress and ammonia is not lowered, cerebral edema, coma, herniation and death [30]. In the acute stages, there is increased blood brain barrier permeability, leading to depletion of intermediates of cell energy metabolism. On an anatomic level, there is disaggregation of microtubules [29].

2.2 Chronic effects of HA

Chronic effects of HA include alterations in axonal development as well as alterations in brain amino acid and neurotransmitter levels. Electrophysiologic effects of HA include direct effects on inhibitory postsynaptic potentiation (IPSP). Neurotransmission is impacted due to increased extracellular glutamate levels and downregulation of AMPA-kainate receptors, enhanced tryptophan uptake, elevated quinolinic acid levels, and enhanced NMDA activity. Activation of NMDA receptors increases calcium in postsynaptic neurons which binds to calmodulin and activates neuronal nitric oxide (NO) synthase, increasing NO, which activates guanylate cyclase, increasing cyclic guanine monophosphate (cGMP), part of which is released to the extracellular space [31, 32]. Activation of this glutamate- NO-cGMP pathway may be involved in some forms of learning.

Recent reports indicate that guanylate cyclase and cGMP are important in learning and memory; induction of LTP is the molecular basis of some forms of learning and memory [33]. Because glial cells also have these receptors, the excessive glutamate leads to glial cell swelling, which seems to protect the neurons from excitotoxic injury. Studies in spf mouse models of OTC and other animal models of HE show neuropathological evidence of excitotoxic neuronal cell death which suggests that overactivation of NMDAR is a feature of urea cycle disorders [34] and may be age dependent [35].

3. What are the cognitive implications of HA on the brain?

A proportion of individuals with OTCD have a wide spectrum of neuropsychological complications including developmental delay, intellectual disability, and executive function deficits [36]. Most adult-onset patients remain asymptomatic, until they present with rapid decline in mental status and subsequently chronic encephalopathy [36, 37].

Fluctuating HA may cause delirium, confusion, and incoherent speech. In addition to subsequent regression, lack of attention leads to unemployment and introverted behavior [37]. Waisbren et al. demonstrated that nearly all asymptomatic 156 women with OTCD attained a full-scale intelligence quotient (IQ) of 102 ± 16. Among 25 men, the full-scale IQ measured was 101 ± 21. No differences were noted between the verbal and performance scores. In addition, in 27% of females and 33% of males, working memory deficiency was observed as a constant finding. The ammonia concentration and its duration appear to be key determinants of the long-term outcome [37].

3.1 Long-term sequelae of HA: Executive function

Executive function (EF) is the ability to control and regulate actions and thoughts [38]. It includes processes such as working memory, self-regulation, and inhibitory control.

Executive functions are a set of cognitive processes and competencies that control behavior and learning. It is an umbrella term which comprises neurologically based skills involving mental control and self-regulation. The general agreement of the core elements of executive functions includes inhibition, working memory, and cognitive flexibility [38, 39]. These elements are highly interrelated, and the interplay of these processes is vital in flexible, goal-directed behaviors. From the core elements of basic executive functions, higher-order executive functions such as reasoning, fluid intelligence, problem-solving, and planning are built [40].

Historically, executive functioning has been thought to be regulated by the prefrontal cortex of the frontal lobes; however reviews found indications for the sensitivity, but not for the specificity, of executive function measures to the frontal lobe [41]. Both the frontal brain regions and other structures of the brain are involved and necessary for successful application of these skills.

Individuals are not born with executive function skills, but rather are born with the potential to develop them. With any genetic or environmental insult to the brain, the executive functions and prefrontal cortex are one of the first to suffer and suffer disproportionately. The disruption of the brain architecture can seriously delay or impair the development of executive functioning [38].

To date, most evaluations of EF rely on parents' reports such as the Behavior Rating Inventory of Executive Function Preschool (BRIEF-P) form, which may not capture the development of executive skills to its full expression [42].

EF has been previously studied using task-based functional MRI (fMRI) scans, which can be difficult to adapt for children [43]. Instead, multiple studies have relied on resting state functional MRI [44]. Reineberg et al. investigated differences in brain connectivity in relation to individual performances during different EF behavioral tasks [45]. However, to our best knowledge, resting state fMRI has not been applied to characterize EF-related brain connectivity differences in children, especially at a young age (2–5 years old).

4. How can neuroimaging help us probe markers of neurological dysfunction in IEMs?

Multiple studies using multimodal MRI suggest its value in the recogni-tion of microscopic anatomic damage that *precedes clinical symptoms* in many inborn errors of metabolism and neurodegenerative disorders. Depending upon the type of imaging study, it may answer a different question regarding the pathology, biochemistry, or physiology. Neuroimaging may detect subtle abnormalities that can be correlated with neurocognitive abnormalities even in asymptomatic OTCD heterozygotes. The neuroimaging/neurocognitive stud-ies we performed as part of the UCDC focused on adolescents and adults with OTCD. Our collective studies demonstrated that OTCD heterozygous females have changes in function of the prefrontal cortex (PFC) in association with an altered neurocognitive profile in working memory, executive functioning, and attention [46].

fMRI can allow us to understand how the brain constructs neural networks to perform cognitive tasks, probe how these networks are altered in brain disorders, and allow us to follow recovery. Magnetic resonance spectroscopy using hydrogen (^{1}H) or carbon (^{13}C) allows us to probe both static and dynamic changes in brain metabolism [47]. The benefits of neuroimaging using MRI are the ability to view the brain in the three orthogonal views, the lack of radiation exposure, and the ability to target the organ or pathology being studied. In addition, high-performance MR hardware is available resulting in faster scans and higher resolution with higher field.

4.1 Use of neuroimaging to assess brain injury in UCDs

Neuroimaging in recent years has come to encompass many different modalities that can be combined in a single imaging session to gain complementary information regarding the brain's structural, functional, and metabolic dimensions.

A typical routine structural MRI protocol includes not only T1- and T2-weighted sequences but also, in most academic and teaching hospitals, fluid attenuation inversion recovery (FLAIR) and voxel-based morphometry (VBM) or other ability to measure tissue volume from acquired structural images on a clinical scanner. Diffusion weighted and diffusion tensor imaging (DWI and DTI) are used to study microstructural variance in WM fiber tracts [48], and proton magnetic resonance spectroscopy is used to measure brain metabolism in static and dynamic models [47]. Multimodal assessment batteries and data fusion give investigators a complex and varied perspective into the structural, functional, and biochemical parameters of the central nervous system in IEMs [49].

4.2 What MRI modalities are available and what do they measure?

4.2.1 Magnetic resonance imaging (MRI)

MRI interrogates tissue water protons via differential populations of proton spins that result when a biological sample is placed in a strong magnetic field. Using MRI, one can define brain anatomy and characterize gray matter and WM micro-structural and macro-structural changes. These are read as signal abnormalities on T1- and T2-weighted images which correspond with the specific tissue pathologies. With MRI one can detect damage at a macroscopic level. One must remember that MRI findings can lag behind clinical changes and stages of disease as well as recovery processes.

4.2.2 Fluid-attenuated inversion recovery

Fluid-attenuated inversion recovery imaging is usually a routine part of most radiology clinical imaging sequences. Diffusion tensor imaging (DTI) is used due to its sensitivity in detecting increases in interstitial water content. Such applications include imaging brain tumors, demyelinating diseases (i.e., multiple sclerosis), metabolic WM disease cerebral infarcts, and gliotic scars.

4.2.3 Diffusion weighted imaging and diffusion tensor imaging

Diffusion MRI is another MRI method which allows fine mapping of the diffusion process of molecules (i.e., water) in biological tissues. DWI and DTI can be used in vivo noninvasively. DWI and DTI techniques are focused on the fact that molecular diffusion in tissues is not free but, rather, is impacted by obstacles, such as macromolecules and cell membranes (myelin). By using DTI, water molecule diffusion patterns can inform about microscopic details regarding myelin integrity and architecture.

DTI relates image intensities to the relative mobility of water molecules in the tissue. It can also imply direction of the motion [48]. In general, areas that have a relatively high mean diffusion appear dark on the diffusion weighted MRI images. Diffusion MRI can also be used to make inferences about WM architecture since the diffusion of water corresponds to cell geometry in axons [48].

Cytotoxic edema in contrast follows sodium/potassium pump failure, often due to energy metabolism failure due to ischemic insult. It is quick and can occur within minutes of the onset of ischemia and can produce increased brain tissue water of up to 3–5%. In HA, cytotoxic edema also results. Therefore, DTI can be used in patients with UCD at baseline to assess whether patients differ with respect to WM integrity. DTI has a role in measuring functional connectivity differences between control groups and patients and to follow up patients over time to monitor disease progres-sion, recovery, or impact of therapies [50]. It is a very good technique to also look at rapid fluxes in water content such as during a HA episode in a patient with UCD and allows follow-up noninvasively during recovery and/or with introduction of a therapeutic agent.

The most commonly used indices for the measurement of anisotropic diffusions by DTI include the relative anisotropy measure, fractional anisotropy (FA), and the volume ratio indices. These indices provide quantitative measurements of the changes of WM integrity in brain regions that are affected by diseases.

We have used DTI techniques together with advanced fiber tracking algorithms, to evaluate the 3D trajectories of neural tracts. This has allowed us to model WM neural connectivity in UCDs. DTI was used to determine whether there are WM microstructural abnormalities in partial OTCD that could underlie the cognitive phenotype. Our focus on WM alterations was based on prior neuropathology studies in HA. These studies have shown WM is almost exclusively affected. There is also a relationship between Gln toxicity and WM damage [51]. Anisotropy was calculated by standard methods, from the eigenvalues of the diffusion tensor by using the FA metric. After comparison between UCD patients and age-matched control groups, we established that FA of the frontal WM was significantly decreased in patients with UCD compared to the age-matched controls. This, in turn, is indicative of changes in WM microstructure (**Figure 2**). Additionally, we found an inverse relationship between FA and disease severity that was not age dependent. Based on this, we could conclude that MR imaging in OTCD may be normal in patients with late-onset disease, heterozygotes, or those not in hyperammonemic crisis at the time of the study.

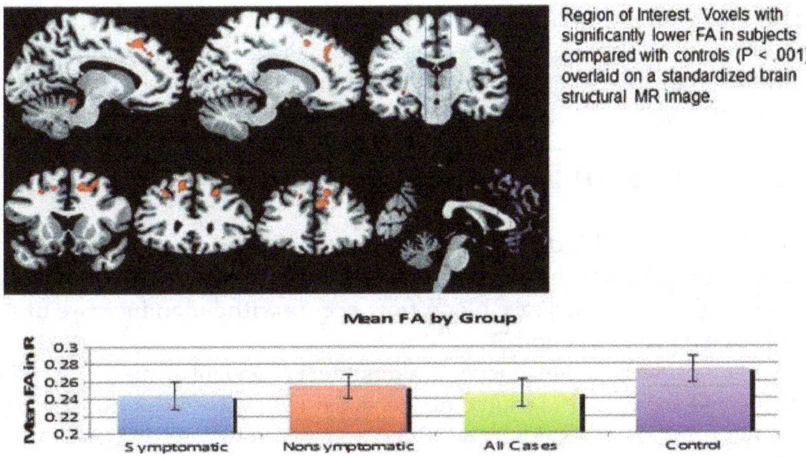

Figure 2.
Decreased FA in the anterior cingulate. The decrease in FA is most significant in patients with OTCD who are symptomatic, but note also FA decreased in those "asymptomatic" patients as compared to controls without any urea cycle disorder.

DTI was much more sensitive to changes in WM microstructural differences than fast spin echo (FSE) T2-weighted imaging for detecting abnormalities in normal-appearing WM. We also found that the degree of the abnormality correlated with degree of cognitive deficits. The location of the deficits in the frontal WM is highly significant as this area is important in the connectivity of fibers vital to executive function, working memory, and attention.

4.3 Progressive WM injury predicts cognitive decline with the most pronounced effects on processing speed and executive function

With our research we have shown DTI evidence of WM injury in motor tracts that subserve executive attention and working memory and can correlate measures of FA with specific working memory tasks. These changes result from WM tract disruption and related cortical disconnection. Quantitative data on the WM microstructure provide a more direct measurement of brain tissue integrity than standard MRI sequences.

4.3.1 WM damage in OTCD

Neuropathological findings have been extensively examined in patients who have died due to urea cycle disorders. These findings shared pathology with other more common conditions such as hepatic encephalopathy as well as hypoxic ischemic encephalopathy. Previous autopsy and, more recently, neuroimaging studies suggest that OTCD results in a predilection for WM injury. And in patients several months prior to death after surviving a neonatal presentation, neuropathological findings consisting of cortical atrophy, ventriculomegaly, gliosis with Alzheimer type II astrocytes, spongiform changes at the gray/white junction, ulegyria, and spongiform changes in the deep gray nucleibasal ganglia and thalamus have been reported in the literature [52–55].

Neuroimaging studies which have been performed several months after a neonatal hyperammonemic event, months later in neonatal coma survivors, are consistent with these pathological findings, correlating with hypomyelination of WM, myelination delay, cystic changes of the WM, and gliosis of the deep gray matter nuclei. The original reports were small case series using clinical CT initially and then, only later, MRI. Survivors of prolonged hyperammonemic coma had severe anatomic abnormalities including ventriculomegaly and cortical atrophy. Today, these severe findings are rarely encountered if patients are diagnosed promptly, and duration of hyperammonemia is shortened.

4.4 Functional MRI and UCD research (fMRI)

The basic premise behind fMRI is the increase in blood flow to the local vasculature that accompanies neural activity in the brain. This leads to a local reduction in deoxyhemoglobin. An increase blood flow occurs without an increase of similar magnitude in oxygen extraction.

Deoxyhemoglobin is paramagnetic; it alters the T2*-weighted magnetic resonance image signal and serves as the source of the signal for fMRI [56]. Coupling between neural activity and changes in blood flow was first reported in 1890 (Roy and Sherrington) [56]. By using fMRI, one can observe how the brain is functioning and what areas of brain are activated while a person is performing a specific task. It can allow unmasking of regional vulnerability, circuitry, and recovery of function after damage or intervention.

4.5 Magnetic resonance spectroscopy

MRS is another clinical sequence that provides noninvasive analytic method of identifying and measuring the individual brain chemicals present in various brain regions. ^{1}H–MR spectroscopy is widely used in clinical practice to provide information on brain metabolites. The major metabolites that can be seen include choline (Cho), creatine (Cr), N-acetyl aspartate (NAA), glutamine (Gln), and the osmolytes: myoinositol (mI) and taurine (Taur). Metabolites that can be detected have a unique frequency resonance that is termed the chemical shift. The chemical shift is reported as parts per million (ppm). The advantage of this measure in ppm is that it is the same at any magnetic field strength. The basis of the signal derives from Larmor frequency and coupling. The frequency of individual nuclei is compared to a reference compound called tetramethylsilane (TMS). The MRS is read from right to left. The metabolites that are disrupted in OTCD include Gln, Cho, and mI [57, 58]. One can quantitate the metabolites by either using in house software or using a commercially available program such as linear combination modeling or LCModel [59].

The magnetic resonance (MR) signal detectable is directly proportional to the concentration of the nuclei in the prescribed voxel. Because the brain is mainly composed of water which has a concentration of 55.5 mmol per gram, this must be subtracted in the analysis as the concentration of other chemicals such as NAA or PCr is on the order of 0.015 and 0.5 mmol per gram of tissue. Common voxel sizes used in MRS are from 1 to 5 mm^3.

Both the size and shape of a peak seen on a spectrum are due to the contribution of five attributes:

1. The concentration of nuclei.

2. The T1 and T2 relaxation times of the metabolite. These are also affected by the TR (relaxation time) and TE (echo time) of the MRS sequence as certain metabolites at low concentration may only be seen best at low TE.

3. Magnetic inhomogeneity across the sample. This can be corrected to some extent by a process called shimming. This implies the process by which the main magnetic field (Bo) is made more homogenous by applying small electrical

currents. This can be done passively, as many vendors have automated shimming packages on the scanners, or manually.

4. Another consideration, especially in the case of ^{1}H MRS in the UCD, is the presence of overlapping peaks since several metabolites in whole or in part may have overlapping peaks at a certain ppm.

5. Whether the line is expected to be single or a multiplet. This is determined by the chemical structure J coupling effects.

Some of these principles are explained below.

MR proton spectroscopy has great utility in evaluation of brain metabolic disturbances. Although a nonspecific pattern (elevated Cho, depressed NAA) is common in many types of brain disease, short echo time (TE) MRS (i.e., TE < 30 msec) can reveal more specific metabolic signatures. It is also useful to focus on the temporal changes of chemicals rather than only what is abnormal. Furthermore, temporal changes on subsequent exams can help support or refute the benefit of ongoing therapeutic measures. A simple single voxel technique boasts better signal-to-noise ratios (SNR) and allows shorter TE options than multivoxel technique.

Voxel size is always a consideration, since there is a balance between signal-to-noise ratio and tissue specificity; ideally, it should be as large as possible to achieve satisfactory SNR but small enough to target the area of interest. Generally, a 2 × 2 × 2 cm (2 cm^3) voxel is sufficient; voxels smaller than 1 cm^3 are unlikely to be worthy of the acquisition time it would require to achieve reasonable SNR. Voxel location and echo times should be selected based on the suspected and/or discovered disease patterns. We typically perform ultrashort (TE 14, TR 1500; STEAM technique), short (TE 35, TR 1500-2000; PRESS technique), and intermediate (TE 144, TR 1500-2000; PRESS) or long (TE 288, TR 1500-2000; PRESS) echo time sequences. Examples of metabolites that are best seen at short echo include gluta-mine and glutamate.

In the case of glycine at 3.55 ppm, it is necessary to obtain at least one MRS data point using an intermediate (i.e., 144 msec) or long (i.e., 288 msec) echo time to remove the spectral contamination of mI that is also around 3.5 ppm.

When are longer echo times preferred? Longer echo times improve diagnostic specificity in disorders such as maple syrup urine disease (MSUD) by eliminating the normal background macromolecular signal that can hide branched-chain amino and ketoacid peaks.

The noninvasive detection of elevated brain glutamine by ^{1}H MRS has also been shown to be a useful biomarker in chronic hepatic encephalopathy [60–62]. We have observed clinically, and it has been shown that glutamine has been implicated in hyperammonemic encephalopathy. A rise in plasma glutamine levels precedes HA [60–62]. The importance of glutamine in this process is further strengthened by the relationship between HA, neurologic dysfunction, and cerebral spinal fluid glutamine concentrations observed in patients with hepatic encephalopathy.

The UCDC presented the largest series of adult patients with OTCD who were imaged using ^{1}H MRS at 3 T and discuss the utility of advanced imaging in understanding the underlying mechanisms of dysfunction [57]. ^{1}H MRS studies have demonstrated elevations in Gln and decreases in mI and Cho in patients who are clinically symptomatic [51]. We showed with ^{1}H MRS decreased mI is also an important biomarker and also seen in females who describe themselves as asymptomatic [57]. We have hypothesized that the decrement of mI might constitute a useful biochemical marker with which to discriminate females with a partial deficiency.

4.5.1 ^{13}C MRS

Although ^{1}H MR spectroscopy is a sensitive tool to detect biochemical abnormalities in individual patients, in vivo ^{13}C MR spectroscopy can reliably be used to quantitate distinct signals from glutamate and glutamine. Unambiguous assignment of these metabolites can contribute to a better understanding of the pathogenesis and treatment of brain dysfunction in UCDs. With the use of carbon 13 (^{13}C) MR spectroscopy, abnormalities in cerebral glutamate metabolism have been noted in patients with chronic hepatic encephalopathy. Therefore, the next step was to use this technique to investigate cerebral glutamate turnover rate in patients with partial OTCD.

This method allows study of glutamate neurotransmission which is carried out by a glial neuronal process that includes the oxidation of glucose and the Gln/Glu cycle [63]. The metabolic model predicts that under conditions of elevated plasma ammonia, the increase in the rate of Gln synthesis is stoichiometrically coupled to increase in the uptake of the anaplerotic substrates CO_2 and ammonia with concurrent efflux of Gln from the brain. Furthermore, studies in hyperammonemic rats suggest that only a fraction of Gln is used to synthesize GABA via Glu. The remainder passes through the neuronal TCA cycle. Bluml et al. have previously shown that there is disturbed neurotransmitter Glu/Gln cycling in chronic hepatic encephalopathy [63, 64]. In their studies, Glu enrichment was decreased and Gln enrichment was increased.

5. Beyond ammonia: relationship of dysregulation of glutamatergic and GABAergic neurons in patients with HA and occurrence of seizures

Despite decades of research, the mechanisms leading to neural injury in HA are still not well understood. Ammonia toxicity is not necessarily an adequate explanation for the degree of cognitive dysfunction seen in patients with argininosuccinate synthase (ASS), argininosuccinate lyase (ASL), and arginase deficiency (ARG). ASS is also referred to as citrullinemia.

Recent studies by our group and others, however, have shown that HA exposure alters several amino acid pathways and neurotransmitter systems, cerebral energy metabolism, nitric oxide synthesis, oxidative stress, and signal transduction pathways which all increase the risk for seizures [30, 65].

Epilepsy had previously been considered an infrequent manifestation in urea cycle disorders, but our longitudinal study (LS) of infants with UCD at a single site (Children's National Health System) found subclinical electrographic seizures (ES) (detected on EEG without clinical manifestations) to be surprisingly common during acute hyperammonemic episodes [66].

This unanticipated finding was particularly identified in neonates with HA. This new finding raises the question of whether seizures play an important role in the etiology of neurocognitive deficits in UCD and whether seizures could afford a biomarker that correlates with brain damage in these disorders.

We have observed ES developing in patients in whom HA rebounded following discontinuation of ammonia scavengers. During HA, the brain is vulnerable to injury as a result of increased permeability and alterations in energy metabolism. In this small cohort, we observed that children with evidence of ES had abnormal MRI scans and/or adverse neurodevelopmental outcomes. This is consistent with what is seen in animal models. The animals develop change in behavior and ataxia and ultimately seizures [67].

HA episodes are critical periods in which to intervene in order to prevent long-term cognitive disability. In neonates and children, HA episodes occur in the context of a developing brain that already has vulnerabilities as a result of normal remodeling, synaptogenesis, and ion channel development. It is also the period in life recognized to be at highest risk for seizures. It is possible that seizures in HA infants and children with UCD are an early biomarker of perturbed metabolism and, left untreated, may contribute to brain injury and subsequent intellectual and other developmental disabilities.

Considerable evidence shows that HA compromises brain energy metabolism which predisposes the patient to seizures due to neuronal depolarization in association with lower energy. The seizures then in turn further lower brain energy, setting in motion a physiologic cycle: HA → lower ATP → depolarization/increased vulnerability to seizures → frank seizures (or, at least, ES) → further lowering of ATP → more seizures (or ES) [68].

5.1 Seizures in UCD

While there is a paucity of investigation of the incidence/prevalence of seizures in UCD, one early study evaluated 11 EEG tracings of 4 infants, irrespective of clinical seizure status [69]. This small study identified epileptiform EEG alterations and hypothesized that they may be a characteristic manifestation of UCD. Later, a retrospective analysis of EEG tracings and head CT scans in 49 UCD patients revealed EEG abnormalities during a clinically stable period that were predominantly observed in patients with abnormal CT scans [70].

In agreement with these reports, our group has identified that patients with distal UCD (where ammonia levels are not as elevated as in proximal disorders) have a high frequency of epilepsy and cognitive dysfunction, raising the possibility that seizures may be associated with other biochemical abnormalities in distal UCD [36, 71]. Additionally, we have shown that disrupted neural networks underlying working memory are a consistent finding in UCD patients with mild as well as severe symptoms [25, 46]. These studies strengthen the importance of understanding the incidence/prevalence of epilepsy in patients with UCDs and if and how this may affect later cognitive function. It further suggests that patients may be undertreated if we focus solely on ammonia-lowering agents and fail to recognize and treat concurrent seizures, as there are clearly other factors, such as disturbed mitochondrial function and oxidative stress, implicated in ammonia-induced neurotoxicity [68, 72–74].

5.1.1 Animal models of UCD and seizures

Both myoclonic and tonic–clonic seizures have been reported in the OTCD spf-ash mouse, an animal model of late-onset UCD with variable phenotype and severity. One study showed that the seizures were linked to the neurotoxic effect of HA on astrocytes, which increased and desynchronized astrocytic Ca^{2+} signaling and compromised the ability to buffer extracellular potassium. Using an animal model of inducible HA, the NAGS knockout (NAGSko) mouse, develops HA within a few hours of withdrawal of effective treatment with N-carbamylglutamate and L-citrulline (NCG + Cit) [75]. Studies in these mice demonstrated that they manifested seizures during HA but unexpectedly also had seizures during baseline recording when blood ammonia levels were normal. In this study, EEG seemed to be a sensitive measure of detecting neuronal dysfunction from HA and suggested its use as an early biomarker of its damaging effects.

The common hypothesis as to how ammonia may lead to seizures is based on the idea that reducing ammonia in the blood will reduce the influx into astrocytes, thereby inhibiting the glutamine synthetase enzyme. This is the basis of ammonia lowering agents in clinical use. However, the work of Rangroo-Thrane et al. challenged that idea by showing that this only may worsen the neurological condition by increasing neuronal exposure to both [NH_4^+] and [K^+]. They have shown that failure of buffering potassium in astrocytes actually is the critical mechanism that contributes to ammonia neurotoxicity. Using awake, intact mice that were induced to have HA, this research group reported that when ammonia was blocked from entering the astrocyte via the potassium transporter Na^+-K^+-$2Cl^-$ cotransporter isoform 1 (NKCC1), seizures developed. They also conclude that a therapeutic intervention should work by blocking this pathway by inhibiting NKCC1 [75].

To determine the incidence of seizures in our UCD cohorts, we conducted a data mining study of six children who were enrolled in the UCDC since 2003 and who had continuous video EEG (cVEEG) during a HA episode and sufficient data to abstract. We accessed their data including clinical information, MRI scans, and metabolic laboratory results including ammonia and glutamine levels. We found that seizures occurred in neonates with UCD even when ammonia and glutamine levels had returned to normal. We further found that interburst interval duration (the time between brain activity and silences) correlated with ammonia levels. During periods of HA, the duration of electrical silence was prolonged, and the EEG pattern could be used to predict elevated ammonia levels (**Figure 3**). A prolonged interval was also correlated with cerebral dysfunction and an abnormal follow-up MRI showing injury (**Figure 4**) [76]. cVEEG therefore can be a useful tool for managing infants with HA and may be essential for seizure management, especially for infants in deep metabolic coma. Features of the EEG appeared predictive of short-term cognitive outcome and structural injury on MRI in this cohort.

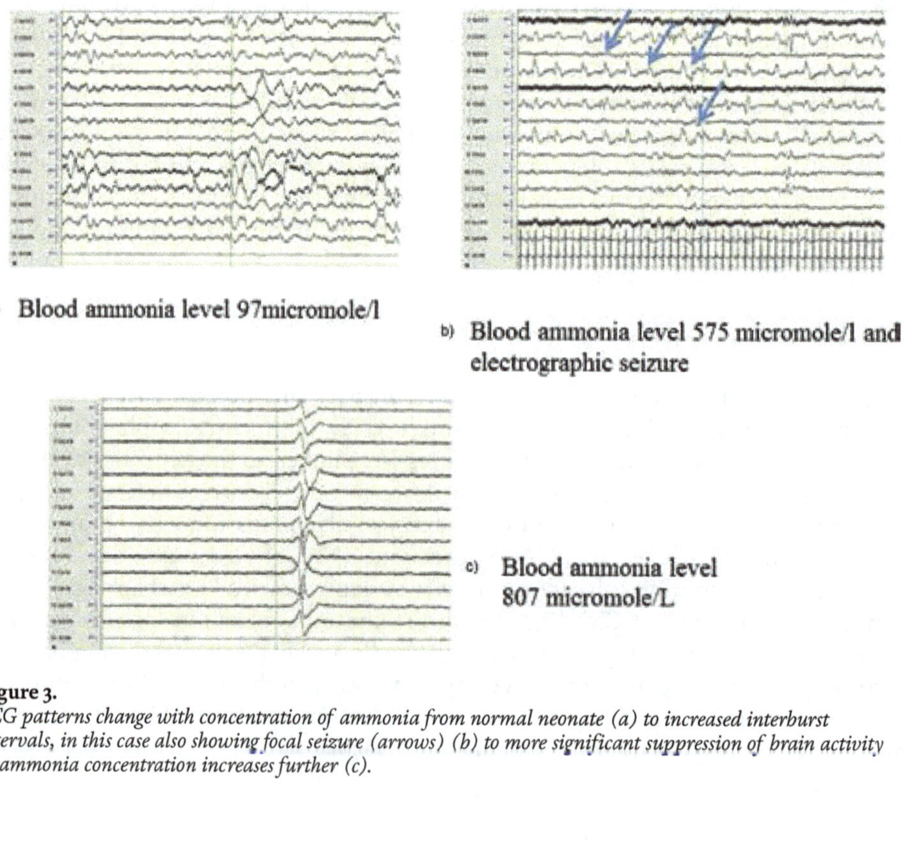

Figure 3.
EEG patterns change with concentration of ammonia from normal neonate (a) to increased interburst intervals, in this case also showing focal seizure (arrows) (b) to more significant suppression of brain activity as ammonia concentration increases further (c).

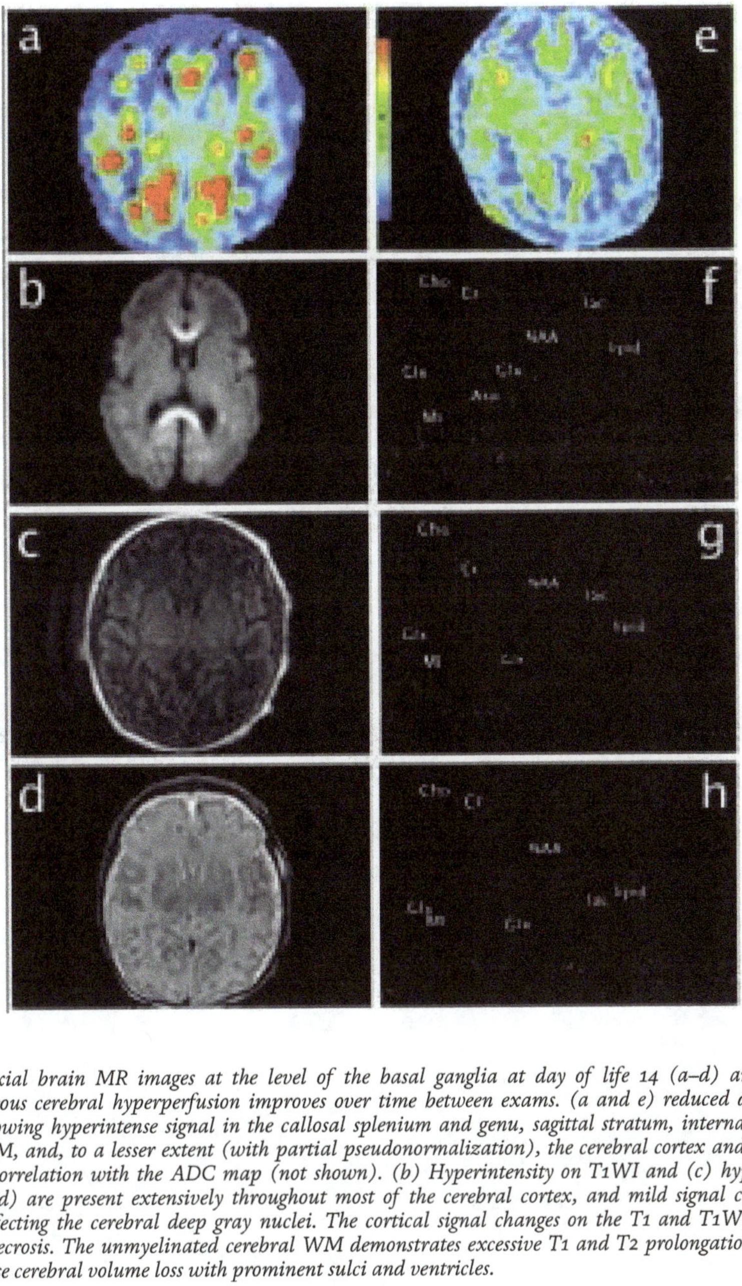

Figure 4.
Selected axial brain MR images at the level of the basal ganglia at day of life 14 (a–d) and 18 (e). Heterogeneous cerebral hyperperfusion improves over time between exams. (a and e) reduced diffusion is present showing hyperintense signal in the callosal splenium and genu, sagittal stratum, internal capsules, frontal WM, and, to a lesser extent (with partial pseudonormalization), the cerebral cortex and deep gray nuclei in correlation with the ADC map (not shown). (b) Hyperintensity on T1WI and (c) hypointensity onT2WI (d) are present extensively throughout most of the cerebral cortex, and mild signal changes are present affecting the cerebral deep gray nuclei. The cortical signal changes on the T1 and T1WI represent laminar necrosis. The unmyelinated cerebral WM demonstrates excessive T1 and T2 prolongation. There is mild diffuse cerebral volume loss with prominent sulci and ventricles.

6. Conclusions

Our understanding of the neurocognitive challenges of OTCD has been improved with the study of advanced MR imaging techniques; however, many issues remain unresolved. We are beginning to understand the neural networks impacted and have been able to correlate imaging findings with specific cognitive outcomes. We now need to scale it back to determine the earliest markers of brain injury. New findings of electrographic seizures in neonates with UCD raise questions of whether seizures play an important role in the etiology of neurocognitive deficits in UCD and whether changes on the EEG could afford a biomarker that correlates with brain changes in these disorders. Future studies are directed towards studying a more diverse group of UCD patients during baseline as well as during HA events. Pattern recognition to inform the structural and biochemical changes on MRI will allow us to move towards precision management where neuromonitoring may inform moment to moment changes in clinical management.

Acknowledgements

We acknowledge grant support from NICHD, NCATS U54 HD061221, and the O'Malley family foundation.

Notes/thanks/other declarations

We would like to thank the National Urea Cycle Foundation (NUCDF) and the families and patients who participated in our studies.

Author details

Andrea L. Gropman
Division of Neurogenetics and Developmental Pediatrics, Neurology and Pediatrics, Children's National Medical Center, N.W. Washington, United States of America

*Address all correspondence to: agropman@childrensnational.org

References

[1] Nagata N, Matsuda I, Oyanagi K. Estimated frequency of urea cycle enzymopathies in Japan. American Journal of Medical Genetics. 1991;**39**(2):228-229

[2] Applegarth DA, Toone JR, Lowry RB. Incidence of inborn errors of metabolism in British Columbia, 1969- 1996. Pediatrics. 2000;**105**(1):e10

[3] Scriver CR, Sly WS, Childs B, Beaudet AL, Valle D, Kinzler KW, editors. The Metabolic and Molecular Bases of Inherited Disease. 8th ed. New York: McGraw-Hill; 1909-1964

[4] Caldovic L, Morizono H, Panglao MG, Cheng SF, Packman S, Tuchman M. Null mutations in the N-acetylglutamate synthase gene associated with acute neonatal disease and hyperammonemia. Human Genetics. 2003;**112**(4):364-368

[5] Dionisi-Vici C, Rizzo C, Burlina AB, Caruso U, Sabetta G, Uziel G, et al. Inborn errors of metabolism in the Italian pediatric population: A national retrospective survey. The Journal of Pediatrics. 2002;**140**(3):321-327

[6] Tuchman M, Jaleel N, Morizono H, Sheehy L, Lynch MG. Mutations and polymorphisms in the human ornithine transcarbamylase gene. Human Mutation. 2002;**19**(2):93-107

[7] Caldovic L, Abdikarim I, Narain S, Tuchman M, Morizono H. Genotype-phenotype correlations in ornithine transcarbamylase deficiency: A mutation update. Journal of Genetics and Genomics. 2015;**42**(5):181-194

[8] Kang ES, Snodgrass PJ, Gerald PS. Ornithine transcarbamylase deficiency in the newborn infant. The Journal of Pediatrics. 1973;**82**(4):642-649

[9] Maestri NE, Clissold D, Brusilow SW. Neonatal onset ornithine transcarbamylase deficiency: A retrospective analysis. The Journal of Pediatrics. 1999;**134**(3):268-272

[10] McCullough BA, Yudkoff M, Batshaw ML, Wilson JM, Raper SE, Tuchman M. Genotype spectrum of ornithine transcarbamylase deficiency: Correlation with the clinical and biochemical phenotype. American Journal of Medical Genetics. 2000;**93**(4):313-319

[11] Msall M, Batshaw ML, Suss R, Brusilow SW, Mellits ED. Neurologic outcome in children with inborn errors of urea synthesis. Outcome of urea-cycle enzymopathies. The New England Journal of Medicine. 1984;**310**(23):1500-1505

[12] Msall M, Monahan PS, Chapanis N, Batshaw ML. Cognitive development in children with inborn errors of urea synthesis. Acta Paediatrica Japonica. 1988;**30**(4):435-441

[13] Campbell AG, Rosenberg LE, Snodgrass PJ, Nuzum CT. Ornithine transcarbamylase deficiency: A cause of lethal neonatal hyperammonemia in males. The New England Journal of Medicine. 1973;**288**(1):1-6

[14] Brusilow SW, Batshaw ML, Waber L. Neonatal hyperammonemic coma. Adv Pediatr. 1982;**29**:69-103

[15] Rowe PC, Newman SL, Brusilow SW. Natural history of symptomatic partial ornithine transcarbamylase deficiency. The New England Journal of Medicine. 1986;**314**(9):541-547

[16] DiMagno EP, Lowe JE, Snodgrass PJ, Jones JD. Ornithine transcarbamylase deficiency--a cause of bizarre behavior

in a man. The New England Journal of Medicine. 1986;**315**(12):744-747

[17] Nicolaides P, Liebsch D, Dale N, Leonard J, Surtees R. Neurological outcome of patients with ornithine carbamoyltransferase deficiency. Archives of Disease in Childhood. 2002;**86**(1):54-56

[18] Brusilow SW, Danney M, Waber LJ, Batshaw M, Burton B, Levitsky L, et al. Treatment of episodic hyperammonemia in children with inborn errors of urea synthesis. The New England Journal of Medicine. 1984;**310**(25):1630-1634

[19] Batshaw ML, Brusilow S, Waber L, Blom W, Brubakk AM, Burton BK, et al. Treatment of inborn errors of urea synthesis: Activation of alternative pathways of waste nitrogen synthesis and excretion. The New England Journal of Medicine. 1982;**306**(23):1387-1392

[20] Batshaw ML, Wachtel RC, Cohen L, Starrett A, Boyd E, Perret YM, et al. Neurologic outcome in premature infants with transient asymptomatic hyperammonemia. The Journal of Pediatrics. 1986;**108**(2):271-275

[21] Christodoulou J, Qureshi IA, McInnes RR, Clarke JT. Ornithine transcarbamylase deficiency presenting with stroke like episodes. The Journal of Pediatrics. 1993;**122**(3):423-425

[22] Maestri NE, Lord C, Glynn M, Bale A, Brusilow SW. The phenotype of ostensibly healthy women who are carriers for ornithine transcarbamylase deficiency. Medicine. 1998;**77**(6):389-397

[23] Pridmore CL, Clarke JT, Blaser S. Ornithine transcarbamylase deficiency in females: An often overlooked cause of treatable encephalopathy. Journal of Child Neurology. 1995;**10**(5):369-374

[24] Yudkoff M, Daikhin Y, Nissim I, Jawad A, Wilson J, Batshaw M. In vivo nitrogen metabolism in ornithine transcarbamylase deficiency. The Journal of Clinical Investigation. 1996;**98**(9):2167-2173

[25] Sprouse C, King J, Helman G, Pacheco-Colón I, Shattuck K, Breeden A, et al. Investigating neurological deficits in carriers and affected patients with ornithine transcarbamylase deficiency. Molecular Genetics and Metabolism. 2014;**113**(1-2):136-141

[26] Dasarathy S, Mookerjee RP, Rackayova V, Rangroo Thrane V, Vairappan B, Ott P, et al. Ammonia toxicity: From head to toe? Metabolic Brain Disease. 2017;**32**(2):529-538

[27] Butterworth RF. Effects of hyperammonaemia on brain function. Journal of Inherited Metabolic Disease. 1998;**21**(Suppl 1):6-20

[28] Brusilow SW, Maestri NE. Urea cycle disorders: Diagnosis, pathophysiology, and therapy. Advances in Pediatrics. 1996;**43**:127-170

[29] Felipo V, Grau E, Miñana GS. Activation of NMDA receptor mediates the toxicity of ammonia and the effects of ammonia on the microtubule-associated protein MAP-2. Advances in Experimental Medicine and Biology. 1993;**341**:83-93

[30] Braissant O, McLin VA, Cudalbu C. Ammonia toxicity to the brain. Journal of Inherited Metabolic Disease. 2013;**36**(4):595-612

[31] Takahashi H, Koehler RC, Brusilow SW, Traystman RJ. Inhibition of brain glutamine accumulation prevents cerebral edema in hyperammonemic rats. The American Journal of Physiology. 1991;**261** (3 Pt 2):H825-H829

[32] Cauli O, González-Usano A, Agustí A, Felipo V. Differential modulation of the glutamate-nitric oxide-cyclic GMP pathway by distinct neurosteroids in cerebellum in vivo. Neuroscience. 2011;**190**:27-36

[33] Monfort P, Muñoz MD, Kosenko E, Llansola M, et al. Sequential activation of soluble guanylate cyclase, protein kinase G and cGMP-degrading phosphodiesterase is necessary for proper induction of long-term potentiation in CA1 of hippocampus. Alterations in hyperammonemia. Neurochemistry International. 2004;**45**(6):895-901

[34] Li MX, Nakajima T, Fukushige T, Kobayashi K, Seiler N, Saheki T. Aberrations of ammonia metabolism in ornithine carbamoyltransferase-deficient spf-ash mice and their prevention by treatment with urea cycle intermediate amino acids and an ornithine aminotransferase inactivator. Biochimica et Biophysica Acta. 1999;**1455**(1):1-11

[35] Allegri G, Deplazes S, Rimann N, Causton B, Scherer T, Leff JW, et al. Comprehensive characterization of ureagenesis in the spfash mouse, a model of human ornithine transcarbamylase deficiency, reveals age-dependency of ammonia detoxification. Journal of Inherited Metabolic Disease. 2019 [Epub ahead of print]

[36] Krivitzky L, Babikian T, Lee HS, et al. Intellectual, adaptive, and behavioral functioning in children with urea cycle disorders. Pediatric Research. 2000;**66**(1):96-101

[37] Waisbren SE, Cuthbertson D, Burgard P, Holbert A, McCarter R, Cederbaum S. Members of the urea cycle disorders consortium. Biochemical markers and neuropsychological functioning in distal urea cycle disorders. Journal of Inherited Metabolic Disease. 2018;**41**(4):657-667

[38] Diamond A. Executive functions. Annual Review of Psychology. 2013;**64**:135-168

[39] Collins A, Koechlin E. Reasoning, learning, and creativity: Frontal lobe function and human decision-making. PLoS Biology. 2012;**10**(3):e1001293

[40] Lunt L, Bramham J, Morris RG, Bullock PR, Selway RP, Xenitidis K, et al. Prefrontal cortex dysfunction and 'jumping to conclusions': Bias or deficit? Journal of Neuropsychology. 2012;**6**(1):65-78

[41] Alvarez J, Emory E. Executive function and the frontal lobes: A meta analytic review. Neuropsychology Review. 2006;**16**(1):17-42

[42] Gioia GA, Isquith PK, Kenworthy L, Barton RM. Profiles of everyday executive function in acquired and developmental disorders. Child Neuropsychology. 2002;**8**(2):121-137

[43] Sachs M, Kaplan J, Der Sarkissian A, Habibi A. Increased engagement of the cognitive control network associated with music training in children during an fMRI Stroop task. PLoS One. 2017;**12**(10):e0187254

[44] Hohenfeld C, Werner CJ, Reetz K. Resting-state connectivity in neurodegenerative disorders: Is there potential for an imaging biomarker? NeuroImage: Clinical. 2018;**18**:849-870

[45] Reineberg AE, Gustavson DE, Benca C, Banich MT, Friedman NP. The relationship between resting state network connectivity and individual differences in executive functions. Frontiers in Psychology. 1600;**2018**(5):9

[46] Pacheco-Colón I, Washington SD, Sprouse C, Helman G, Gropman AL, VanMeter JW. Reduced functional connectivity of default mode and set-maintenance networks in ornithine

transcarbamylase deficiency. PLoS One. 2015;**10**(6):e0129595

[47] Gropman AL. Expanding the diagnostic and research toolbox for inborn errors of metabolism: The role of magnetic resonance spectroscopy. Molecular Genetics and Metabolism. 2005;**86**:2-9

[48] Basser PJ, Mattiello J, LeBihan D. MR diffusion tensor spectroscopy and imaging. Biophysical Journal. 1994;**66**:259-267

[49] Hirjak D, Rashidi M, Kubera KM, Northoff G, Fritze S, Schmitgen MM, et al. Multimodal magnetic resonance imaging data fusion reveals distinct patterns of abnormal brain structure and function in catatonia. Schizophrenia Bulletin. 2019 [Epub ahead of print]

[50] Meng L, Chen Y, Xu X, Chen T, et al. The neurobiology of brain recovery from traumatic stress: A longitudinal DTI study. Journal of Affective Disorders. 2018;**225**:577-584

[51] Gropman AL, Gertz B, Shattuck K, Kahn IL, et al. Diffusion tensor imaging detects areas of abnormal white matter microstructure in patients with partial ornithine transcarbamylase deficiency. AJNR. American Journal of Neuroradiology. 2010;**31**(9):1719-1723

[52] Dolman CL, Clasen RA, Dorovini-Zis K. Severe cerebral damage in ornithine transcarbamylase deficiency. Clinical Neuropathology. 1988;**7**(1):10-15

[53] Harding BN, Leonard JV, Erdohazi M. Ornithine carbamoyl transferase deficiency: A neuropathological study. European Journal of Pediatrics. 1984;**141**(4):215-220

[54] Krieger I, Snodgrass PJ, Roskamp R. Atypical clinical course of ornithine transcarbamylase deficiency due to a new mutant (comparison with Reye's disease). The Journal of Clinical Endocrinology and Metabolism. 1979;**48**(3):388-392

[55] Kornfeld M, Woodfin BM, Papile L, Davis LE, Bernard LR. Neuropathology of ornithine carbamyl transferase deficiency. Acta Neuropathologica. 1985;**65**(3-4):261-264

[56] Roy CS, Sherrington CS. On the regulation of the blood-supply of the brain. The Journal of Physiology. 1890;**11**(1-2):85-158.17

[57] Gropman AL, Fricke ST, Seltzer RR, Hailu A, Adeyemo A, Sawyer A, et al. Urea cycle disorders consortium 1H MRS identifies symptomatic and asymptomatic subjects with partial ornithine transcarbamylase deficiency. Molecular Genetics and Metabolism. 2008;**95**:21-30

[58] Connelly A, Cross JH, Gadian DG, Hunter JV, Kirkham FJ, Leonard JV. Magnetic resonance spectroscopy shows increased brain glutamine in ornithine carbamoyl transferase deficiency. Pediatric Research. 1993;**33**:77-81

[59] Provencher SW. Estimation of metabolite concentrations from localized in vivo proton NMR spectra. Magnetic Resonance in Medicine. 1993;**30**:672-679

[60] Kreis R, Ross BD, Farrow NA, Ackerman Z. Metabolic disorders of the brain in chronic hepatic encephalopathy detected with H-1 MR spectroscopy. Radiology. 1992;**182**(1):19-27

[61] Bates TE, Williams SR, Kauppinen RA, Gadian DG. Observation of cerebral metabolites in an animal model of acute liver failure in vivo: A 1H and 31P nuclear magnetic resonance study. Journal of Neurochemistry. 1989;**53**(1):102-110

[62] Ross BD, Danielsen ER, Blüml S. Proton magnetic resonance spectroscopy: The new gold standard for diagnosis of clinical and subclinical hepatic encephalopathy? Digestive Diseases. 1996;**14**(Suppl 1):30-39

[63] Westergaard N, Sonnewald U, Schousboe A. Metabolic trafficking between neurons and astrocytes: The glutamate/glutamine cycle revisited. Developmental Neuroscience. 1995;**17**:203-211

[64] Blüml S, Moreno-Torres A, Ross BD. [1-13C] glucose MRS in chronic hepatic encephalopathy in man. Magnetic Resonance in Medicine. 2001;**45**:981-993

[65] Cagnon L, Braissant O. Hyperammonemia-induced toxicity for the developing central nervous system. Brain Research Reviews. 2007;**56**(1):183-197

[66] Wiwattanadittakul N, Prust M, Gaillard WD, et al. The utility of EEG monitoring in neonates with hyperammonemia due to inborn errors of metabolism. Molecular Genetics and Metabolism. 2018 Nov;**125**(3):235-240

[67] Jang Y, Smith NA, Liu J, et al. Neurological defects in the animal model of inducible hyperammonemia. Molecular Genetics and Metabolism. 2018;**123**(3):240-241

[68] Bosoi CR, Rose CF. Identifying the direct effects of ammonia on the brain. Metabolic Brain Disease. 2008;**24**:95-102

[69] Verma NP, Hart ZH, Kooi KA. Electroencephalographic findings in urea-cycle disorders. Electroencephalography and Clinical Neurophysiology. 1984;**57**(2):105-112

[70] Nagata N, Matsuda I, Matsuura T, Oyanagi K, Tada K, Narisawa K, et al. Retrospective survey of urea cycle disorders: Part 2. Neurological outcome in forty-nine Japanese patients with urea cycle enzymopathies. American Journal of Medical Genetics. 1991;**40**(4):477-481

[71] Waisbren SE, Gropman AL, Members of the Urea Cycle Disorders Consortium (UCDC), Batshaw ML. Improving long term outcomes in urea cycle disorders-report from the urea cycle disorders consortium. Journal of Inherited Metabolic Disease. 2016;**39**(4):573-584

[72] Norenberg M. Oxidative and nitrosative stress in ammonia neurotoxicity. Hepatology. 2003;**37**:245-248

[73] Felipo V, Butterworth RF. Mitochondrial dysfunction in acute hyperammonemia. Neurochemistry International. 2002;**40**:487-491

[74] Lai JCK, Cooper AJL. Neurotoxicity of ammonia and fatty acids: Differential inhibition of mitochondrial dehydrogenases by ammonia and fatty acyl coenzyme a derivatives. Neurochemical Research. 1991;**16**:795-803

[75] Rangroo Thrane V, Thrane AS, Wang F, Cotrina ML, Smith NA, Chen M, et al. Ammonia triggers neuronal disinhibition and seizures by impairing astrocyte potassium buffering. Nature Medicine. 2013;**19**(12):1643-1648

[76] Wiwattanadittakul N, Prust M, Gaillard WD, Massaro A, Vezina G, Tsuchida TN, et al. The utility of EEG monitoring in neonates with hyperammonemia due to inborn errors of metabolism. Molecular Genetics and Metabolism. 2018;**125**(3):235-240

12

Cardiovascular Risk/Disease in Type 2 Diabetes Mellitus

Gabriela Roman and Anca Pantea Stoian

Abstract

People with Type 2 diabetes mellitus (T2DM) have a 2–3 times higher cardiovascular risk (CVR) than people without diabetes. Atherosclerotic cardiovascular disease (ASCVD) is the major cause of morbidity and mortality in T2DM. Over 30% of those with T2DM have CVD (cardiovascular disease), and over half die from it, mainly from coronary heart disease. The presence of T2DM reduces life expectancy by 10–14 years. The European Society of Cardiology stratifies the CVR into *moderate* (young patients, with a short duration of diabetes, no risk factors), *high* (duration of diabetes >10 years, no target organ damage, plus any additional risk factor) and *very high* (patients with established CVD, target organ injury three CVD risk factors: age, hypertension, dyslipidemia, obesity, or Type 1 diabetes mellitus (T1DM) over 20 years duration). The American Association of Clinical Endocrinologists (AACE) considers that diabetes per se involves high risk. Heart failure (HF) is the second most common complication after obstructive peripheral arterial disease. T2DM associates a 75% higher risk of CV mortality or hospitalization for HF. A multifactorial approach is required to reduce CV morbidity and mortality.

Keywords: cardiovascular risk, cardiovascular disease, type 2 diabetes mellitus

1. Introduction

Diabetes mellitus (DM), a very complex and heterogeneous chronic disease, is associated with chronic complications generated by alteration of the endothelium at all arterial vascular territory levels. The micro-and macro-vascular complications affect the entire body; their pathogenic mechanisms are intricate and involve multiple pathways and risk factors. The main risk factors are insulin resistance/hyperinsulinemia, hyperglycemia, dyslipidemia, hypercholesterolemia, hypertension, smoking, obesity/overweight, all of which cause endothelial dysfunction. Consequently, they appear vasoconstriction, oxidative stress, subclinical inflammation, vascular calcification and thrombosis, essential pathogenic elements in the process of atherogenesis/macroangiopathy and microangiopathy [1].

Cardiovascular disease is considered a major cause of morbidity and mortality among people with diabetes, with a significantly increased prevalence compare with people without diabetes. The risk increases with the level of glycaemia, the duration of diabetes and the number of risk factors. People with T2DM have much earlier and more extensive process of accelerated atherosclerosis, more vulnerable and larger volume atherosclerotic plaques and coronary artery lumen with a smaller diameter, compared to people without DM. The pathogenic complex is influenced by genetic factors, age, personal history and a pro-risk lifestyle (unhealthy eating, sedentary lifestyle, smoking, sleep disorders, psychosocial stress and depression) [1].

The macrovascular complications mainly refer to the atherosclerotic cardiovascular disease, represented by coronary artery disease (acute and chronic coronary syndrome) (CAD), chronic peripheral artery disease (PAD) and cerebrovascular disease (CBV). A particular complication is heart failure (HF). The pathogenic mechanisms of micro-and macro-angiopathy are extremely complex, with multiple interactions at the molecular-cellular and vascular-organic level.

Although all complications declined in the last decades, the most significant decreases in diabetes-related complications occurred for heart attack and stroke, especially for people aged 75 years and older. The CVD risk and mortality rate has declined in both the general population [2] and the people with diabetes [3]. However, diabetes, mainly T2DM, continues to be an important generator of cardiovascular disease. As the number of patients with diabetes is predicted to increase, reaching 700 million in 2045 [4], it is expected that the number of people with CVD will also increase. Thus, these major diabetes complications continue to place a heavy burden on health care systems.

2. Cardiovascular risk

2.1 Cardiovascular risk factors in diabetes

The concept of cardiovascular risk began to be of interest in the 1930s. Studies of cardiovascular disease epidemiology and its causes began in the 1950s, with Framingham Heart Study being one of the first. This extended program has demonstrated the existence of multiple CVR factors over the years, introducing in the literature and medical concept the term "risk factor". Among the study's first results, high blood pressure, hypercholesterolemia, and smoking are considered traditional, classic, risk factors. Subsequently, Framingham study and other epidemiological studies, have identified other risk factors, including obesity, diabetes, dyslipidemia, sedentary lifestyle. Thus diabetes is associated with a 2- to 3-fold increase in the risk of developing CVD and glucose intolerance with a 1.5-fold increase [5–7]. In addition to hyperglycemia, diabetes, mainly T2DM, is accompanied by other cardiovascular risk factors, within the metabolic syndrome: insulin resistance, abdominal obesity, atherogenic dyslipidemia (hypertriglyceridemia, low HDL-C (High-density lipoprotein cholesterol), LDL-C (low density lipoprotein cholesterol particles), remnant lipoproteins, postprandial hyperlipidemia), high blood pressure, prothrombotic, proinflammatory and oxidative stress state, microalbuminuria, non-alcoholic fatty liver disease [8].

High blood pressure increases 2–4 times the risk of CVD, kidney and death, atherogenic dyslipidemia induces a residual CVR, even under statin treatment and LDL-C control and abdominal obesity, as a component of metabolic syndrome, significantly increases the risk of coronary heart disease, stroke and death [9–11].

2.2 Cardiovascular risk in T2DM

Overall, people with T2DM have 2 to 4 times increased risk of cardiovascular morbidity and mortality than individuals without diabetes. This risk increases with the increase of fasting glycaemia since the stages of prediabetes. The risk for CAD is rising by 160%, for ischemic heart disease by 127%, for stroke by 56%, for CVD death by 132% [4], and for HF is 2–4 times higher in men and five times higher in women, compared to those without diabetes [5].

Epidemiological studies have revealed that the excess relative risk of vascular events is more significant in women, at younger ages, in long-standing DM and in the presence of microvascular complications, mainly renal disease or proteinuria [12].

Although T2DM was initially considered a "cardiovascular risk equivalent" [12–14], it has since been shown that CVR, mainly CAD risk, is not similar for all persons with diabetes, but is highly heterogeneous. Thus, the CVR should be differentiated based on the presence of other CVR factors or overt CVD.

For T1DM, The Swedish National Diabetes Register has shown actual results in CVD and CV death prevalence. [13] For T1DM, 27,195 patients were stratified by age and sex. Early-onset at 1–10 years of age was correlated with an HR (hazard ratio) of 7.38 for CV mortality, 30.95 for acute myocardial infarction (MI), and 12.9 for heart failure (HF). Progress of T1DM between 1 and 10 years of age resulted in a loss of 17.7 years of life in women and 14.2 years in men. [13] Notwithstanding, T2DM is more common than T1DM.These results confirm the loss of years of life in both populations, which is more severe in the younger patients and in young-onset female individuals with T2DM, highlighting the need for early and intensive risk-factor interventions in these clusters of patients. [12]

The European Society of Cardiology stratifies the CVR into three risk categories, including T1DM, considering that presence of DM overall represents a significant risk factor for CVD:

- **moderate:** young patients (T1DM aged <35 years or T2DM aged <50 years), with DM duration <10 years, without other risk factors,
- **high:** patients with DM duration ≥10 years without target organ damage plus any other additional risk factor, and
- **very high:** patients with DM and established CVD, or further target organ damage (proteinuria, renal impairment defined as eGFR <30 mL/min/1.73 m^2, left ventricular hypertrophy, or retinopathy), or three or more major risk factors (age, hypertension, dyslipidemia, smoking, obesity), or early-onset T1DM of >20-year duration [12].

The American Association of Clinical Endocrinologists (AACE) stratifies the CVR into three levels: **high** (DM without other CVR factors), **very high** (DM with at least one CVR factor), and **extreme risk** (overt CVD in those with DM, chronic kidney disease) [15].

T2DM is reputed as an independent risk factor for the development of HF. Higher levels of glycated hemoglobin A1c (HbA1c) in T2DM patients have been associated with significantly more incident HF cases than in patients with T2DM and lower HbA1c levels. The incidence is even higher in patients with established CAD, in which each 1% increase in HbA1c level was associated with a 36% increased risk for HF hospitalization [16].

2.3 Cardiovascular risk assessment

A rational and successful approach to reducing the CVR involves stratifying risk and the periodic evaluation of the outcomes. Accurate CVD risk estimation in people with T2DM without established CVD can identify patients at high risk of developing CVD and can thus be used to adapt the intensity and complexity of appropriate treatment. For practice, CVR assessment scores which can be applied in diabetes are beneficial. Several CVR calculation scores in DM have been developed. The first is the result of the United Kingdom Prospective Diabetes Study (UKPDS), in which, in people newly diagnosed with T2DM, the effect of intensive treatment on the evolution of chronic complications compared to conventional treatment was followed. UKPDS Risk Engine estimates the risk of fatal and non-fatal coronary events and fatal and non-fatal stroke at 15 and 30 years, in people with T2DM without CV disease, considering the duration of DM, age, gender, ethnicity, smoking, presence of atrial fibrillation, the level of HbA1c, systolic blood pressure, total cholesterol and HDL-C (https://www.dtu.ox.ac.uk/riskengine/) [17].

Another score (Advance Risk Engine) is based on ADVANCE and ADVANCE-ON studies. It refers to patients with T2DM without CVD, is based on the usual parameters and estimates the risk of major CV events at four years, the risk of renal events at five years and the risk of major vascular disease at ten years (www.advanceriskengine.com) [18].

The American Heart Association (AHA) and American College of Cardiology (ACC), developed a risk score to estimate the ten-year risk for the first ASCVD event (non-fatal myocardial infarction or CHD death, or fatal/non-fatal stroke) (available online at tools.acc.org/ASCVD-Risk-Estimator-Plus) [19].

Based on the ACCORD trial population, the BRAVO risk engine has been recently developed. It contains three separate modules addressed to events (stroke, MI, HF, angina, revascularization surgery, renal events, blindness, hypoglycemia), risk factors and mortality [20].

None of these CVR estimation scores is perfect, so clinical judgment and consideration of CVR factors are important for setting and selecting appropriate therapeutic goals and interventions.

3. Cardiovascular disease

The high impact that T2DM has on the CVD has generated numerous studies, and population analyzes in order to determine the prevalence of the cardiovascular pathologies in people with diabetes. Although the description and the diagnostic criteria used to define the different manifestations of CVD were different across the epidemiologic studies, overall, the results show that CVD is a major cause of comorbidity and mortality among patients with T2DM. Thus, the CVD, including myocardial infarction, stroke, angina, heart failure, atherosclerosis and coronary artery disease, is present in 32.2% of people with T2DM. The most frequent type of CVD seems to be the CAD (21.2%), males having higher rates (18.7%) than females (14.3%) [21]. A large cohort of 1,921,260 individuals, 1.8% with T2DM, followed 5.5 years, has been analyzed in terms of the most common initial manifestations of CVD. In T2DM individuals, peripheral artery disease was the most frequent first presentation (16.2%) followed by HF (14.1%), significantly higher compared with those without diabetes [22]. The prevalence of CVD and the incidence of primary adverse outcomes is higher in women with T2DM than their male counterparts ($RR = 9.29$; $P < 0.0001$ for CVD and $RR = 5.25$; $P < 0.0001$ for incident major adverse outcomes [23]. Data from the UK Biobank showed that a cardiovascular event's excess risk was approximately 50% higher in women ($HR = 1.96$) than in men, and more importantly, the incidence of myocardial infarction (MI) [A1] was higher in men than in women(28.8%). This observation was exciting since the sex differences were the same in all age-related groups and were attenuated with increasing age, from 0.27 (0.18 to 0.41), in 45 years age cluster versus 0.45 (0.40 to 0.50) in 65 years age cluster. [24].

Although diabetes-related excess mortality is lower in the contemporary era than previously, T2DM is associated with a two to six times increased risk of CVD mortality than people without diabetes. Cardiovascular disease accounts for 52% of deaths in type 2 DM [25]. MI is considered to be the main cause of death for individuals with diabetes mellitus.

T2DM is a significant predictor of HF, independent of the simultaneous presence of hypertension and coronary heart disease. In diabetes, HF has multiple risk factors: hyperglycemia, insulin resistance, age, ischemic heart disease, high blood pressure, left ventricular hypertrophy, diffuse, accelerated and severe atherosclerosis. Even in the absence of CAD, microvascular complications, arterial thickening and fibrosis, endothelial and vasomotor dysfunction, increase the risk of HF. The risk of HF is 2–4 times higher in men and five times higher in women compared to those without diabetes, according to Framingham Heart Study [5, 16].

The presence of DM in those hospitalized for HF worsens the prognosis, prolongs hospitalization, and increases the number of hospitalizations and the risk of death. In stable forms of HF, T2DM is an independent predictor of hospitalization and mortality, regardless of the ejection fraction [26]. Patients with T2DM have a 2.5-fold increased risk of developing HF and a 75% higher risk of CV mortality or hospitalization for HF than those without DM [27].

HF with preserved ejection fraction (HFpEF) is a frequent phenotype in T2DM and is related to an important risk of morbidity and mortality than those without diabetes, with multiple comorbidities, reduced exercise capacity, increased markers of inflammation, fibrosis, endothelial dysfunction, congestion, increased left ventricular pressure [28].

At similar ejection fraction values, patients with T2DM have a higher NYHA functional class, and more expressed symptoms than those without diabetes. Left ventricular (LV) diastolic dysfunction can be detected in approximately 75% of patients with T2DM, even in the early stages. The degree of hyperglycemia and its duration correlate with LV diastolic dysfunction's severity, with the increased risk of HF and CV mortality [16].

A retrospective cohort study that included 208,792 adults with diabetes has analyzed the impact of varying combinations of heart disease, stroke, moderate chronic kidney disease (CKD) on mortality, over a median follow-up of 8.5 years [29]. The effect of heart diseases, stroke, CKD, and the combination of these conditions on all-cause mortality has been found to be independent and cumulative. The mortality risks were 1.75 times, 2.63 times, and 3.58 times greater for patients with one, two, and all three above mentioned conditions, and life expectancy for a 40-year-old with one, two, and three conditions was reduced with 20, 25, and 30 years for men and 25, 30, and 35 years, respectively, for women, compared with patients without these diseases [29].

Diabetes is the most known cause of CKD, and end-stage renal disease, more than 50% of people with DM are likely to develop CKD. CKD, especially severe CKD, has a significant impact on life expectancy and mortality risk in patients with diabetes. The risk of CV death increases as renal function declines (eGFR <60 ml/min/1.73 m^2 and/ or ACR (albumin-creatinine ratio) ≥10 mg/g) [30]. Rates of HF are approximately 3-times higher in patients with eGFR <60 ml/min/1.7 m^2 [31].

The CardioRenal Metabolic Syndrome (CaReMe) has been introduced as terminology to describe the ongoing relationship between obesity, diabetes, kidney disease and heart failure with preserved systolic function, with significant mortality implications rate and therapeutic interventions [32].

Within CVD, diabetic cardiomyopathy or "cardiac microvascular disease", is a phenotype with distinct manifestations, described as a "structural and functional alteration of the myocardium, in the absence of hypertension, ischemic coronary heart disease, valvulopathies, or other FRCV, in patients with diabetes, especially with long-term diabetes and poor control." Fibrosis, stiffness, and cardiac hypertrophy are basic changes, which are initially associated with diastolic dysfunction, LV hypertrophy and reduced compliance with preserving the ejection fraction, later evolving into systolic dysfunction clinically manifest HF [16, 33]. As risk factors of diabetic cardiomyopathy, hyperglycemia, dyslipidemia, altered energy metabolism, dysregulated insulin signaling, inflammation, endoplasmic reticular stress, mitochondrial dysfunction, oxidative stress and accumulation of advanced glycation end-products (AGEs) and activation of the renin-angiotensin-aldosterone system (RAAS) are described [33].

4. Hyperglycemia and CVD relationship

Compared with individuals without diabetes, patients with T2DM are disproportionately affected by CVD morbidity and mortality. Most of this excess risk is associated with an increased prevalence of risk factors such as hypertension, dyslipidemia, and obesity in these patients. However, hyperglycemia, as a distinct characteristic of DM, appears to be an independent risk factor for all-cause and CVD mortality independent of other modifiable CVD risk factors:

- Increased HbA1c is associated with coronary heart disease [34],
- Long-term intraindividual variability of HbA1c or basal blood glucose is associated with micro-and macro-vascular complications and with an increased risk of important adverse CV events [35–37],
- 24-hour glycemic variability, greater than 35.9 mg/dl, is independently associated with an increased risk of left ventricular diastolic dysfunction, even in asymptomatic patients and with a preserved ejection fraction (odds ratio: 3.67; $p < 0.05$) [38],
- Basal glycemia >126 mg/dl was associated with a risk of fatal/non-fatal coronary heart disease, fatal/non-fatal stroke of 39% in women and 48% in men [39],
- Each 1-point increase in HbA1c is associated with an 8% increase in the risk of HF and a 36% increased risk for HF hospitalization [16, 39],
- The onset of diabetes in young people and the long duration of diabetes, are associated with an increased risk of mortality, mainly CV, due to ischemic disease and stroke [40, 41],
- The risk of coronary artery disease increases in diabetic patients by 11% for each 1% increment in HbA1c greater than 6.5%. In adults with diabetes, but without baseline CVD, an HbA1c of 9% is correlated with an increased risk of myocardial infarction or acute coronary syndrome (odds ratio [OR] = 1.18), stroke (OR = 1.29) and heart failure (OR = 1.37) [42],
- Severe hypoglycemia is linked with an increased risk of recurrence (especially in T1DM) and an increased risk of CV events, including death. In the Veteran Affairs Diabetes Trial, severe hypoglycemia was associated in the following 3 months with CV events (HR = 1.9; $p = 0.03$), CV mortality (HR = 3.7; $p = 0.01$) and mortality of any cause (HR = 2.4; $p = 0.02$) [43].

5. Therapeutic approach

The major objectives of clinical management in T2DM are preventing or delaying chronic complications, increasing life expectancy, and quality. The basic principle of clinical management is the "patient-centred" approach, respectively, the intervention's individualisation [19]. The significant reduction of cardiovascular morbidity/mortality implies a multifactorial approach, addressed simultaneously to all CV risk factors. The 7.8-year STENO-2 study, which included patients with T2DM with increased CV risk (microalbuminuria), showed that the multifactorial approach significantly reduced mortality from any cause (20% reduction in absolute risk), 13% in CV mortality and 29% in the absolute risk of CV events [44, 45]. The 21-year assessment showed that life expectancy was extended by eight years in the intensive care group, and the relative risk of HF was reduced by 76% [45]. Analysis of data from an extensive registry has shown that in people with T2DM, simultaneous control of major CV risk factors (blood glucose, blood pressure, cholesterol and lifestyle, no smoking), can reduce over 60% of cardiovascular and coronary atherosclerotic events [46]. Patients with diabetes who have five risk-factor variables within the target ranges (HbA1c, LDL-cholesterol, blood pressure, no albuminuria and no smoking) seem to have lower or no excess risk of overall death or myocardial infarction and/or stroke, as similar to the general population, as shown in a study that included 271,174 patients with T2DM registered in the Swedish National Diabetes Register [47].

5.1 Specific therapeutic targets

The specific targets of clinical management are presented in **Table 1**.

5.2 Lifestyle optimization

Lifestyle interventions addressed to optimize nutrition, increase physical activity, control body weight, stop smoking, are the cornerstone for T2DM therapy, for both glycemic and other CV risk factors [19, 50]. In terms of CV risk control, lifestyle management aims to, and it is proven to achieve [51]:

- Control of body weight and improvements of obesity-related complications,
- Improvement of glycemic control,
- Lowering of blood pressure,
- Improvement of the lipid profile
- Improvement of fitness, well-being and mental health.

Although the Look AHEAD (Action for Health in Diabetes) study did not show. that intensive lifestyle optimization reduces CV events in people with diabetes and obesity, it showed significant control of CV risk factors, with fewer medication requirements and long-term weight loss maintenance (4.7% at eight years) [52].

5.2.1 Medical nutrition therapy

Medical nutrition therapy is based on healthy eating principles. Its goals are to promote and support healthful eating patterns, with various nutrient-dense foods

Risk factors	Therapeutic targets
HbA1c [19]	< 7% (53 mmol/mol), for most patients < 6,5% (48 mmol/mol), if the target can be achieved without hypoglycemia† (usually younger patients, short duration of diabetes, longer life expectancy, no complications) < 8% (64 mmol/mol), for patients with long-standing diabetes, complex glucose-lowering treatment including insulin, history of severe hypoglycemia, limited life expectancy, advanced microvascular or macrovascular, multiple complications and comorbidities
Fasting glycemia, postprandial glycemia, glycemic variability [19]	• Fasting/pre-meal glycemia = 80–130 mg/dl • Postprandial glycemia (1–2 hours) < 180 mg/dl • Reduced glycemic variability (< 36%) • Avoidance of hypoglycemia: > 70% time in range (70–180 mg/dL/3.9–10.0 mmol/L) < 4% glucose range below 70 mg (3.9 mmol/L)
Blood pressure [18, 19]	< 130/80 mmHg (≥ 65 years = 130–139/70–79 mmHg)
Lipids [12]	• LDL-cholesterol: < 100 mg/dl - moderate CV risk < 70 mg/dl and > 50% reduction - high CV risk < 55 mg/dl and > 50% reduction – very high CV risk • Non-HDL-col <85 mg/dl in patients at very high CVR, and < 100 mg/dl in those at high CVR • HDL-cholesterol >40 mg/dl in men and > 45 mg/dl in women * • Triglycerides <150 mg/dl *
Body weight [19]	> 5–10% weight loss and • long term weight loss maintenance
Lifestyle [19]	• Physical activity • Healthy nutrition, caloric adjusted • No smoking • No/moderate alcohol intake

†Glucose range below 70 mg/dl in less then 4% of the measurements determined by continuous glucose monitoring or ambulatory glycemic profile per day [19, 49].
**Not a goal but indicates a lower risk.*

Table 1.
Therapeutic targets in T2DM (adapted by [12, 19, 48, 49]).

in appropriate portion sizes. To ensure adherence and effectiveness, nutritional interventions should be individualized by meeting individual's needs, personal and cultural preferences, access to healthy foods, the pleasure of eating, and providing practical tools to implement the recommendations [50].

Dietary interventions can reduce HbA1c with up to 2% [50]. Caloric intake should be adapted to maintain body weight control. General recommendations are to avoid saturated lipids and foods with a high glycemic index. The Mediterranean diet, the DASH diet (Dietary Approaches to Stop Hypertension), plant-based diets, are recommended for their proven benefits. Carbohydrates should come from foods rich in fibers: vegetables, legumes, whole grains, fruits. Lipids should be mainly mono- and polyunsaturated and omega-3. Supplementation with minerals or vitamins is only recommended in case of deficiency. Hydration is important, with the selection of non-caloric drinks. Consumption of alcohol in moderate amounts is acceptable, but special attention should be paid to the risk of hypoglycemia, hyperglycemia and additional caloric intake. Sodium intake should be <2,300 mg/day [50].

5.2.2 Physical activity

Most patients' recommended physical activity is at least 150 minutes/week, at least three times/week, moderate/high intensity, aerobic, 2–3 sessions/week of endurance and flexibility exercises. Reducing the time spent in sedentary lifestyle is also an important recommendation [50].

5.3 Pharmacotherapy

5.3.1 Pharmacotherapy of hyperglycemia

The pharmacotherapy of hyperglycemia should be patient-centred, addressed to reduce glycemia and the overall CVR [53]. Metformin remains the first step of the treatment. It is initiated simultaneously with lifestyle optimization, from the diagnosis, and is continued throughout the treatment, associated with the other therapeutic classes. It is stopped in case of intolerance and at an eGFR <30 mL/min/1.73 m^2. For patients with established ASCVD or indicators of high ASCVD risk, established kidney disease, or heart failure, the guidelines recommend the medication with demonstrated CVD benefit, independent of the HbA1c value: SGLT-2 inhibitor (sodium-glucose cotransporter 2 inhibitor – empagliflozin, canagliflozin, dapagliflozin) or GLP-1RA (glucagon-like peptide 1 receptor ago-nist – semaglutide, liraglutide, dulaglutide, long-acting exenatide, lixisenatide) (**Table 2**) [12, 19, 53–59].

For patients without the conditions mentioned above, a second or third agent's choice as an add-on to metformin is based on CV safety, the effect on body weight, and avoidance of hypoglycemia. Sulfonylureas have controversial CV effects. To date, the ADVANCE study (Action in Diabetes and Vascular Disease: Preterax

GLP-1 RA/Study				
	MACE	CV mortality	hHF	Renal effects
Liraglutide (LEADER) [54]	HR = 0.87 (0.78–0.97)	HR = 0.78 (0.66–0.93)	NS	HR = 0.78 (0.67–0.92)
Semaglutide (SUSTAIN) [55]	HR = 0.74 (0.58–0.95) *	NS	NS	HR = 0.64 (0.46–0.88)
Dulaglutide (REWIND) [56]	HR = 0·88 (0·79–0·99)	NS	NS	HR = 0·85 (0·77–0·93)
SGLT2-inh/Study				
Empagliflozin (EMPAREG-OUTCOME) [57]	HR = 0.86 (0.74–0.99)	HR = 0.62 (0.49–0.77)	HR = 0.65 (0.50–0.85)	HR = 0.61 (0.53–0.70)
Canagliflozin (CANVAS) [58]	HR = 0.86 (0.75–0.97)	HR = 0.78 (0.67–0.91) (CV mortality and hHF)	HR = 0.67 (0.52–0.87)	HR = 0.60 (0.47–0.77)
Dapagliflozin (DECLARE-TIMI 58) [59]	NS	HR = 0.83 (0.73–0.95) (CV mortality and hHF)	HR = 0.73 (0.61–0.88)	HR = 0.53 (0.43–0.66)

Table 2.
Antihyperglycemic agents with proven CVD/CKD/hHF benefits [12, 19, 53–59].

and Diamicron MR Controlled Evaluation) is the only one that has shown that gliclazide is CV neutral and has beneficial effects in reducing kidney disease [60]. The PROactive study (PROspective pioglitAzone Clinical Trial In macroVascular Events) which included patients with T2DM treated with pioglitazone in secondary prevention, demonstrated benefits of reducing events (myocardial infarction, stroke), but also increased risk for heart failure [61]. DPP-IV-inhibitors (dipeptidyl peptidase IV), are CV safe, body weight neutral, and with low risk for hypogly-cemia. Insulin therapy is initiated at HbA1c values above 10%, or in the presence of symptoms of hyperglycemia. New generations of basal insulin analogues are recommended, preferably in association with GLP-1 RA [53].

Mention: only the statistically significant results are included in the table; HR-hazard ratio; NS-non statistically significant; MACE-major adverse cardiovascular events: CV mortality, non-fatal myocardial infarction and non-fatal stroke; hHF- hospitalization for heart failure; Renal effects: new or worsening nephropathy (persistent macroalbuminuria, persistent doubling of the serum creatinine level and a creatinine clearance of less than 45 ml per minute per 1.73 m2 of body-surface area, or the need for continuous renal-replacement therapy); LEADER- Liraglutide Effect and Action in Diabetes: Evaluation of Cardiovascular Outcome Results; SUSTAIN-Efficacy and Safety of Semaglutide Once-weekly Versus Placebo in Drug-naïve Subjects With Type 2 Diabetes; EXSCEL- Exenatide Study of Cardiovascular Event Lowering Trial; EMPA-REG-OUTCOME-The Empagliflozin Cardiovascular Outcome Event Trial in Type 2 Diabetes Mellitus Patients–Removing Excess Glucose; CANVAS- Canagliflozin Cardiovascular Assessment Study; DECLARE-TIMI 58-Dapagliflozin Effect on Cardiovascular Events–Thrombolysis in Myocardial Infarction 58.

Unfortunately, the Exenatide Study of Cardiovascular Event Lowering (EXSCEL) was unable to demonstrate that once-weekly exenatide can achieve a statistically significant reduction in the incidence of major adverse cardiovascular events, compare to placebo [62].

DAPA-HF study ("Dapagliflozin and Prevention of Adverse Outcomes in Heart Failure"), which included patients with heart failure (ejection fraction <40% and NYHA class II-IV or symptoms) has demonstrated, after a median of 18.2 months of treatment with dapagliflozin vs. placebo, a significant reduction (HR = 0.74) of the primary composite outcome (worsening heart failure or death from cardiovascular causes) and the secondary outcomes (cardiovascular death or heart-failure hospitalization) (HR = 0.75) [63]. A similar design has been used in EMPEROR-Reduced trial (Empagliflozin Outcome Trial in Patients with Chronic Heart Failure and a Reduced Ejection Fraction). Compare to placebo, empagliflozin has significantly reduced the primary composite outcome of death from cardiovascular causes or hospitalization for HF (HR = 0.75) and the total number of hospitalizations for heart failure (HR = 0.70) [64].

5.3.2 Pharmacotherapy of hypertension

At a systolic blood pressure ≥ 140 mmHg and/or diastolic blood pressure ≥ 90 mmHg, drug therapy is necessary in combination with non-pharmacological treatment [12, 19]. An association of antihypertensive classes is often needed, with a renin-angiotensin-aldosterone system (RAAS) blocker, and a calcium channel blocker or diuretic. A RAAS blocker is recommended particularly in the presence of microalbuminuria, albuminuria, proteinuria, or LV hypertrophy. Dual therapy is recommended as first-line treatment, with a combination of a RAAS blocker with a calcium channel blocker or thiazide/thiazide-like diuretic. It is recommended that at least one antihypertensive drug to be administered in the evening [12]. In the case of resistant hypertension, if therapeutic targets are not met with three classes of antihypertensive drugs (including a diuretic), a mineralocorticoid receptor antagonist (spironolactone) is added. Fixed combinations are preferred to increase adherence [12, 19].

5.3.3 Pharmacotherapy of dyslipidemia

LDL-cholesterol is the first therapeutic target. Depending on the risk class, statins are used in moderate doses (Atorvastatin 10–20 mg, Rosuvastatin 5–10 mg, Simvastatin 20–40 mg) or high doses (Atorvastatin 40–80 mg, Rosuvastatin 20–40 mg) [12, 19, 48]. If LDL-C target is not reached, ezetimibe or PCSK9 inhibitors (evolocumab or alirocumab) are added. In the presence of atherogenic dyslipidemia (persistent triglycerides >200 mg/dl), the fenofibrate can be added to statin. The lipid arm of ACCORD study showed a further 31% reduction in CV events' relative risk in this combination. Omega-3 fatty acids can further reduce the level of triglycerides in a dose of >2 g/day. The "Reduction of Cardiovascular Events with EPA-Intervention Trial (REDUCE-IT) study", showed that 2 grams icosapent ethyl (EPA), administered twice daily, was associated with 25% relative risk reduction ($P < 0.001$) in major adverse CV events (MACE), compared to placebo [65].

5.3.4 Antithrombotic/anticoagulant pharmacotherapy

If there is no contraindication, in individuals with T2DM over the age of 50, at high/very high CV risk, in the presence of the family history of premature atherosclerotic disease, hypertension, dyslipidemia, smoking, chronic kidney disease, aspirin (75–100 mg/day) may be considered in primary prevention. Aspirin (75–162 mg/day), or clopidogrel (75 mg/day) are recommended for secondary prevention. Dual antiplatelet therapy (low-dose aspirin and P2Y12 inhibitors: ticagrelor, clopidogrel, prasugrel) is recommended after an acute coronary syndrome, for a period of one year, possibly with benefits and longer use [12, 19].

Rivaroxaban 2.5 mg administered twice daily in combination with 100 mg aspirin, in T2DM patients with stable atherosclerotic vascular disease, has been shown to significantly reduce the primary outcome of CV death, stroke, myocardial infarction and major adverse limb events including amputation, but with more major bleeding events than those assigned to aspirin alone [66, 67].

5.3.5 Surgical procedures

a. Bariatric surgery is associated with the most important and sustained weight loss, with significant improvements in CV risk factors, including remission of T2DM, cardiac functional parameters and coronary events. The criteria for recommending bariatric surgery in people with T2DM are [68, 69]:

   - BMI $\geq$ 40 kg/m^2
   - BMI = 35.0–39.9 kg/m^2 if long-term weight loss and improvement of comorbidities by non-surgical methods could not be achieved
   - BMI = 30.0–34.9 kg/m^2 in certain situations [69]

b. Myocardial revascularization strategies, either percutaneous coronary intervention (PCI) with drug-eluting stents (DES), preferably the newer-generation everolimus-eluting stents, or coronary artery bypass graft surgery (CABG), are strongly recommended in patients with T2DM. Based on the coronary anatomy complexity, PCI may represent an alternative to CABG for lower complexity, while CABG is recommended for intermediate-to-high anatomical complexity [19].

6. Conclusions

T2DM is a major cardiovascular risk factor. CVD is frequent, associated with high mortality.

The clinical management of T2DM must be early, multifactorial, intensive, and patient-centred. Lifestyle intervention and a combination of several classes of drugs should be addressed to all cardiovascular risk factors.

Author details

Gabriela Roman[1*] and Anca Pantea Stoian[2]

1 Clinical Center of Diabetes, Nutrition and Metabolic Diseases, "Iuliu Haţieganu" University of Medicine and Pharmacy, Cluj-Napoca, Romania

2 Department of Diabetes, Nutrition and Metabolic Diseases, "Carol Davila" University of Medicine and Pharmacy, Bucharest, Romania

*Address all correspondence to: groman@umfcluj.ro

References

[1] Low Wang CC, Hess CN, Hiatt WR, et al. Clinical Update: Cardiovascular Disease in Diabetes Mellitus Atherosclerotic Cardiovascular Disease and Heart Failure in Type 2 Diabetes Mellitus – Mechanisms, Management, and Clinical Considerations. Circulation. 2016;133:2459. DOI: 10.1161/ CIRCULATIONAHA. 116.022194

[2] Mensah GA., Wei GS., Sorlie PD, et al. Decline in Cardiovascular Mortality: Possible Causes and Implications. Circ Res. 2017, 120(2): 366-380. DOI: 10.1161/CIRCRESAHA.116.309115

[3] Harding JL, Pavkov ME, Magliano DJ, et al. Global trends in diabetes complications: a review of current evidenceDiabetologia. 2019; 62:3-16. DOI:10.1007/s00125-018-4711-2

[4] IDF DIABETES ATLAS. Ninth edition, 2019. www. https://www.diabetesatlas.org/upload/resources/2019/IDF_Atlas_9th_ Edition_2019.pdf. [Accessed: 2020-09-15]

[5] Kannel WB, McGee DL. Diabetes and cardiovascular disease. The Framingham study. JAMA.1979;241: 2035-2038.DOI:10.1001/ jama.1979.03290450033020

[6] Kannel WB, McGee DL. Diabetes and glucose intolerance as risk factors for cardiovascular disease: the Framingham study. Diabetes Care. 1979;2:120-126. DOI: 10.2337/ diacare.2.2.120

[7] O'Donnella CJ, Elosua R. Cardiovascular Risk Factors. Insights From Framingham Heart Study. Rev Esp Cardiol. 2008;61(3):299-310. DOI:10.1016/S1885-5857(08)60118-8

[8] Alberti K.G.M.M., Eckel RH, Grundy SM., et al. Harmonizing the Metabolic Syndrome. A Joint Interim Statement of the International Diabetes Federation Task Force on Epidemiology and Prevention; National Heart, Lung, and Blood Institute; American Heart Association; World Heart Federation; International Atherosclerosis Society; and International Association for the Study of Obesity. Circulation. 2009;120:1640-1645. DOI:10.1161/ CIRCULATIONAHA.109.192644

[9] Ferrannini E, Cushman WC. Diabetes and hypertension: The bad companions. The Lancet, 2012;380: 601-610. DOI: 10.1016/S0140-6736(12) 60987-8

[10] Xiao C, Dash S, Morgantini C et al. Pharmacological targeting of the atherogenic dyslipidemia complex: The next frontier in CVD prevention beyond lowering LDL cholesterol. Diabetes 2016;65:1767-1778. DOI: 10.2337/db16-0046

[11] Eeg-Olofsson K., Cederholm J, Nilsson PM, et al. Risk of cardiovascular disease and mortality in overweight and obese patients with type 2 diabetes: an observational study in 13,087 patients. Diabetologia, 2009, 52: 65. DOI: 10.1007/s00125-008-1190-x

[12] Cosentino F, Grant PJ, Aboyans V, et al. 2019 ESC Guidelines on diabetes, pre-diabetes, and cardiovascular diseases developed in collaboration with the EASD: The Task Force for diabetes, pre-diabetes, and cardiovascular diseases of the European Society of Cardiology (ESC) and the European Association for the Study of Diabetes (EASD). European Heart Journal, 2020 (41), 2: 255-323. DOI: 10.1093/eurheartj/ ehz486

[13] Rawshani A, Sattar N, Franzen S, Rawshani A, Hattersley AT, Svensson AM, Eliasson B, Gudbjornsdottir S. Excess mortality and cardiovascular

disease in young adults with type 1 diabetes in relation to age at onset: a nationwide, register-based cohort study. Lancet 2018;392:477-486.

[14] Haffner SM, Lehto S, Rönnemaa T, et al. Mortality from Coronary Heart Disease in Subjects with Type 2 Diabetes and in Nondiabetic Subjects with and without Prior Myocardial Infarction. N Engl J Med 1998; 339:229-234. DOI: 10.1056/NEJM199807233390404

[15] Garber AJ, Abrahamson MJ, Barzilay JI, et al. Consensus Statement by the American Association of Clinical Endocrinologists and American College of Endocrinology on the comprehensive type 2 diabetes management algorithm – 2018 Executive summary. Endocrine practice, 2018, 24(1). DOI: 10.4158/ CS-2018-0535

[16] Seferović PM, Petrie MC, Filippatos GS, et al. Type 2 diabetes mellitus and heart failure: a position statement from the Heart Failure Association of the European Society of Cardiology. Eur J Heart Fail, 2018, 20: 853-872. DOI:10.1002/ejhf.1170

[17] Stevens RJ, Kothari V, Adler AI, et al on behalf of the United Kingdom Prospective Diabetes Study (UKPDS) Group. The UKPDS Risk Engine: a model for the risk of coronary heart disease in type 2 diabetes Clinical Science 2001;101: 671-679. DOI:10.1042/cs1010671

[18] Kengne AP. The ADVANCE cardiovascular risk model and current strategies for cardiovascular disease risk evaluation in people with diabetes. Cardiovasc J Afr. 2013;24 (9-10):376-381. DOI: 10.5830/ CVJA-2013-078

[19] American Diabetes Association. Cardiovascular Disease and Risk Management: Standards of Medical Care in Diabetes, 2020. Diabetes Care 2020;43(Suppl. 1):S111–S134. DOI: 10.2337/dc20-s010

[20] Shao H, Fonseca V, Stoecker C, et al. Novel risk engine for diabetes progression and mortality in USA: Building, Relating, Assessing, and Validating Outcomes (BRAVO). Pharmacoeconomics, 2018;36:1125- 1134. DOI: 10.1007/s40273-019-00807-3

[21] Einarson TR, Acs A, Ludwig C, Panton UH. Prevalence of cardiovascular disease in type 2 diabetes: a systematic literature review of scientific evidence from across the world in 2007-2017. Cardiovasc Diabetol (2018) 17:83. DOI: 10.1186/ s12933-018-0728-6

[22] Shah AD, Langenberg C, Rapsomaniki E, et al. Type 2 diabetes and incidence of cardiovascular diseases:a cohort study in 1·9 million people. Lancet Diabetes Endocrinol. 2015; 3: 105-113. DOI: 10.1016/ S2213-8587(14)70219-0

[23] Succurro E, Fiorentino TV, Miceli S, et al. Relative Risk of cardiovascular Disease Is Higher in Women With Type 2 Diabetes, but Not in Those With Prediabetes, as Compared With Men. Diabetes Care, 2020, dc201401; DOI: 10.2337/dc20-1401

[24] Millett ERC, Peters SAE, Woodward M. Sex differences in risk factors for myocardial infarction: cohort study of UK Biobank participants. BMJ, 2018, 363:k4247. DOI:10.1136/bmj.k4247

[25] Morrish NJ, Wang SL, Stevens LK, et al. Mortality and causes of death in the WHO Multinational Study of Vascular Disease in Diabetes. Diabetologia. 2001;44 Suppl 2:S14-S21. DOI: 10.1007/pl00002934

[26] Low Wang CC, Hess CN, Hiatt W.R., et al. Clinical Update: Cardiovascular Disease in Diabetes Mellitus Atherosclerotic Cardiovascular Disease and Heart Failure in Type 2 Diabetes Mellitus – Mechanisms, Management, and Clinical

Considerations. Circulation. 2016; 133:2459. DOI: 10.1161/CIRCULATIONAHA.116.022194

[27] Maack C., Lehrke M., Backs J., et al. Heart failure and diabetes: metabolic alterations and therapeutic interventions: a state-of-the-art review from the Translational Research Committee of the Heart Failure Association–European Society of Cardiology. European Heart Journal, 2018; 39, 4243-4254. DOI: 10.1093/eurheartj/ehy596

[28] Kelly McHugh, DeVore A.D., Wu J. Heart Failure With Preserved Ejection Fraction and Diabetes. JACC State-of-the-Art Review. J Am Coll Cardiol. 2019;73:602-11. DOI: 10.1016/j.jacc.2018.11.033

[29] Wan EYF, Chin WY, Yu YET, et al. The Impact of Cardiovascular Disease and Chronic Kidney Disease on Life Expectancy and Direct Medical Cost in a 10-year Diabetes Cohort Study. Diabetes Care 2020 May; dc192137. DOI: 10.2337/figshare.12194913.

[30] Chronic Kidney Disease Prognosis Consortium, Matsushita K, van der Velde M, Astor BC et al. Association of estimated glomerular filtration rate and albuminuria with all-cause and cardiovascular mortality in general population cohorts: a collaborative meta-analysis. Lancet. 2010,12;375(9731): 2073-2081. DOI: 10.1016/S0140-6736(10)60674-5

[31] Kottgen A, Russell SD, Loehr LR, et al. Reduced Kidney Function as a Risk Factor for Incident Heart Failure: The Atherosclerosis Risk in Communities (ARIC) Study. JASN 1307-1315; DOI: 10.1681/ASN.200610 1159 Apr 2007, 18 (4)

[32] Connell AW, Sowers JR. The CardioRenal metabolic syndrome. J Am Soc Hypertens. 2014;8(8):604-606. DOI: 10.1016/j.jash.2014.07.003

[33] Thomas MC. Type 2 Diabetes and Heart Failure: Challenges and Solutions. Current Cardiology Reviews, 2016, 12,249 - 255. DOI: 10.2174/157 3403X1266616060 6120254

[34] Yeung SLA, Luo S, Schooling M. The Impact of Glycated Hemoglobin (HbA1c) on Cardiovascular Disease Risk: A Mendelian Randomization Study Using UK Biobank. Diabetes Care. 2018 Sep;41(9):1991-1997. DOI: 10.2337/dc18-0289

[35] Takao T, Matsuyama Y, Yanagisawa H, et al. Association between HbA1c variability and mortality in patients with type 2 diabetes. J Diabetes Complications, 2014;28:494-499. DOI: 10.1016/j.jdiacomp.2014.02.006

[36] Sun B., He F., Gao Y, et al. Prognostic impact of visit-to-visit glycemic variability on the risks of major adverse cardiovascular outcomes and hypoglycemia in patients with different glycemic control and type 2 diabetes. Endocrine, 2019, https://doi.org/10.1007/s12020-019-01893-1

[37] Zhou JJ, Schwenke DC, Bahn G, Reaven P. Glycemic Variation and Cardiovascular Risk in the Veteran's Affairs Diabetes Trial. Diabetes Care. 2018Oct;41(10):2187-2194.DOI:10.2337/dc18-0548

[38] Yokota S, Tanaka H, Mochizuki Y. et al. Association of glycemic variability with left ventricular diastolic function in type 2 diabetes mellitus. JACC, 2019 (73) 9:764. DOI:10.1186/s12933-019-0971-5

[39] Dunlay SM, Givertz MM, Aguilar D, et al. Type 2 Diabetes Mellitus and Heart Failure: A Scientific Statement From the American Heart Association and the Heart Failure Society of AmericaCirculation. 2019, 140:e294–e324. DOI:10.1161/ CIR.0000 000000000691

[40] Huo L, Magliano DJ, Rancière F, et al. Impact of age at diagnosis and duration of type 2 diabetes on mortality in Australia 1997-2011. Diabetologia, 2018, 61:1055-1063. DOI: 10.1007/s00125-018-4544-z

[41] Sattar N., Rawshani A, Franzén S, et al. Age at Diagnosis of Type 2 Diabetes Mellitus and Associations With Cardiovascular and Mortality Risks. Findings From the Swedish National Diabetes Registry. Circulation. 2019; 2019 May 7;139(19):2228-2237. DOI: 10.1161/CIRCULATIONAHA.118.037885

[42] Vazquez-Benitez G, Desai JR, Xu S et al. Preventable major cardiovascular events associated with uncontrolled glucose, blood pressure, and lipids and active smoking in adults with diabetes with and without cardiovascular disease: a contemporary analysis. Diabetes Care, 2015, 38(5): 905-912. DOI: 10.2337/dc14-1877

[43] Davis SN, Duckworth W, Emanuele N, et al. Effects of Severe Hypoglycemia on Cardiovascular Outcomes and Death in the Veterans Affairs Diabetes Trial. Diabetes Care 2019;42:157-163. DOI:10.2337/dc18-1144

[44] Gæde P, Lund-Andersen H, Parving H-H, Pedersen O. Effect of a Multifactorial Intervention on Mortality in Type 2 Diabetes. N Engl J Med. 2008; 358:580-591. DOI: 10.1056/NEJMoa0706245

[45] Gæde, P, Oellgaard J, Carstensen B, et al. Years of life gained by multifactorial intervention in patients with type 2 diabetes mellitus and microalbuminuria: 21 years follow-up on the Steno-2 randomised trial. Diabetologia, 2016, 59: 2298. DOI: 10.1007/s00125-016-4065-6

[46] Fan W, Song Y, Inzucchi SE, et al. Composite cardiovascular risk factor target achievement and its predictors in US adults with diabetes: The Diabetes Collaborative Registry. Diabetes Obes Metab, 2019;21:1121-1112. DOI: 10.1111/dom.13625

[47] Rawshani A, Rawshani A, Franzén S, et al. Risk Factors, Mortality, and Cardiovascular Outcomes in Patients with Type 2 Diabetes. N Engl J Med 2018; 379:633-644. DOI: 10.1056/NEJMoa1800256

[48] Mach F, Baigent C, Catapano AL, et al. 2019 ESC/EAS Guidelines for the management of dyslipidaemias: lipid modification to reduce cardiovascular risk. European Heart Journal (2020) 41, 111188. DOI:10.1093/eurheartj/ehz455

[49] American Diabetes Association. Glycemic Targets: Standards of Medical Care in Diabetes – 2021. Diabetes Care 2021;44(Suppl. 1):S73–S84. DOI: 10.2337/dc21-S006

[50] American Diabetes Association. Facilitating Behavior Change and Well-being to Improve Health Outcomes: Standards of Medical Care in Diabetes—2020. Diabetes Care 2020 Jan; 43(Supplement 1): S48-S65. DOI: 10.2337/dc20-S005

[51] Sun Y, You W, Almeida F, Estabrooks P, Davy B. The Effectiveness and Cost of Lifestyle Interventions Including Nutrition Education for Diabetes Prevention: A Systematic Review and Meta-Analysis. J Acad Nutr Diet. 2017 Mar;117(3):404-421.e36. doi: 10.1016/j.jand.2016.11.016. PMID: 28236962; PMCID: PMC5330213.

[52] Wing RR, Bolin P, Brancati FL, et al.; Look AHEAD Research Group. Cardiovascular effects of intensive lifestyle intervention in type 2 diabetes. N Engl J Med 2013;369:145-154. DOI: 10.1056/NEJMoa1212914

[53] American Diabetes Association. Pharmacologic Approaches to Glycemic Treatment: Standards of Medical Care in Diabetes—2020. Diabetes Care Jan

2020, 43 (Supplement 1) S98-S110; DOI: 10.2337/dc20-S009

[54] Marso SP., Daniels GH., Kirstine Brown-Frandsen, et al. Liraglutide and Cardiovascular Outcomes in Type 2 Diabetes. N Engl J Med. 2016, 28;375(4):311-22. DOI: 10.1056/NEJMoa1603827

[55] Marso SP., Bain SC., Consoli A., et al. Semaglutide and Cardiovascular Outcomes in Patients with Type 2 Diabetes. N Engl J Med, 2016; 375:1834- 1844. DOI: 10.1056/NEJMoa1607141

[56] Gerstein H.C., Colhoun HM., Dagenais GR., et al. Dulaglutide and cardiovascular outcomes in type 2 diabetes (REWIND): a double-blind, randomised placebo-controlled Trial. Lancet 2019; 394: 121-130. DOI: 10.1016/S0140-6736(19)31149-3

[57] Zinman B., Wanner C., Lachin JM., et al. Empagliflozin, Cardiovascular Outcomes, and Mortality in Type 2 Diabetes. N Engl J Med 2015; 373:2117- 2128. DOI: 10.1056/NEJMoa1504720

[58] Neal B., Perkovic V., Mahaffey KW., et al. Canagliflozin and Cardiovascular and Renal Events in Type 2 Diabetes. N Engl J Med 2017; 377: 644-657. DOI: 10.1056/NEJMoa1611925

[59] Wiviott SD., Raz I., Bonaca MP., et al. Dapagliflozin and Cardiovascular Outcomes in Type 2 Diabetes. N Engl J Med 2019; 380:347-357. DOI: 10.1056/NEJMoa1812389

[60] The ADVANCE Collaborative Group. Intensive Blood Glucose Control and Vascular Outc⊠⊠omes in Patients with Type 2 Diabetes. N Engl J Med 2008; 358:2560-2572. DOI: 10.1056/NEJMoa0802987

[61] Erdmann E., Dormandy J., Wilcox R., et al. PROactive 07: pioglitazone in the treatment of type 2 diabetes: results of the PROactive study. Vasc Health Risk Manag. 2007,3(4): 355-370. PMID: 17969365; PMCID: PMC2291341.

[62] Holman RR, Bethel MA, Mentz RJ, et al. Effects of Once-Weekly Exenatide on Cardiovascular Outcomes in Type 2 Diabetes. N Engl J Med. 2017, 28;377(13):1228-1239. DOI: 10.1056/NEJMoa1612917

[63] McMurray J.J.V., Solomon S.D., Inzucchi S.E. et al. Dapagliflozin in Patients with Heart Failure and Reduced Ejection Fraction. N Engl J Med 2019;381:1995-2008. DOI: 10.1056/NEJMoa1911303

[64] Packer M., Anker S.D., Butler J., et al. Cardiovascular and renal outcomes with empagliflozin in heart failure. N Engl J Med. Published online August 29, 2020.DOI: 10.1056/NEJMoa2022190

[65] Bhatt DL, Steg PG, Miller M, et al. Cardiovascular risk reduction with icosapent ethyl for hypertriglyce ridemia. N Engl J Med 2019;380:11-22. DOI: doi: 10.1056/NEJMoa1812792

[66] Connolly SJ, Eikelboom JW, Bosch J, et al. Rivaroxaban with or without aspirin in patients with stable coronary artery disease: an international, randomised, double-blind, placebo-controlled trial. Lancet2018;391: 205-218. DOI: 10.1016/S0140-6736(17)32458-3.

[67] Anand SS, Bosch J, Eikelboom JW, et al. Rivaroxaban with or without aspirin in patients with stable peripheral or carotid artery disease: an international, randomised, double-blind, placebo-controlled trial. Lancet 2018; 391:219-229. DOI: 10.1016/S0140 6736-(17)32409-1

[68] Fisher DP, Johnson E, Haneuse S, et al. Association Between Bariatric Surgery and Macrovascular Disease Outcomes in Patients With Type 2

Diabetes and Severe Obesity. JAMA. 2018;320(15):1570-1582.DOI:10.1001/jama.2018.14619

[69] American Diabetes Association. Obesity Management for the Treatment of Type 2 Diabetes: Standards of Medical Care in Diabetes 2020. Diabetes Care 2020;43(Suppl. 1):S89–S97. DOI:10.2337/dc20-S008

13

Hypothyroidism Therapy

Wissal Abassi, Nejmeddine Ouerghi and Anissa Bouassida

Abstract

Hypothyroidism refers to the common pathological disorder of thyroid hormone deficiency. The successful therapy for hypothyroidism is levothyroxine (LT4) administration, which is the same as thyroxine but produced synthetically. Serum thyrotropin (TSH) normalization with LT4 replacement therapy in hypothyroidism is generally needed to restore a euthyroid state. The daily dose of thyroxine therapy depends on various factors, such as body weight, age, and severity. It also differs from hypothyroidism during pregnancy to congenital hypothyroidism. The pres-ence of various comorbidities may exist such as myxoedema coma, coronary artery disease, obesity, anemia and COVID-19 which necessitate individualized treatment. LT4 intolerance manifested with sympathetic hyperactivity may appear during the first hours after the LT4 administration. It requires starting with very low doses of LT4 that should be increased gradually, and reaching normal TSH may take several months. The sympathetic hyperactivity may be attributable to the presence of uncorrected iron-deficiency anemia that worsens by the use of thyroid hormone.
Keywords: Levothyroxine, dose, hypothyroidism, anemia, myxoedema coma, treatement

1. Introduction

Hypothyroidism refers to the common disorder of thyroid hormone deficiency. According to the time of onset, it may be present at birth (congenital) or develop later in life (acquired). Treatment of hypothyroidism consists of replacing the deficiency of thyroid hormone to improve symptoms and prevent the adverse consequences of hypothyroidism. Levothyroxine (LT4) that is identical to thyroxine but synthetically produced, is the most effective way to treat hypothyroidism. Treating hypothyroidism strategies are based on adopting an adequate dose of LT4, selecting a patient-appropriate serum thyrotropin (TSH) goal, and ensuring maintenance of that desired goal [1, 2]. The serum TSH normalization with LT4 replacement is gen-erally needed to restore a euthyroid state. The daily LT4 dose must be individualized taking into account various factors such as body weight, age and severity of disease. Special situations may be coexisting requiring special treatement [2]. The purpose of this chapter is to provide the different therapeutic strategies of hypothyroidism management in various circumstances.

2. Congenital hypothyroidism

The guidelines of the American Academy of Pediatrics [3] and the European Society of Pediatric Endocrinology [4] recommends keeping TSH values below 5 μU/dl (optimally 0.5–2.0 μU/dl) for the first 3 years of life. LT4 is the treatment for hypothyroidism in children. Infants with suspected hypothyroidism require rapid evaluation and treatment, ideally within the first 2 weeks of life since delays in treatment of congenital hypothyroidism are correlated with poorer outcomes. Congenital hypothyroidism treatment should be initiated with 10–15 lg/kg of LT4 daily. Early starting treatment and adequate LT4 initial doses are associated with rapid normalization of thyroid function and improved neurodevelopmental outcomes [5].

Tablets might be crushed, suspended in a small amount of water, breast milk, or non–soy-based formula, and administered orally via a syringe. LT4 should not be administered with substances that may interfere with its absorption such as multivitamins containing calcium or iron, aluminum hydroxide and soy formula. The medication is administered once a day. The same dose should be repeated if there is immediate vomiting. If a dose is missed, it may be doubled the following day [6].

Serum thyroid function testing should be checked every 1–2 weeks until normal, and then every 1–2 months during the first year of life and every 2–4 months during the 2nd and 3rd years of life with an appropriate adjustment of LT4 dose [7]. The needed dose of LT4 to restore euthyroidism varied with the age and the hypothyroidism severity. The dose required per kilogram body mass to fully replace thyroid function is significantly higher in children than in adults and decreases with age (**Table 1**). Early and adequate prenatal treatment with LT4 prevents severe neurocognitive deficits and results in normal brain development [1].

3. Acquired hypothyroidism

The approach to treat acquired hypothyroidism is similar to that of congenital, consisting of replacing the deficiency of thyroid hormones to improve symptoms and prevent the adverse consequences of hypothyroidism. The target is to keep serum TSH <5 mIU/L (optimally 0.5–2 mIU/L) [8].

The recommended treatment is LT4 in tablets form administered once daily. Adherence to LT4 may be influenced by multiple factors such as the time of day when the medication is administered. Although the majority of experts recommend administration of LT4 prior to food consumption, but bedtime administration may be more convenient and does not appear to be associated with worse control of hypothyroidism in children and adolescents [9].

The dose of LT4 required to restore euthyroidism depends on patient age, body surface area, weight and severity of hypothyroidism (**Table 1**) [1]. The dose required to fully replace thyroid function is significantly higher in children than in

Age	LT4 dose (lg/kg/day)
0–3 months	10–15
3–12 months	6–10
1–3 years	4–6
3–10 years	3–5
10–16 years	2–4
16 years	1.6

Modified from reference [1].

Table 1.
Levothyroxine (LT4) weight-based dose in pediatric hypothyroidism.

adults and decreases with age. Thyroid function should be controlled 6 to 8 weeks after initiating therapy and then every 4 to 6 months until the child achieves final height or every 6 to 8 weeks following a change in LT4 dose [10].

4. Subclinical hypothyroidism

Subclinical hypothyroidism (SCH) is considered in two categories according to the elevation in serum TSH level: mildly increased TSH levels (4.0–10.0 mU/l), and more severely increased TSH value (>10 mU/l) [11]. For the American Thyroid Association recommendation, the goal of therapy is to normalize TSH levels within 0.5–3.5 mIU/L. The starting thyroid hormone treatment is recommended when the serum TSH is above 10 mIU/L. The treatment is generally not recommended when the TSH is 5–10 mIU/L except in patients with increased cardiovascular risk.

These guidelines suggest that LT4 is the treatment of choice with a daily dose of 25–75 mcg, depending on the degree of TSH elevation and to be adjusted based on clinical response and serum thyroid function test monitoring. Addition with liothyronine (T3) is not routinely recommended [1, 12].

For the European Thyroid Association, even in the absence of symptoms, replacement therapy with LT4 is recommended for younger patients (<65–70 years) who have serum TSH levels greater than 10 mIU/L and in younger SCH patients (serum TSH <10 mU/l) with symptoms suggestive of hypothyroidism. For older individuals (>80–85 years) with elevated serum TSH ≤ 10 mU/l, observa-tion without hormonal treatment should be the strategy of choice. The aim is to reach a stable serum TSH in the lower half of the reference range (0.4–2.5 mU/l). A daily dose of LT4 between 50 and 100 µg for most SCH patients and between 25 and 50 mcg for those with cardiac disease and/or older age has been advised to normalize serum TSH. The use of T3 or combined LT4/T3 is no supported in the treatment [11].

According to the Latin American Thyroid Society, LT4 is started in individuals with a serum TSH greater than 10 mIU/L and considered to start in those with a serum TSH 4.5–10 mIU/L especially greater than 7 mIU/L, who are younger than 65 years with increased cardiovascular risk. Patients with persistent mild elevations in TSH with positive serum thyroid peroxidase (TPO) antibodies and ultrasound findings consistent with autoimmune thyroiditis could be considered for treatment. These guidelines recommends against treatment of elderly (65–80 years) patients with TSH levels less than 10 mIU/L [13].

The British Thyroid Association recommends starting treatment in patients with positive antibodies or serum TSH >10 mU/L. The guideline advise to start with LT4 doses between 50 and 125 mcg/day. Treatment should aim for TSH levels within the normal reference range. Older individuals or with cardiac disease, require smaller starting doses to avoid inducing cardiac ischaemia. Combined therapy with LT4 and liothyronine (LT3) is not routinely recommended [14].

The American Association of Clinical Endocrinology recommends starting treatment with LT4 in patients with TSH > 10 mIU / L or those with symptoms, cardiovascular risk factors or positive TPO antibody. The goal of LT4 therapy is to normalize TSH. The starting dose of LT4 is 1.5 mcg/kg in the absence of cardiovascular disease. In patients with cardiovascular disease, the starting dose of LT4 is 25 mcg and is up-titrated as needed [12]. The American Endocrine Society recommends against routine treatment of SCH patients with serum TSH levels of 4.5–10 mU/liter, but indicated that treatment was reasonable for patients with TSH levels greater than 10 mU/liter [15].

The American Association of Clinical Endocrinology recommends starting treatment with LT4 in patients with TSH > 10 mIU / L or those with symptoms, cardiovascular risk factors or positive TPO antibody. The goal of LT4 therapy is to normalize TSH. The starting dose of LT4 is 1.5 mcg/kg in the absence of cardiovascular disease. In patients with cardiovascular disease, the starting dose of LT4 is 25 mcg and is up-titrated as needed [12].

5. Myxoedema coma

Myxedema coma is the most extreme, life-threatening expression of severe hypothyroidism. Usual precipitating factors are discontinuation of therapy, infections and exposure to cold. Its clinical manifestations include hypothermia, respira-tory depression, cardiovascular instability and altered mental status [1]. According to the American Thyroid Association guidelines, patients with myxedema coma should first receive empiric glucocorticoid coverage with intravenous glucocorti-coid administration (at appropriate doses for the stressed state). These guidelines recommends a daily intravenous dose of 200 to 400 mcg of LT4, with lower doses given for patients who are of smaller stature, older, or who have a history of coro-nary disease or arrhythmia. Subsequently, daily dosing of 1.6 mcg/kg/day reduced to 75%, should be given.

Additional administration of intravenous T3 in a loading dose of 5–20 μg, followed by a maintenance dose of 2.5 to 10 mcg every 8 hours is recommended [1, 16–18]. For the Latin American Thyroid Society guidelines, LT4 replacement is the backbone of myxedema coma therapy. The daily starting dose is 300–500 μg/day followed by maintenance dose 50–100 μg/day. T3 can be added to therapy at 10–20 mg bolus followed by 10 μg every 4–6 hours [19]. The necessity of intravenous LT4 replacement is apparent, but administering LT4 alone or with a combination of LT4 and LT3 remains controversial [1, 16–18]. Carter et al. [20] prouved that combinated treatment with LT3 and LT4 are more beneficial than LT4 monotherapy to improve symptoms and to reverse the biochemical abnormalities in patients with myxedema coma. At the same Ono et al. [21] showed that patients receiving LT4 combined with LT3 had a lower mortality rate than those who received LT4 alone. T3 is an active hormone in the body with an immediate onset of action, whereas LT4 has a slow onset of action with relatively few adverse events. Also T3 affinity for the nuclear receptor is 10- to 20- fold higher than that of T4 and it reaches a peak level in 2–4 h after administration [22]. The beneficial effect of this combined treatment on neuropsychiatric symptoms has been confirmed because of the capacity of T3 to cross the blood brain barrier [23].

6. Pregnancy

6.1 Treatment of overt hypothyroidism in pregnancy

Women with overt hypothyroidism who contemplate pregnancy must first normalize TSH and thyroid hormone levels. LT4 is the first choice of treatment. American Thyroid Association guidelines specifies a goal of TSH less than 2·5 mIU/L in women planning to become pregnant [24]. Women with a previous diagnosis of hypothyroidism, should checked their serum TSH concentration and increased LT4 dose made as soon as possible after pregnancy is confirmed [25]. The LT4 replacement dose may be as high as 2.0–2.4 μ g/kg body weight per day, which is 25–50% higher than levels used in the general population (1.6–1.8 μg/kg body weight) [26].

Once pregnant, women should increased the dose of LT4 by about 30–50%, beginning in the first 4–8 weeks of gestation and gradually increasing the dose through 16–20 weeks of gestation [27]. After initiation of therapy, thyroid function should be retested within 30–40 days, and then every 4–6 weeks. Maternal TSH concentrations should be maintained between 0·1–2·5 mIU/L in the first trimester, 0·2–3·0 mIU/L in the second, and 0·3–3·0 mIU/L in the third [27, 28].

6.2 Treatment of SCH in pregnancy

According to the American Thyroid Association guidelines, TSH levels should be keept below 2.5 mU/L and 3.0 mU/L in the first and second/third trimesters, respectively [28]. Autoantibodies to TPO should be mesured in women with SCH, since TPO positivity acts in synergy with SCH and it is an independent risk factor for pregnancy outcomes [29–31]. The rise of maternal TSH level increase the risk of miscarriage. This risk becomes even greater at a TSH level > 2.5 mU/L, accompanied by TPO posi-tivity [24]. Pregnant women with SCH have a higher risk of several side effects such as pregnancy loss and premature birth [32]. In women receiving LT4 treatment for SCH, the risk of miscarriage in early pregnancy increases at a TSH level > 4.5 mU/L.

It was showed that LT4 treatment is capable to reduce the rate of pregnancy loss in TPO positive euthyroid women, therefore starting treatment with low doses of LT4 may be considered in TPO positive women with a history of recurrent pregnancy loss [33, 34]. Starting doses of LT4 were based on TSH levels, either 1 μg/kg/day (TSH 1.5–2.5 mU/L) or 0.5 μg/kg/day (TSH 0.5–1.5 mU/L), raising or lowering amounts by 0.5 μg/day at TSH levels >3 mU/L or < 0.5 mU/L, respectively [34].

7. Coronary artery disease

If untreated, hypothyroidism in patients with coronary heart disease might lead to ventricular arrhythmias, congestive heart failure or acute myocardial infarction.

Therefore, patients with coronary artery disease and hypothyroidism should be treated cautiously with thyroid hormone replacement [35]. Therapy is performed with the administration of LT4,that induced significant improvement in myocardial function, cardiac pump function and resting cardiac output [36]. However, in subjects with an already compromised myocardial blood supply due to coronary atherosclerosis, LT4 treatment may lead to anginal symptoms. A previous study showed that 2% among 1503 hypothyroid subjects presented new onset angina dur-ing LT4 therapy [37]. The recommended initial therapeutic dose for patients with pre-existing angina is 25 mg / day or even less and it should not be increased faster than at 4 week intervals [38].

8. Elderly

The serum TSH levels reference range is 4–5 mIU/L in the elderly patients. The hormone replacement with LT4 is the specific therapy for old individuals with hypothyroidism [39]. A normal starting dose of LT4 in an elderly subject is around 1 mg/kg/day, which is maintained for 4–6 weeks [40]. Dosage adjustments are guided by the TSH response and the clinical state, with emphasis on possible cardiac adverse effects [38]. Some experts recommended to increase the dosage by 25 μg in accordance with the patient's complaints [41]. While this therapeutic dose should be reduced of about 50 μg/day, or 25 μg/day when the patient was underweight or developed cardiovascular disease [39].

Hormone replacement with LT4 in patients with hypothyroidism requires specific therapy. LT4 is available in the form of tablets at various doses, but also in liquid form as an oral solution, thus enabling a customized dose titration [41]. The European Thyroid Association recommends treatment in patients above 70 years old who have a TSH >10 mUI/L and signs and/or symptoms of hypothyroidism [11]. However, the LT4 dose needs to be tailored, aiming for a TSH level between 4 and 10 mIU/L, while the patient's health condition and the potential presence of dyslipidemia and other metabolic derangements should be considered [42].

Likewise, van den Beld et al. [43] showed that low serum FT4 levels are associated with a longer 4-year survival, reflecting a possible adaptive mechanism to prevent excessive catabolism in the elderly. Hypothyroidism is prevalent among the elderly population, that's why efforts should be made to maintain optimal thyroid function with individualized treatment based on severity and comorbidities [43].

9. COVID-19

Hypothyroidism is not a risk factor associated with worse outcomes in COVID-19 positive patients, therefore no particular changes or consultations are envisaged relating to the diagnosis and treatment of hypothyroidism during the COVID-19 crisis [44]. It is recommended to pursue the same form, frequence and dosage of thyroid hormone replacement therapy. No particular therapeutic dosage was addressed. Regular blood test monitoring may be difficult, but when patients on thyroid hormone replacement feel significantly unwell or if there are significant weight changes, thyroid function testing with measurement of serum TSH and free thyroxine, is recommended to adjust medication if needed [45].

10. Obesity

LT4 is the treatment of patients developed SCH associated with hyperlipidae. T3 can be added for patients with adequately substituted hypothyroidism with obesity resistant to a lifestyle intervention or suspected of thyroid hormone conversion disorders [46].

An individual approach of treatment is necessary for obese patients with higher TSH levels. The clinical condition of elevated TSH without symptoms, thyroid antibodies, goiter, or associated thyroidal illness, was defined as obesity-associated hyperthyrotropinemia. The efficacy of thyroid hormone therapy in obese subjects with hyperthyrotropinemia has not yet been evaluated [47]. LT4 treatment of obese with hyperthyrotropinemia were performed only in children demonstrating that the increase TSH levels causes impaired glucose metabolism and dyslipidemia [48].

However, no evidence regarding a favorable effect of LT4 treatment on body weight in obese subjects with TSH < 10 mU/L and normal free thyroxine level has been demonstrated, so such treatment is not recommended. In such subjects dietary-behavioral intervention contributed to weight loss irrespective of LT4 use. However, in normal-weight patients with hypothyroidism LT4 treatment is an effec-tive strategy to decrease total and LDL cholesterol, this effect is more pronounced in patients with TSH > 10 mU/L [11]. Obese individuals with isolated hyperthyrotro-pinemia without symptoms or other signs of thyroid disease should not be treated with thyroid hormone replacement [47].

The role of thyroid hormone in treating obesity was confirmed. So that T3 and T4 administration at varying doses and durations has been shown to enhance weight loss in obese euthyroid subjects [49]. Such therapy increased fat loss without decreasing skeletal muscle mass and strength or inducing cardiac dys-function [50]. However, a systematic review by Kaptein et al. [51] describing the effects of T3 or T3/T4 treatment on weight loss in euthyroid individuals during caloric deprivation showed that T3 is associated with decreases in serum T4 concentrations, indicating pituitary suppression of TSH, resulting in subclinical hyperthyroidism.

11. Adrenal insufficiency

Glucocorticoid deficiency can accompany hypothyroidism and if it is not detected, it may be exacerbated by thyroid hormone replacement, raising the risk of adrenal crisis that can be fatal. The signs and symptoms of coexisting adrenal insufficiency are manifold and can be summarized as hypoglycemia, anemia, nausea, vomiting, alopecia areata, weight loss, abdominal pain, hypo-tension and hyperpigmentation [52]. If so, it is important to start with replacing glucocorticoids first before starting LT4 and then approaching LT4 replacement 5–7 days later. Starting by LT4 therapy can contribuate to an increased metabo-lism which raise demand of cortisol, leading then to an increased risk of an adrenal crisis [38, 53].

Not to be confused, the modest increase of serum TSH levels in cases of untreated adrenal insufficiency, might return to normal without for thyroid hormone replacement but only by the correct hydrocortisone replacement treatment. The diurnal variation in TSH levels is influenced by cortisol, it manifests higher concentrations at night and lower in the morning, which explains the increase in TSH levels in untreated adrenal insufficiency [54].

12. Differentiated thyroid cancer

Differentiated thyroid cancer should be treated with an appropriate expertise in order to ensure an optimal long-term treatment quality. The therapeutic approach is individualized and risk-adapted. Surgery and radioiodine therapy followed by LT4 substitution is the established therapeutic procedures. For widely invasive follicular thyroid carcinomas and follicular thyroid carcinoma with vascular infiltration, thyreoidectomy is recommended. Radioiodine therapy has been established for more than 60 years, consisted on systemic administration of radioiodine (I-131) to irradiate thyroid remnants as well as non-resectable or incompletely resected differentiated thyroid cancer.

The benefit of I-131 therapy was confirmed in individuals with differentiated thyroid cancer at high risk for recurrence, however in subjects with very low-risk differentiated thyroid cancer the positive effect of radioiodine on tumor-free and overall survival has not been proven [55]. Patients who undergone thyroidectomy for differentiated thyroid cancer, with or without additional treatment with I-131, need to take LT4 not only for treatment of hypothyroidism but also to minimize potential TSH stimulation of tumor growth [56]. The recommended levels of thyroid-stimulating hormone (mIU/L) targets in patients with differentiated thyroid cancer by American Thyroid Association are;<0.1 for high and intermediate risk (initial therapy) and persistent disease, 0.1–0.5 for high-risk disease free (follow-up) and low risk (initial therapy) and 0.3–2.0 for low-risk disease free (follow-up) [57].

13. Anemia

Anemia seems to be associated with thyroid dysfunction particularly hypothyroidism. Anemia was found in 65% of children and adolescents with hypothyroidism [58]. At the same, 10% of the individuals with thyroiditis manifested anemia. These patients may be particularly sensitive at the beginning of LT4 replacement therapy [59]. Lack of stimulation of erythroid colony development by thyroid hormones, reduction in oxygen distribution to tissues and diminution of erythropoietin level in the absence of thyroid hormones leads to anemia [60]. The correction of hypothyroidism promotes an adequate therapeutic response to iron salts. Iron may interfere with the absorption of the thyroid hormone, that's why LT4 should be taken at least four hours apart from the iron intake [52].

14. LT4 intolerance

A sympathetic hyperactivity may appear during the first hours after the LT4 administration manifested by precordialgia or palpitations. This condition requires starting with very low doses of LT4 that should be increased gradually, and reaching normal TSH may took several months. The sympathetic hyperactivity may be attributable to the presence of uncorrected iron-deficiency anemia that worsens by the use of thyroid hormone. The anemia should be corrected and thyroid hormone therapy should be stoped temporarily and then restarted at low doses. Occasionally, beta blockers can be used in these cases to control the symptoms during the first few weeks of thyroid replacement treatment [59, 61].

15. Conclusion

Hypothyroidism is a common disorder, diagnosed by the measurement of blood levels of thyroid hormones. In the face of this challenge, efforts should be made to maintain optimal thyroid function. Therapy of choise is the administration of LT4. Treatment should be individualized in accordance with the subject's age, weight, severity and comorbidities. Special situations can coexist such as adrenal insufficiency, pregnant women, elderly patients and patients with differentiated thyroid cancer, needing special considerations to ensure the patient's welfare and prevent therapeutic complications. Under- or over-treatment is common in clinical practice and should be avoided and lifelong follow-up is strongly indicated.

Author details

Wissal Abassi[1,2]*, Nejmeddine Ouerghi[1,3] and Anissa Bouassida[1]

1 Research Unit, Sportive Performance and Physical Rehabilitation, High Institute of Sports and Physical Education of Kef, University of Jendouba, El Kef, Tunisia

2 Higher Institute of Sport and Physical Education of Ksar Said, University of "La Manouba", Manouba, Tunisia

3 University of Tunis El Manar, Faculty of Medicine of Tunis, Rabta Hospital, Tunis, Tunisia

*Address all correspondence to: wissalabassi93@gmail.com

References

[1] Jonklaas J, Bianco AC, Bauer AJ, Burman KD, Cappola AR, Celi FS, Cooper DS, Kim BW, Peeters RP, Rosenthal MS, Sawka AM. American Thyroid Association Task Force on Thyroid Hormone Replacement. Guidelines for the treatment of hypothyroidism: prepared by the american thyroid association task force on thyroid hormone replacement. Thyroid. 2014;24:1670-1751. DOI: 10.1089/thy.2014.0028.

[2] Jonklaas J. Update on the treatment of hypothyroidism. CurrOpin Oncol. 2016;28:18-25. DOI: 10.1097/CCO.0000000000000242.

[3] American Academy of Pediatrics, Rose SR; Section on Endocrinology and Committee on Genetics, Ameri-can Thyroid Association, Brown RS; Public Health Committee, Lawson Wilkins Pediatric Endocrine Society, Foley T, Kaplowitz PB, Kaye CI, Sundararajan S, Varma SK. Update of newborn screening and therapy for congenital hypothyroidism. Pediatrics. 2006 ;117: 2290-2303. DOI: 10.1542/peds.2006-0915.

[4] European Society for Pediatric Endocrinology. Revised guidelines for neonatal screening programmes for primary congenital hypothyroidism. Horm Res. 1999;52:49-52. DOI: 10.1159/000023433.

[5] Rose SR, Brown RS, Foley T, Kaplowitz PB, Kaye CI, Sundararajan S, Varma SK. Update of newborn screening and therapy for congenital hypothyroidism. Pediatrics.2006;117: 2290-22303. DOI: 10.1542/peds.2006-0915.

[6] Diaz A, Lipman Diaz EG. Hypothyroidism. Pediatr Rev. 2014; 35:336-337. DOI: 10.1542/pir.35-8-336.

[7] Léger J, Olivieri A, Donaldson M, Torresani T, Krude H, van Vliet G, Polak M, Butler G. Congenital Hypothyroidism Consensus Conference Group. European Society for Paediatric Endocrinology consensus guidelines on screening, diagnosis, and management of congenital hypothyroidism. Horm Res Paediatr. 2014;81:80-103. DOI: 10.1159/000358198.

[8] LaFranchi S. Acquired hypothyroidism in childhood and adolescence. In: Post TW, editor. UpToDate. Waltham. 2019.

[9] Akin O. Morning vs. bedtime levothyroxine administration: what is the ideal choice for children? J Pediatr Endocrinol Metab. 2018 ;31 : 1249-1255. DOI: 10.1515/jpem-2018-0168.

[10] Hanley P, Lord K, Bauer AJ. Thyroid Disorders in Children and Adolescents: A Review. JAMA Pediatr. 2016;170:1008- 1019. DOI: 10.1001/jamapediatrics.2016.0486.

[11] Pearce SH, Brabant G, Duntas LH, Monzani F, Peeters RP, Razvi S, Wemeau JL. 2013 ETA Guideline: Management of Subclinical Hypothyroidism. Eur Thyroid J. 2013 ; 2:215-228. DOI: 10.1159/000356507.

[12] Garber JR, Cobin RH, Gharib H, Hennessey JV, Klein I, Mechanick JI, Pessah-Pollack R, Singer PA, Woeber KA. Clinical practice guidelines for hypothyroidism in adults: cosponsored by the American Association of Clinical Endocrinologists and the American Thyroid Association. EndocrPract. 2012; 18:988-1028. DOI: 10.4158/EP12280.GL.

[13] Brenta G, Vaisman M, Sgarbi JA, Bergoglio LM, Andrada NC, Bravo PP, Orlandi AM, Graf H; Task Force on Hypothyroidism of the Latin American Thyroid Society (LATS). Clinical

practice guidelines for the management of hypothyroidism. Arq Bras Endocrinol Metabol. 2013;57:265-291. DOI: 10.1590/s0004-27302013000400003.

[14] Parretti H, Okosieme O, Vanderpump M. Current recommendations in the management of hypothyroidism: developed from a statement by the British Thyroid Association Executive. Br J Gen Pract. 2016;66: 538-540.DOI:10.3399/bjgp16X687493.

[15] Gharib H, Tuttle RM, Baskin HJ, Fish LH, Singer PA, McDermott MT. Subclinical thyroid dysfunction: a joint statement on management from the American Association of Clinical Endocrinologists, the American Thyroid Association, and the Endocrine Society. J Clin Endocrinol Metab. 2005;90:581-585. DOI: 10.1210/jc.2004-1231.

[16] Dubbs SB, Spangler R. Hypothyroidism: causes, killers, and life-saving treatments. Emerg Med Clin North Am.2014; 32: 303-317. DOI: 10.1016/j.emc.2013.12.003.

[17] Wartofsky L. Myxedema coma. Endocrinol Metab Clin North Am.2006;35: 687-698. DOI: 10.1016/j.ecl.2006.09.003.

[18] Mathew V, Misgar RA, Ghosh S, Mukhopadhyay P, Roychowdhury P, Pandit K, Mukhopadhyay S, Chowdhury S. Myxedema coma: a new look into an old crisis. J Thyroid Res. 2011:493462. DOI: 10.4061/2011/493462.

[19] Dutta P, Bhansali A, Masoodi SR, Bhadada S, Sharma N, Rajput R. Predictors of outcome in myxoedema coma: a study from a tertiary care centre. Crit Care. 2008;12:R1. DOI: 10.1186/cc6211.

[20] Carter JN, Eastmen CJ, Corcoran JM, Lazarus L. Inhibition of conversion of thyroxine to triiodothyronine in patients with severe chronic illness. Clin Endocrinol (Oxf). 1976 ; 5: 587-594. DOI: 10.1111/j.1365-2265.1976.tb03861.x.

[21] Ono Y, Ono S, Yasunaga H, Matsui H, Fushimi K, Tanaka Y.Clinical characteristics and outcomes of myxedema coma: analysis of a national inpatient database in Japan. J Epidemiol.2017 ; 27: 117-122. DOI: 10.1016/j.je.2016.04.002.

[22] Biondi B, Wartofsky L. Combination treatment with T4 and T3: toward personalized replacement therapy in hypothyroidism? J Clin Endocrinol Metab.2012 ; 97: 2256– 2271. DOI: 10.1210/jc.2011-3399.

[23] Chernow B, Burman KD, Johnson DL, McGuire RA, O'Brian JT, Wartofsky L, Georges LP.T3 may be a better agent than T4 in the critically ill hypothyroid patient: evaluation of trans- port across the blood-brain barrier in a primate model. Crit Care Med.1983 ; 11: 99-104. DOI: 10.1097/00003246-198302000-00009.

[24] Alexander EK, Pearce EN, Brent GA, Brown RS, Chen H, Dosiou C, Grobman WA, Laurberg P, Lazarus JH, Mandel SJ, Peeters RP, Sullivan S. Guidelines of the American Thyroid Association for the Diagnosis and Management of Thyroid Disease During Pregnancy and the Postpartum. Thyroid. 2017;27:315-389. DOI: 10.1089/thy.2016.0457.

[25] Alexander EK, Marqusee E, Lawrence J, Jarolim P, Fischer GA, Larsen PR. Timing and magnitude of increases in levothyroxine requirements during pregnancy in women with hypothyroidism. N Engl J Med. 2004;351:241-249. DOI: 10.1056/NEJMoa040079.

[26] Abalovich M, Vázquez A, Alcaraz G, Kitaigrodsky A, Szuman G, Calabrese C, Astarita G, Frydman M, Gutiérrez S.

Adequate levothyroxine doses for the treatment of hypothyroidism newly discovered during pregnancy. Thyroid. 2013 ;23:1479-1483. DOI: 10.1089/thy.2013.0024.

[27] De Groot L, Abalovich M, Alexander EK, Amino N, Barbour L, Cobin RH, Eastman CJ, Lazarus JH, Luton D, Mandel SJ, Mestman J, Rovet J, Sullivan S. Management of thyroid dysfunction during pregnancy and postpartum: an Endocrine Society clinical practice guideline. J Clin Endocrinol Metab. 2012;97:2543-2565. DOI: 10.1210/jc.2011-2803.

[28] Stagnaro-Green A, Abalovich M, Alexander E, Azizi F, Mestman J, Negro R, Nixon A, Pearce EN, Soldin OP, Sullivan S, Wiersinga W; American Thyroid Association Taskforce on Thyroid Disease During Pregnancy and Postpartum. Guidelines of the American Thyroid Association for the diagnosis and management of thyroid disease during pregnancy and postpartum. Thyroid. 2011;21:1081-1125. DOI: 10.1089/thy.2011.0087.

[29] Karakosta P, Alegakis D, Georgiou V, Roumeliotaki T, Fthenou E, Vassilaki M, Boumpas D, Castanas E, Kogevinas M, Chatzi L. Thyroid dysfunction and autoantibodies in early pregnancy are associated with increased risk of gestational diabetes and adverse birth outcomes. J Clin Endocrinol Metab. 2012;97:4464-4472. DOI:10.1210/jc.2012-2540.

[30] Liu H, Shan Z, Li C, Mao J, Xie X, Wang W, Fan C, Wang H, Zhang H, Han C, Wang X, Liu X, Fan Y, Bao S, Teng W. Maternal subclinical hypothyroidism, thyroid autoimmunity, and the risk of miscarriage: a prospective cohort study. Thyroid. 2014 ;24:1642-1649. DOI: 10.1089/thy.2014.0029.

[31] Ying H, Tang YP, Bao YR, Su XJ, Cai X, Li YH, Wang DF. Maternal TSH level and TPOAb status in early pregnancy and their relationship to the risk of gestational diabetes mellitus. Endocrine. 2016;54:742-750. DOI: 10.1007/s12020-016-1022-6.

[32] Maraka S, Ospina NM, O'Keeffe DT, Espinosa De Ycaza AE, Gionfriddo MR, Erwin PJ, Coddington CC 3rd, Stan MN, Murad MH, Montori VM. Subclinical Hypothyroidism in Pregnancy: A Systematic Review and Meta-Analysis. Thyroid. 2016;26:580-590. DOI: 10.1089/thy.2015.0418.

[33] Lepoutre T, Debieve F, Gruson D, Daumerie C. Reduction of miscarriages through universal screening and treatment of thyroid autoimmune diseases. Gynecologic and obstetric investigation.2012;74:265-273. DOI:10.1159/000343759.

[34] Negro R, Formoso G, Mangieri T, Pezzarossa A, Dazzi D, Hassan H. Levothyroxine treatment in euthyroid pregnant women with autoimmune thyroid disease: effects on obstetrical complications. J Clin Endocrinol Metab. 2006 ;91:2587-2591. DOI: 10.1210/jc.2005-1603.

[35] Aronow WS. The heart and thyroid disease.ClinGeriatrMed1995;11:219-229.

[36] Mourouzis I, Forini F, Pantos C, Iervasi G. Thyroid hormone and cardiac disease: from basic concepts to clinical application. J Thyroid Res. 2011;2011: 958626. DOI: 10.4061/2011/958626.

[37] Keating FR, Parkin TW, Selby JB, Dickinson LS. Treatment of heart disease associated with myxedema. Prog Cardiovasc Dis. 1961; 3: 364-381. DOI: 10.1016/s0033-0620(61)90004-4

[38] Khandelwal D, Tandon N. Overt and subclinical hypothyroidism: who to treat and how. Drugs. 2012;72:17-33. DOI: 10.2165/11598070- 000000000-00000.

[39] Cooper DS. Thyroid disease in the oldest old: the exception to the rule. JAMA. 2004;292:2651-2654. DOI: 10.1001/jama.292.21.2651.

[40] Laurberg P, Andersen S, Bülow Pedersen I, Carlé A. Hypothyroidism in the elderly: pathophysiology, diagnosis and treatment. Drugs Aging. 2005;22:23-38. DOI:10.2165/00002512-200522010-00002

[41] Duntas LH, Yen PM. Diagnosis and treatment of hypothyroidism in the elderly. Endocrine. 2019;66:63-69. DOI: 10.1007/s12020-019-02067-9.

[42] Jonklaas J, Razvi S. Reference intervals in the diagnosis of thyroid dysfunction: treating patients not numbers. Lancet Diabetes Endocrinol. 2019;7:473-483. DOI: 10.1016/S2213-8587(18)30371-1.

[43] Van den Beld AW, Visser TJ, Feelders RA, Grobbee DE, Lamberts SW. Thyroid hormone concentrations, disease, physical function, and mortality in elderly men. J Clin Endocrinol Metab. 2005;90: 6403-6409. DOI: 10.1210/jc.2005-0872.

[44] Chaker L, Bianco AC, Jonklaas J, Peeters RP. Hypothyroidism. Lancet. 2017;390:1550-1562. DOI: 10.1016/S0140-6736(17)30703-1.

[45] Boelaert K, Visser WE, Taylor PN, Moran C, Léger J, Persani L. ENDOCRINOLOGY IN THE TIME OF COVID-19: Management of hyperthyroidism and hypothyroidism. Eur J Endocrinol. 2020;183:G33-G39. DOI: 10.1530/EJE-20-0445.

[46] Krotkiewski M. Thyroid hormones in the pathogenesis and treatment of obesity. Eur J Pharmacol. 2002; 440: 85-98. DOI: 10.1016/s0014-2999(02)01420-6.

[47] Gajda SN, Kuryłowicz A, Żach M, Bednarczuk T, Wyleżoł M. Diagnosis and treatment of thyroid disorders in obese patients - what do we know? Endokrynol Pol. 2019;70:271-276. DOI: 10.5603/EP.a2018.0089.

[48] Radhakishun NN, van Vliet M, von Rosenstiel IA, Weijer O, Beijnen JH, Brandjes DP, Diamant M. Increasing thyroid-stimulating hormone is associated with impaired glucose metabolism in euthyroid obese children and adolescents. J Pediatr Endocrinol Metab. 2013;26:531-537. DOI: 10.1515/jpem-2012-0302.

[49] Adler SM, Wartofsky L.The nonthyroidal illness syndrome. Endocrinol Metab Clin North Am.2007 ; 36:657– 672. DOI: 10.1016/j.ecl.2007.04.007.

[50] Burman KD, Wartofsky L, Dinterman RE, Kesler P, Wannemacher RW Jr. The effect of T3 and reverse T3 administration on muscle protein catabolism during fasting as measured by 3-methylhistidine excretion. Metabolism. 1979;28:805-813. DOI: 10.1016/0026-0495(79)90206-3.

[51] Kaptein EM, Beale E, Chan LS. Thyroid hormone therapy for obesity and nonthyroidal illnesses: a systematic review. J Clin Endocrinol Metab. 2009;94:3663-3675. DOI: 10.1210/jc.2009-0899.

[52] Rizzo LFL, Mana DL. Treatment of hypothyroidism in special situations. Medicina (B Aires). 2020;80:83-93.

[53] Arima H, Iwama S, Inaba H, Ariyasu H, Makita N, Otsuki M, Kageyama K, Imagawa A, Akamizu T. Management of immune-related adverse events in endocrine organs induced by immune checkpoint inhibitors: clinical guidelines of the Japan Endocrine Society. Endocr J. 2019;66:581-586. DOI: 10.1507/endocrj.EJ19-0163.

[54] Samuels MH. Effects of variations in physiological cortisol levels on

thyrotropin secretion in subjects with adrenal insufficiency: a clinical research center study. J Clin Endocrinol Metab 2000; 85: 1388-1393. DOI:10.1210/jcem.85.4.6540

[55] Schmidbauer B, Menhart K, Hellwig D, Grosse J. Differentiated Thyroid Cancer-Treatment: State of the Art. Int J Mol Sci. 2017;18:1292. DOI: 10.3390/ijms18061292.

[56] McGriff NJ, Csako G, Gourgiotis L, Lori C G, Pucino F, Sarlis NJ. Effects of thyroid hormone suppression therapy on adverse clinical outcomes in thyroid cancer. Ann Med. 2002;34:554-564. DOI: 10.1080/078538902321117760.

[57] Cooper DS, Doherty GM, Haugen BR, Kloos RT, Lee SL, Mandel SJ, Mazzaferri EL, McIver B, Pacini F, Schlumberger M, Sherman SI, Steward DL, Tuttle RM. American Thyroid Association (ATA) Guidelines Taskforce on Thyroid Nodules and Differentiated Thyroid Cancer, Revised American Thyroid Association management guidelines for patients with thyroid nodules and differentiated thyroid cancer. Thyroid. 2009;19:1167-1214. DOI: 10.1089/thy.2009.0110.

[58] Chu JY, Monteleone JA, Peden VH, Graviss ER, Vernava AM. Anemia in children and adolescents with hypothyroidism. Clin Pediatr (Phila). 1981; 20:696-699. DOI: 10.1177/000992288102001102.

[59] Cinemre H, Bilir C, Gokosmanoglu F, Bahcebasi T. Hematologic effects of levothyroxine in iron-deficient subclinical hypothyroid patients: a randomized, double-blind, controlled study. J Clin Endocrinol Metab. 2009;94:151-156. DOI: 10.1210/jc.2008-1440.

[60] Soliman AT, De Sanctis V, Yassin M, Wagdy M, Soliman N. Chronic anemia and thyroid function. Acta Biomed. 2017 28;88:119-127. DOI: 10.23750/abm. v88i1.6048.

[61] Roberts CG, Ladenson PW. Hypothyroidism. Lancet. 2004;363:793-803. DOI: 10.1016/ S0140-6736(04) 156 96-1.

14

Epigenetic: New Insight in Gestational Diabetes Mellitus

Maria Grazia Dalfrà, Silvia Burlina and Annunziara Lapolla

Abstract

Gestational diabetes mellitus (GDM) is the more frequent metabolic complication of pregnancy with a prevalence that is significantly increased in the last decade accounting for 12–18% of all pregnancies. Recent evidences strongly suggests that epigenetic profile changes could be involved in the onset of GDM and its related maternal and fetal complications. In particular, the unfavorable intrauterine environment related to hyperglycemia, a feature of GDM, has been evidenced to exert a negative impact on the establishment of the epigenome of the offspring. Furthermore the adverse in utero environment could be one of the mechanisms engaged in the development of adult chronic diseases. The purpose of this article is to review a number of published studies to fill the gap in our understanding of how maternal lifestyle and intrauterine environment influence molecular modifications in the offspring, with an emphasis on epigenetic alterations.

Keywords: gestational diabetes, epigenetic, maternal complications, fetal complications

1. Introduction

Gestational diabetes mellitus (GDM), defined as a glucose intolerance developing or first recognized during pregnancy that is not clearly overt diabetes [1], is increasingly worldwide due mainly to a rising rates of obesity [2–7].

GDM, if not properly diagnosed and/or treated can lead to adverse outcomes for the mother and the child both during and after pregnancy [8–10]. Of note women experiencing GDM and their children are at high risk to develop cardiometabolic diseases (type 2 diabetes, obesity, hyperlipemia, metabolic syndrome, hyperten-sion, cardiovascular disease) later in life [8–10].

Insulin resistance and beta-cell disfunction are the main physiopathological mechanisms involved in GDM development [4–7]. However all the actors involved are not completely understood as an intricate network of metabolic pathways work in pregnancy complicated by GDM, that includes an abnormal expression of proteins involved in glucose and lipid metabolism, inflammation, oxidative stress, immune response, organ development, and cell death regulation. In this context recent studies have suggested that genetic, epigenetic and environmental factors contributes in GDM development [11–14], (**Figure 1**). In addition, the adverse intrauterine environment in patients with GDM could also have a negative impact on the establishment of the epigenomes of the offspring [15, 16].

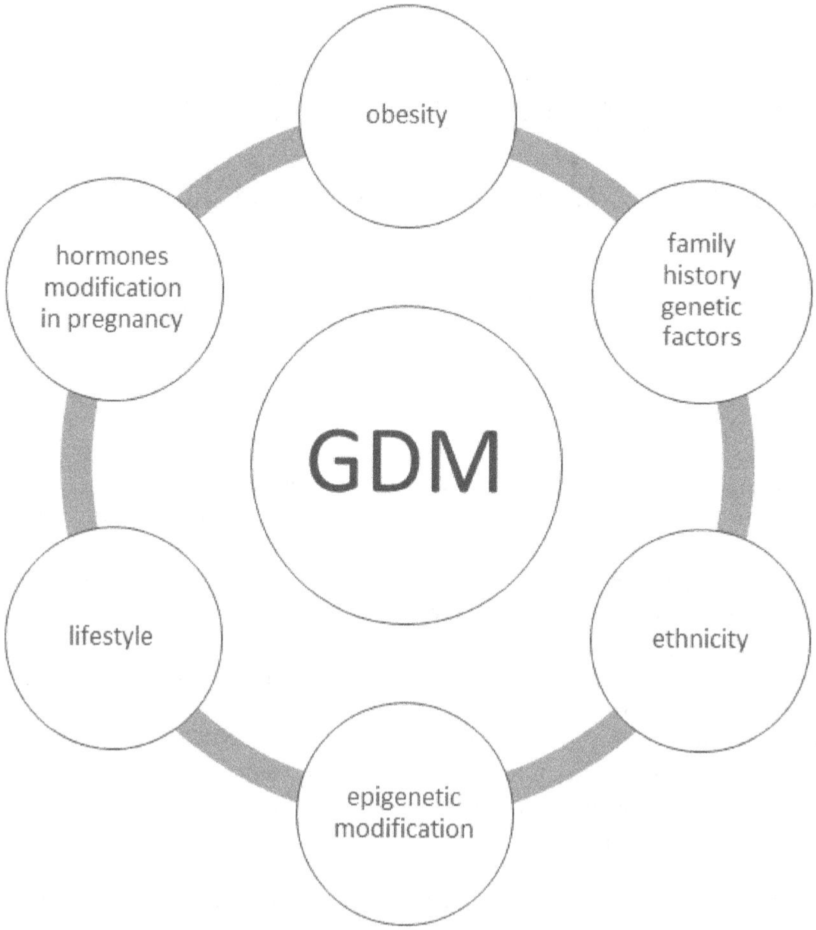

Figure 1.
Factors contributing to GDM development.

The purpose of this article is to review a number of published studies to fill the gap in our understanding how the intrauterine environment can determine molecular modifications in the offspring, with an emphasis on epigenetic alterations.

2. Epigenetic: the meaning

Epigenetic is the study of changes in gene expression caused by mechanisms not involving variations in DNA sequences but determining changes as DNA methyla-tion, histone modifications, and messenger RNA (mRNA) binding by microRNAs (miRNAs).

The study of epigenetic modifications can therefore be useful in deepening and clarifying the pathogenesis of GDM as well as in the use of markers for diagnosis, risk prediction and follow up of different types of pathologies as GDM.

Methylation of cytosine on CpG in the DNA so determining the formation of methylcytosine (5-mc) is the first studied DNA modification. Methylation can determine an increased gene expression by silencing some repressor elements, but, in some regions of DNA as the promoter ones, it can reduce gene expression by inhibiting the activity of enhancer elements [17].

Histone modification can influence the gene expression by modifying the chromatin packing [18].

Micro RNAas are small non-coding single stranded RNAs of about 22 nucleotides that are involved in post-trancriptional regulation of gene expression. It has been evidenced that miRNAs can affect the stability and translation of RNA [19]. Interestingly, some recently-identified miRNAs have been associated with insulin secretion, insulin resistance, and inflammation in patients affected by type 2 diabetes [14].

3. Epigenetic and GDM

It has been demonstrated that even slight increases in glycemia can be associated with epigenetic adaptations via the so-called "metabolic memory", and in this context few studies have examined the association between methylation and GDM development. Of note Whu and coworkers firstly identified two differently methylated genes in plasma, umbilical cord and placenta samples of pregnant women that develop GDM, the Hook Microtubule Tethering Protein 2 (HOOK2), and Retinol Dehydrogenase 12 (RDH12). HOOK2 is a protein that mediates binding to organelles, and is involved in cilia morphogenesis and endocytosis. RDH12 encodes a retinal reductase involved in short-chain aldehyde metabolism [20] (**Tables 1** and **2**).

In this frame more studies have been performed evaluating microRNA: Zhao and coworkers [21], evidenced that the expression of miR222, miR-132 and miR-29a was significantly lower in women who were diagnosed as affected by GDM at 24–28 WG with respect to non GDM control pregnant women. MiR-29 has a role in glucose homeostasis, in particular when overexpressed reduce the insulin-stimulated glu-cose uptake and the gluconeogenesis [22]. MiR-132 is involved in the regulation of cytochrome P450,mediated by insulin, furthermore when its expression is reduced impairs the correct development of trophoblast, [14, 22].

Successively, as omental adipose tissue is known to play a role in insulin resistance in GDM, the differential expression patterns of miRNAs in omental adipose tissues from GDM patients and pregnant women with normal glucose tolerance was studied [23]. MiR-222 was found to be significantly up-regulated in GDM by quantitative real-time PCR and its expression was related with serum estradiol levels, whereas the expressions of estrogen receptor (ER)-α protein and insulin-sensitive membrane transporter glucose transporter 4 (GLUT4) protein were markedly reduced. Then in order to silence miR-222 in 3 T3-L1 adipocytes the antisense transfection oligonucleotides of miR-222 was applied. An important increase of the expressions of ERα and GLUT4, the insulin-stimulated translocation of GLUT4 from the cytoplasm to the cell membrane and of the uptake of glucose was evidenced in mature adipocites. On the basis of their results the authors conclude that: "miR-222 is a potential regulator of ERα expression in estrogen-induced insulin resistance in GDM and could be a candidate biomarker and therapeutic target for GDM".

Cao and coworkers [24], in 85 pregnant women with GDM found that the relative and absolute expression of plasma microRNA-16-5p, −17-5p, −20a-5p were significantly upregulated, with respect to 72 pregnant women without GDM. During pregnancy, the expression of those microRNAs from GDM women were also positively correlated with insulin resistance. Furthermore, significative differences were found in GDM women with respect to normal pregnant ones in the plasma levels of microRNA-16-5p, −17-5p, −20a-5p and in the areas under the curve (0.92, 0.88, and 0.74, respectively). The authors conclude that plasma microRNA-16-5p, −17-5p and -20a-5p are potential diagnostic biomarkers in GDM. MiR16.5

	Gene	Authors
Mothers	**HOOK2**	WHU, 2016
	RDH12	Whu,2016
	H3K27	Michialczy 2016
	H3K4	Michialczy 2016
Placenta	**ADIPOQ**	Bouchard 2010
	TNFRSF1B	Cardenas 2018
	LDLR	Cardenas2018
	BLM	Cardenas 2018
	PDE4B	Cardenas 2018
	ABCA1	Houde 2013
	MEST	Hajj 2018
	NR3C1	Hajj 2018
Offspring	**NR3C1**	Hajj 2018
	PYGO1	Allard 2015
	CLN8	Allard 2015
	PRDM16	Côté 2016
	BMP7	Côté 2016
	PPARGC1a	Côté 2016
	MEST	Hajj 2018
	ATPSA1	Haerle 2017
	NFAP4	Haerle 2017
	PRKCH	Haerle 2017
	SLC17A	Haerle 2017

Table 1.
Gene methylation in gestational diabetes mellitus.

is implicated in the insulin sensitivity regulation and it is upregulated in type 2 diabetes. MiR17–5 has a role is the proliferation of smooth muscle cell. MiR20a-5p is upregulated in preeclampsia, a well known complication of diabetic pregnancy.

Wander and coworkers [25] analyzed the role of miRNA in women affected by GDM and different body mass index. MiR155-5p, and 21–3p were found positively associated with GDM. The miR-21-3p and miR-210-3p were positively associ-ate only in GDM overweight/obese women. MiR-155 and MiR21–3 have a role in pathways that regulate cell survival, and inflammation. MiR210-3p is associated with angiogenesis [14].

As for histone modification, Michalczyk and coworkers [26], analyzed several epigenetic markers during and after pregnancy in a small, multiethnic popula-tion. The evaluation of the proportion of total H3 histone methylated GDM women who developed type 2 diabetes after pregnancy showed a significantly lower H3K27 (50%)with respect to non-diabetic women; furthermore type2 dia-betic women with previous GDM had also significantly lower H3K4 (75%) with respect to GDM with normal glucose tolerance after pregnancy. A study evaluat-ing a large sample size for a longer post partum follow up is however necessary to confirm that histone methylation could be a useful predictor of type 2 diabetes in women with GDM.

	miRNA	miRNA impaired	Author
Mothers	miR-132	Reduced expression	Zhou et al. 2019 Zhao et al. 2011
	miR29a	Reduced expression	Zhao et al. 2011
	miR222	Reduced expression	Zhao et al. 2011
	miR16-5p	Up-regulation	Cao et al. 2017
	miR17–5p	Up-regulation	Cao et al. 2017
	miR20a-5p	Up-regulation	Cao et al. 2017
	miR 155-5p	Overexpression	Wander et al. 2017
	miR 21–3p	Overexpression	Wander et al. 2017
	miR210-3p	Overexpressiion	Wander et al. 2017
Placenta	miR98	Up-regulation	Cao et al. 2016

Table 2.
Studies assessing the role of mRNAs in gestational diabetes.

4. Epigenetic and placenta

The placenta undergoes a number of structural and functional changes in pregnant women affected by diabetes due to the increased production of inflammatory cytokines determined by the high levels of maternal glucose [27]. In this frame, utilizing different mass spectrometry approaches - such as MALDI-MS and LC-MS^E – in the evaluation of placental samples from women with and without GDM, it has been showed that if well controlled, GDM induces only minor changes in the placental proteome [28]. So it is of interest to verify if epigenetic modifications can however occur at the placental level even with relatively low maternal glucose levels and if the extent of these modifications is in some way related to glycemic levels (**Tables 1** and **2**).

Lesseur and coworkers investigate the relations between prepregnancy obesity and GDM and placental leptin DNA methylation on 535 mother-neonate enrolled in the Rhode Island Child health Study. The results of the study showed that neonates of mothers affected by GDM had higher placenta leptin methylation levels similar to those of the mothers with prepregnancy obesity. So maternal metabolic milieu before and during pregnancy can determine impairment of placenta methylation so contributing to the metabolic fetal programming of obesity [29]. These data well fit with those reported by Bouchard et al. [30]. In a subsequent paper Bouchard et al. [31], evaluated the possible association between the methylation of adiponectin gene (ADIPOQ) in plasma cord blood and pla-centa tissue and plasma glucose levels of pregnant women. They found low DNA methylation levels in the ADIPOQ promoter on the fetal side of the placenta that were positively related with high maternal glucose levels in the second trimester of pregnancy. Furthermore, the low DNA methylation levels on the maternal side of the placenta were also positively related to insulin resistance, assessed with the homeostasis model assessment method (HOMA), and to high circulating adipo-nectin levels during pregnancy.

Furthermore, a negative correlation between DNA methylation of the ATP-binding cassette transporter A1 (ABCA1) gene on the placenta maternal site and HDL and 2 hour OGTT plasma glucose was found in 26 GDM women. When looking at the placenta fetal site, DNA methylation of ABCA1 was negatively associated with cord blood tryglicerides [32].

In a well conducted study, Cao et al. aim to verify the role of miRNA-98 in placental tissues from GDM patients, considering that MiRNA-98 is implicated in the correct embryo implantation [33]. They showed that, in the placentas of GDM patients miR-98 is upregulated and total DNA methylation levels are reduced with respect to normal pregnant women. These results,considering that MiRNA-98 regulates the Mecp2 target gene a key protein for embrio development, coud have important consequences for fetal growth.

More recently Cardenas and coworkers [34] conducted an elegant epigenome-wide association study (involving 850,000 CpG sites) on samples of placenta and plasma glucose, and related them to 2 h post-OGTT plasma glucose levels in 448 mother-and-infant pairs at 24–30 weeks of gestation. They found a lower DNA methylation of 4 CpG sites within the phosphodiesterase 4b gene that are positively correlated with plasma glucose at 2 h OGTT. Furthermore, a differentially methyla-tion behavior in relation with maternal glucose was found for 3 CpG sites in the TNFRSF1B, LDLR and BLM.

DNA methylation correlated with expression of its respective genes in placental tissue at three out of four independent identified loci:PDE4B, TNFRSF1B, and LDLR. TNFRSF1B is involved in apoptosis, LDLR encodes a lipoprotein receptor that mediates LDL endocytosis in the cells, and is also expressed in the placenta. BLM is associated with genome stability and maintenance. So maternal glycemic levels during pregnancy were associated with placental DNA methylation of inflam-matory genes, the expression of which depends on epigenetic changes.

5. Epigenetic and offspring

The Developmental Origins of Health and Disease, largely derived from the Barker hypothesis [16], strongly suggests that not only undernutrition but also overnutrition, maternal obesity and diabetes can determine chronic diseases in the offspring through an early exposure to a suboptimal fetal environment; in this context epigenetic modifications have been showed to contribute mainly to this (**Table 1**).

Hajj et al. [35], have evaluated the effect of GDM on the epigenome of the offspring. To reach this aim they analyzed cord blood and placental tissue from the newborn of GDM patients 88 of them treated with diet and 98 with insulin. The results of the study show meaningful lower methylation levels in GDM compared with pregnant women without GDM in the levels of the maternal imprinting MEST gene and the non-imprinting glucocorticoid receptor NR3C1 gene. It is to notice that these genes are associated with placental and fetal growth. Low levels of MEST methylation have also been found in plasma of adults with obesity with respect to normal-weight controls. So the intrauterine exposure to GDM has effects on the epigenome of the offspring, and epigenetic malprogramming of MEST can contribute to predisposing individuals to obesity later in life.

The effect of the exposure to maternal diabetes in utero has been investigated by a genome-wide methylation analysis on peripheral mononuclear cell's DNA in 21 healthy children of GDM mothers, by utilizing a mediation analysis [36]. A series of genes have been identified to be associated with cardiometabolic risk among that the ubiquitin proteasome system (UPS) was the most important. An increased methylation of PYGO1 and CLN 8 showed the most important mediation effect on VCAM-1 levels of the children. TheVCAM-1 protein mediates the adhesion of lymphocytes, monocytes, eosinophils, and basophils to vascular endothelium. It also functions in leukocyte-endothelial cell signal transduction, and it may play a role in the development of atherosclerosis.

A 2 step epigenetic Mendelian randomization approach was used by Allard et al. on data of 485 mothers and their children [37]. To take into consideration maternal glycemia, a genetic risk score, based on 10 known genetic variant related to glycemia, was firstly developed (GRs 10). The results of the study showed that an high GRs 10 was associated with a lower methylation of cg 12083122 that is located near the leptin gene. The low methylation levels at cg12083122 was associated with high cord leptin levels, so evidencing that maternal glycemia can influence offspring leptin epigenetic modulation. In this frame, to evaluate the possible relation of maternal hyperglycemia and DNA methylation of genes involved in brown adipose tissue activation, the DNA methylation levels were measured in placenta samples from normal and GDM women and compared to results of maternal plasma glucose levels. The values of maternal plasma glucose, at the second and third trimester of pregnancy, resulted correlated with the methylation levels of PRDM16, BMP 7 and PPARGC1a and with cord blood leptin levels. These results suggest that maternal glycemia can determine modification in genes related to obesity development in the offsring [38]. More recently, an Illumina 450 K methylation arrays was utilized to analyze genome-wide methylation patterns in fetal cord blood of pregnant women with and without GDM. Significant differences in methylation were found between the GDM patients and the normal pregnant women; furthermore, these differences were more significant in GDM women treated with insulin. A series of genes were found modified by methylation and in particular: ATPSA1, which encodes a subunit of mitochondrial ATP synthetase that acts also reducing mitochondrial oxidation; MFAP4, which is engaged in the process of cell adhesion and intercellular interaction; PRKCH, a component of the protein C family engaged in numerous signaling pathways; and SLC17A, or sodium/phosphate cotransporter involved in hypoxia events. It is to emphasize that these methylation modifications even if had a small effect size, affects many genes/loci [39]. Furthermore, methylation that affects a series of genes that can impair insulin secretion and increase the risk of diabetes and obesity has been reported in offspring of mother affected by type 2 diabetes a condition that shares the same physiopathological mechanisms of GDM [40].

6. Conclusions

The studies taken into consideration made a significant contribution to the knowledge of the physiopathological basis of GDM and of its complications, however methodological problems, small sample size, different GDM diagnostic criteria, make difficult to have final conclusion.

Further researches with high study power need to be undertaken in order to be more confident on the role of epigenetic in GDM disease, bearing also in mind that epigenetic expression in pregnancy varies with weeks of gestation, sex of the fetus, ethnicity, type of sample considered. These studies must be able to determine new road for intervention so to reduce in GDM patients and their children the development of the chronic metabolic diseases [41, 42].

Abbreviations

GDM	Gestational diabetes mellitus
miRNAs	microRNAs
CpG	C-phosphate-G sites
5-mc	methylcytosine
HOOK2	Hook Microtubule Tethering Protein 2

RDH12	Retinol Dehydrogenase 12
ER	estrogen receptor (ER)-α protein
GLUT4	sodium glucose transporter 4
H3K27	tri-methylation of lysine 27 on histone H3 protein.
H3K4	H3K4 histone
MALDI-MS	Matrix-Assisted Laser Desorption/Ionization Mass Spectrometry
LC-MS^E	Liquid Chromatography / Mass Spectrometry
ADIPOQ	Human Adiponectin, C1Q and Collagen Domain
HOMA	Homeostasis Model Assessment Method
ABCA1	ATP-Binding Cassette Transporter A1
Mecp2	Methyl-CpG Binding Protein 2
TNFRSF1B	TNF Receptor Superfamily Member 1B
LDLR	Low Density Lipoprotein Receptor
BLM	BLM RecQ Like Helicase
MEST	Mesodermic Specific Transcript Gene
NR3C1	Nuclear Receptor Subfamily 3 Group C Member 1
UPS	Ubiquitin Proteasome System
PYGO1	Pygopus Family PHD Finger 1
CLN 8	CLN8 Transmembrane ER and ERGIC Protein
VCAM	Human Vascular Cell Adhesion Molecule 1
GRs 10	Genetic Risk Score
PRDM16	PR Domain Containing 16
BMP 7	Bone Morphogenic Protein 7
PPARGC1a	Peroxisome Proliferator Activated Receptor Gamma
ATPSA1	Subunit of Mitocondrial ATP Synthetase
MFAP4	Microfibril Associated Protein 4
PRKCH	Protein Kinase C family
CTBP2	C-Terminal Binding Protein 2

Author details

Maria Grazia Dalfrà, Silvia Burlina and Annunziara Lapolla*
DPT Medicine-Padova University, Padova, Italy

*Address all correspondence to: annunziata.lapolla@unipd.it

References

[1] International Association of Diabetes and Pregnancy Study Groups Consensus Panel. International Association of Diabetes and Pregnancy Study Groups: Recommendations on the diagnosis and classification of hyperglycemia in pregnancy. Diabetes Care. 2010; 33: 676-668.

[2] Koning SH, Hoogenberg K, Lutgers HL, Van den Berg PP, Wolffenbuttel BHR. Gestational diabetes mellitus: current knowledge and unmet needs. J Diabetes. 2016; 8(6): 770-781.

[3] Lapolla A, Metzger BE .The post-HAPO situation with gestational diabetes: the bright and dark sides. Acta Diabetol. 2018; 55(9):885-892.

[4] Ryan EA, Enns L. Role of gestational hormones in the induction of insulin resistance. J Clin Endocrinol Metab. 1988; 67:341-347.

[5] Kampmann U, Knorr S, Fuglsang J, Ovesen P. Determinants of Maternal Insulin Resistance during Pregnancy: An Updated Overview. J Diabetes Res. 2019 Nov 19;2019:5320156. doi: 10.1155/2019/5320156. PMID: 31828 161; PMCID: PMC6885766.

[6] Lain K, Catalano PM. Metabolic changes in pregnancy. Clin Obstet Gynaecol.2007; 50:938- 948.

[7] Lapolla A, Dalfrà MG, Mello G, Parretti E, Cioni R, Marzari C et al. Early detection of insulin sensitivity and beta-cell function with simple tests indicates future derangements in late pregnancy. J Clin Endocrinol Metab. 2008; 93:876-880.

[8] Burlina S, Dalfrà MG, Lapolla A. Short- and long-term consequences for offspring exposed to maternal diabetes: a review. J Matern Fetal Neonatal Med. 2019; 32(4):687-694.

[9] Lowe WL, Scholtens DM, Lowe LP, Kuang A, Nodzenky M, Talbot O et al. Association of gestational diabetes with maternal disorders of glucose metabolism and childhood adiposity. JAMA. 2018;11;320(10):1005-1016.

[10] Burlina S, Dalfrà MG, Lapolla A. Clinical and biochemical approach to predicting post-pregnancy metabolic decompensation. Diabetes Res Clin Pract. 2018;145:178-183.

[11] Whu L, Cui L, Tam WH, Ma RC, Wang CC. Genetic variants associated with gestational diabetes mellitus: a meta-analysis and subgroup analysis. Scientific Reports. 2016; 6: 30539.

[12] Pantham P, Aye IL, Powell T. Inflammation in maternal obesity and gestational diabetes mellitus. Placenta. 2015; 36:709-715.

[13] Lapolla A, Dalfrà MG, Sanzari M, Fedele D, Betterle C, Masin M et al. Lymphocyte subsets and cytokines in women with gestational diabetes mellitus and their newborn. Cytochine. 2005; 31:280-287.

[14] Zhu H, Leung SW.Identification of microRNA biomarkers in type 2 diabetes: a meta-analysis of controlled profiling studies. Diabetologia. 2015; 58: 900-911.

[15] Agarwal P, Morriseau TS, Kereliuk SM, Doucette CA, Wicklow BA, Dolinsky VW. Maternal obesity, diabetes during pregnancy and epigenetic mechanisms that influence the developmental origins of cardiometabolic disease in the offspring. Crit Rev Clin Lab Sci. 2018; Mar;55(2):71-101. doi: 10.1080/10408363.2017.1422109. Epub 2018 Jan 8. PMID: 29308692.

[16] Barker DJ. The fetal and infant origins of adult disease. BMJ. 1990; 301:1111-1122.

[17] Hernando-Herraez I, Garcia- Perez R, Sharp AJ, Marques-Bonet T. DNA Methylation: Insights into Human Evolution. PLoS Genet. 2015; Dec 10;11(12):e1005661. doi: 10.1371/journal.pgen.1005661. PMID: 26658498; PMCID: PMC4684328.

[18] Martin C, Zhang Y. The diverse functions of histone lysine methylation. Nat Rev Mol Cell Biol. 2005 Nov;6(11):838-849. doi: 10.1038/nrm1761. PMID: 16261189.

[19] Zhao S, Liu MF. Mechanisms of microRNA-mediated gene regulation. Sci China C Life Sci. 2009 Dec;52(12):1111-1116. doi: 10.1007/s11427-009-0152-y. Epub 2009 Dec 17. PMID: 20016967.

[20] Whu P, Farrel W E, Haworth K E, Emes RD, Kitchen MO, Glossop JR, Fryer AA. Maternal genome-wide DNA methylation profile in gestational diabetes shows distinctive disease-associated changes relative to matched healthy pregnancies. Epigenetics.2016; 13: 122-128.

[21] Zhao C, Dong J, Jiang T, Shi Z, Yu B, Zhu Y et al. Early second trimester serum miRNA profiling predicts gestational diabetes mellitus. PLoS ONE. 2011: 6:23925-23923.

[22] Zhou X, Xiang C, Zheng X. MiR-132 serves as a diagnostic biomarker in gestational diabetes mellitus and its regulatory effect on trophoblast cell viability. Diagnostic Pathology. 2019; 14:119-128.

[23] Shi Z, Zhao C, Guo X, Ding H, Cui Y, Shen R, Liu J. Differential expression of microRNAs in omental adipose tissue from gestational diabetes mellitus subjects reveals miR-222 as a regulator of ERα expression in estrogen-induced insulin resistance. Endocrinology. 2014; May;155(5): 1982-90. doi: 10.1210/en.2013-2046. Epub 2014 Mar 6. PMID: 24601884.

[24] Cao YL, Jia YJ, Xing BH, Shi DD, Dong XJ. Plasma microRNA-16-5p, 17-5p and 20a-5p: novel diagnostic biomarkers for gestational diabetes mellitus. Obstet Gynecol Res 2017; 43:974-981.

[25] Wander PL, Boyko EJ, Hevner K, Parikin VJ, Tadesse MG, Sorensen TK et al. Circulating early- and mid-pregnancy microRNAs and risk of gestational diabetes. Diab Res Clin Pract. 2017; 132:1-9.

[26] Michalczyk AA, Dunbar JA, Janus ED, Best JD, Ebeling PR, Ackland MJ et al. Epigenetic markers to predict conversion from gestational diabetes to type 2 diabetes. J Clin Endocrinol Metab. 2016; 101: 2396-2404.

[27] Hauguel-de Mouzon S, Desoye G The placenta in diabetic pregnancy: new methodological approaches. In Gestational Diabetes: a decade after the HAPO study. Lapolla A, Metzger BE eds, Frontier in Diabetes Karger 2020; pp 145-154.

[28] Burlina S, Banfi C, Brioschi M, Visentin S, Dalfrà MG, Traldi P, Lapolla A. Is the placenta proteome impaired in well-controlled gestational diabetes? J Mass Spectrom. 2019; 54:359-365.

[29] Lesseur C, Armstrong DA, Paquette AG, Li Z, Padbury JF, Marsit CJ. Maternal obesity and gestational diabetes are associated with placental leptin DNA methylation. Am J Obstet Gynecol. 2014; Dec;211(6):654.e1-9. doi: 10.1016/j.ajog.2014.06.037. Epub 2014 Jun 19. PMID: 24954653; PMCID: PMC4254188.

[30] Bouchard L, Thibault S, Guay SP, Santure M, Monpetit A, St-Pierre J et al. Leptin gene epigenetic adaptation to impaired glucose metabolism during pregnancy. Diabetes Care. 2010; Nov;33(11):2436-41. doi: 10.2337/dc10-1024. Epub 2010 Aug 19. PMID: 20724651; PMCID: PMC2963508.

[31] Bouchard L, Hivert M, Guay SP, StPierre J, Perron P, Brisson D.Placental adiponectin gene DNA methylation levels are associated with mother's blood glucose concentration. Diabetes. 2012; 61:1272-1280.

[32] Houde AA, Guay SP, Desgagné V, Hivert MF, Baillargeon JP, St-Pierre J et al. Adaptations of placental and cord blood ABCA1 DNA methylation profile to maternal metabolic status. Epigenetics. 2013; Dec;8(12):1289-302. doi: 10.4161/epi.26554. Epub 2013 Oct 10. PMID: 24113149; PMCID: PMC3933490.

[33] Cao JL, Zhang L, Lin J, Tian S, Lv XD, Wang XQ et al. Up-regulation of miR-98 and unravelling regulatory mechanisms in gestational diabetes mellitus. Sci Rep.2016; 6:32268-32274.

[34] Cardenas A, Gagnè-Quellet V, Allard C, Brisson D, Perron P, Bouchard L et al. Placental DNA methylation adaptation to maternal glycemic response in pregnancy. Diabetes. 2018; 67: 1673-1683.

[35] Hajj NE, Pliushch G, Schneider E, Dittrich M, Muller T, Korenov M et al. Metabolic programming of MEST DNA methylation by intrauterine exposure to gestational diabetes mellitus. Diabetes. 2018; 62: 1320-1328.

[36] West NA, Kechris K, Dabelea D. Exposure to Maternal Diabetes in Utero and DNA Methylation Patterns in the Offspring. Immunometabolism. 2013; Mar;1:1-9. doi: 10.2478/immun-2013-0001. PMID: 23741625; PMCID: PMC3670583.

[37] Allard C, Desgagné V, Patenaude J, Lacroix M, Guillemette L, Battista MC, et al.Mendelian randomization supports causality between maternal hyperglycemia and epigenetic regulation of leptin gene in newborns. Epigenetics. 2015;10(4):342-51. doi: 10.1080/1559 2294.2015.1029700. PMID: 25800063; PMCID: PMC4622547.

[38] Côté S, Gagné-Ouellet V, Guay SP, Allard C, Houde AA, Perron P et al.. PPARGC1α gene DNA methylation variations in human placenta mediate the link between maternal hyperglycemia and leptin levels in newborns. Clin Epigenetics. 2016; Jun 22;8:72. doi: 10.1186/s13148-016-0239-9. PMID: 27340502; PMCID: PMC4918074.

[39] Haerle L, Haj N, Dittrich M, Muller T, Nanda I, Lehnen H. Epigenetic signatures of gestational diabetes mellitus on cord blood methylation. Clinical Epigenetics.2017; 9:28-38.

[40] Chen P, Piaggi P, Traurig M, Bogardus C, Knowler WC, Baier LJ, Hanson RL. Differential methylation of genes in individuals exposed to maternal diabetes in utero. Diabetologia. 2017;Apr;60(4):645-655. doi: 10.1007/s00125-016-4203-1. Epub 2017 Jan 26. PMID: 28127622; PMCID: PMC7194355.

[41] Michels K. Epigenetics epidemiology. 2012. Dondrecht New York Springer Verlag.

[42] Ibarra A, Vega-Guedes B, Brito-Casillas Y, Wagner A Diabetes in pregnancy and microRNAs :promises and limitations in their clinical applications. Non-coding RNA; 2018; 4:1-250.

15

Morphology Aspects of Hypothyroidism

Fernando Candanedo-Gonzalez, Javier Rios-Valencia, Dafne Noemi Pacheco-Garcilazo, Wilfredo Valenzuela-Gonzalez and Armando Gamboa-Dominguez

Abstract

Hypothyroidism is a common endocrine disorder resulting of low levels of thyroid circulating hormones. The prevalence in the general population varies between 0.3% and 3.7%. Presents as clinical or subclinical disease based on presence of symptoms and levels of serum TSH and free thyroxine and T4, respectively. Hypothyroidism has numerous etiologies, some of them are originated on the thyroid itself and some others are of extrathyroid origin, with variable manifesta-tions. Classified as primary, secondary, tertiary and peripheral. Thyroid autoim-mune disease is the principal cause. A new class of drugs against cancer, like the anti-CTLA-4 and anti-PD-L1/PD1 therapies have been associated with primary or secondary hypothyroidism. Endocrine disorders can be difficult to diagnose based only on morphological features because endocrine manifestations are caused primarily by a hormonal imbalance. Hypothyroidism may have a higher risk of morbidity and mortality. Finally, myxedematous coma is the main complication of terminal stages hypothyroidism.

Keywords: hypothyroidism, epidemiology, pathophysiology, etiology, pathology, anti-CTLA-4 and anti-PD-L1/PD1 therapies, treatment, prognosis, complications

1. Introduction

1.1 Anatomy

The thyroid gland is a butterfly-shaped organ formed by a right and left lobe connected at the midline by a thin structure called isthmus. Located in the neck, the thyroid covers the anterior side of the trachea underneath the larynx at the vertebral levels of C5 to T1 (**Figure 1A**). The average size of a thyroid gland is of 5 cm height and 5 cm wide and it weighs between 20 and 30 grams in adults (**Figure 1B**), being a little more heavy in women. Is a highly vascular organ, receiving blood supply from two main sources, the superior thyroid artery, branch of the external carotid artery irrigates the superior half of the thyroid in more than 95% of the popula-tion, the inferior half is irrigated by the inferior thyroid artery that branches from the thyrocervical trunk which is a branch of the subclavian artery. Furthermore, the thyroid gland has extensive lymphatic drainage that involves multiple levels of

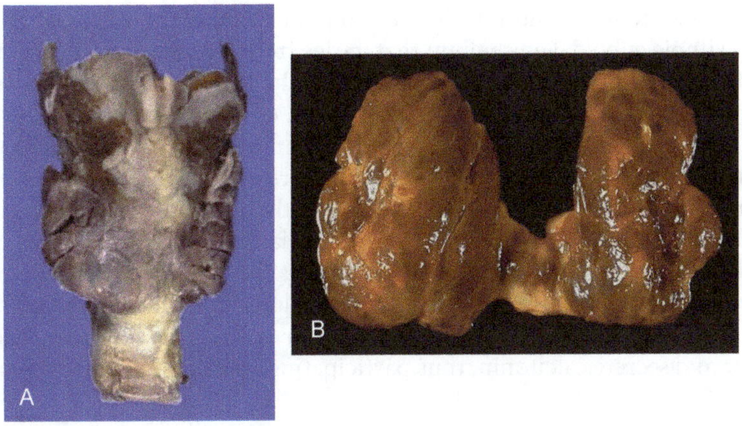

Figure 1.
Thyroid gland: A) butterfly-shaped, located in the anterior side of the trachea underneath the larynx; B) formed by a right and left lobes connected at the midline by a thin structure called isthmus.

lymphatic nodes, including the prelaryngeal, pre and paratracheal, retropharyngeal, retroesophageal and the internal jugular nodes [1].

1.2 Embryology

The thyroid gland is the first endocrine organ that develops during fetal development [2]. It begins to develop during the fourth week of gestation as an epithelial diverticulum arising from the endoderm of the foregut near the base of the primi-tive tongue, it progressively extending downward starting from week fifth as the fetus develops [2, 3]. It reaches its final shape and size at the end of the seventh week of gestation [2].

1.3 Normal histology

The normal thyroid gland is composed of numerous follicles surrounded by a fibrous capsule that forms septa that divide the parenchyma in multiple lobules. These septa contain nerves and blood vessels that irrigate each lobule. Each lobule

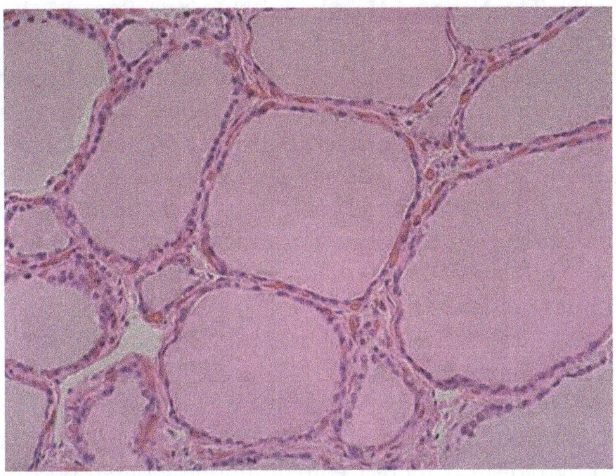

Figure 2.
Normal thyroid gland histology.

contain from 20 to 40 round follicles of 200 μm of diameter on average, these are coated by simple cuboidal epithelium that varies from plane to low according to the current functional activity, when more active the follicle is, taller the follicular epithelium will be. The follicular cells have small, dark and uniform nuclei that are localized at the center of the cell and some of them have an abundant granular and eosinophil cytoplasm, a variant known as Hürthle cells. The follicles contain colloid, a viscous material composed predominantly by the precursor protein of the thyroglobulin (**Figure 2**). The normal thyroid gland contains up to 3 months of thyroglobulin stored in the colloids. Alternating, the parafollicular cells or the C cells, derived from the neural crest through the ultimobranchial body, are found in a higher concentration in the middle and superior portions of the lobes. These cells synthesize and secrete calcitonin, thus participating on calcium homeostasis [4].

2. Definition

Described in 1850, hypothyroidism was the first disorder of endocrine deficiency ever reported [5]. Hypothyroidism is the result of low levels of thyroid circulating hormones. Due to the wide variety of clinical presentations and the lack of specific symptoms, the definition of hypothyroidism is mainly biochemi-cal [6]. Therefore, hormonal levels in overt hypothyroidism are: TSH (Thyroid Stimulating Hormone) >4.8 UI/l, FT4 < 13 pmol/l, and in subclinical hypothy-roidism are: TSH >4.8 UI/l, FT4: 13–23 pmol/l. [7] Recent research suggests that the superior reference values for serum TSH varies among different age groups [8]. Nevertheless, up to this present day there is no exact definition of a cut point for serum TSH values regarding age in our population [9–12]. According to the moment of clinical presentation, hypothyroidism is divided in congenital or acquired, according to the level of endocrine dysfunction is divided in primary or secondary or central and according to the severity of hypothyroidism is divided in severe or clinic hypothyroidism or in mild or subclinical hypothyroidism [13].

3. Epidemiology

The prevalence of overt hypothyroidism in the general population varies between 0.3% and 3.7% in the US and between 0.2% and 5.3% in Europe, according to the used definition [6, 10–15]. The National Health and Nutrition Examination Survey found that the prevalence of overt hypothyroidism between people older than 12 years old in the United States is of 0.3% and of subclinical hypothyroidism is of 4.3% [12]. The difference in iodine status affects the prevalence of hypothyroidism, which occurs in population with a relatively high intake of iodine as well as in populations with deficient intake of iodine. The most common cause of thyroid dysfunction is iodine deficiency and it is estimated that 2 thousand millions of people have an insufficient iodine intake [16]. Hypothyroidism is more common in women and the incidence increases with age (>65 years old) and in Caucasian individuals, although data regarding ethnical difference are scarce [6]. Female gender and older age individuals are related to an increase of TSH and prevalence of anti-thyroid antibodies [12]. Among women in reproductive age (12–49 year old), the prevalence of hypothyroidism is of 3.1%. While women older than 80 years old or more, have 5 times more probabilities to suffer from hypothyroidism, compared to the 12–49 year old women population. Hypothyroidism is more frequent among women born with low height and of low body mass index at childhood. However, in countries with good iodine supply, autoimmune disorders are the most common causes of hypothyroidism [12].

3.1 Genetic epidemiology

It is estimated that the heritability of serum levels of TSH and of free thyroxin levels is of 65% and of 23–65% respectively [17, 18]. The results of studies of genome association of all the genome, up to this present day have now explained only but a small proportion of the variability in thyroid function and only three studies have focused on hypothyroidism [19]. The loci that are more consistently implicated in hypothyroidism include genes related to autoimmunity and regulating genes specific to thyroid. The majority of these loci are also related to serum concentrations of TSH within the reference rank [19–23]. Monogenetic disorders that cause congenital hypothyroidism are rare and include TSH resistance (due to an inactivating mutation on the TSH receptor), thyroid digenesis and thyroid dyshormonogenesis.

4. Pathophysiology

To understand better hypothyroidism and its consequences it is important to remember the normal physiology of the thyroid gland. The main function of the thyroid follicular cells is the synthesis of thyroid hormones, tetraiodothyronine or (T4; 3,5,3′,5′-L-tetraiodothyronine) and triiodothyronine (T3; 3,5,3′-L-triiodothyronine). Iodine is essential for thyroid hormone synthesis. Food and water are the main sources for iodine intake, with a daily supply that ranges from 50 to 300 μg being absorbed in the small intestine. Both thyroid hormones are synthesized by the iodination and condensation of two tyrosine molecules and differ by an iodine atom. The production and release of thyroid hormones is stimulated by the hypothalamus-pituitary axis. The thyrotropin-releasing hormone (TRH) released from the hypothalamus stimulates the anterior pituitary gland to release thyrotropin, also called TSH (**Figure 3**) [24]. In response to the stimuli of TSH, the thyroid follicular cells produce thyroglobulin, an inactive protein that is then released from the apical surface into the follicle as a colloid [4]. TSH is released into the bloodstream and it then binds to the thyroid stimulating hormone receptor (TSH-R) in the basolateral surface of the follicular cell of the thyroid gland. The TSH-R is a G-protein coupled receptor and its triggering yields to the activation of the Adenylate Cyclase and of increased levels of intracellular cAMP. An increased cAMP activates the protein kinase A (PKA). PKA phosphorylates different proteins in order to change their functions. The thyroid hormone biosynthesis is made by steps, regulated by enzymes that are stimulated by TSH, these steps are: 1) thyro-globulin synthesis (TG): the thyrocites in the thyroid follicles produce a protein called thyroglobulin. Thyroglobulin does not contain iodine and is a precursor pro-tein storaged in the follicle lumen. Thyroglobulin is produced in the rough endoplas-mic reticulum, then the Golgi apparatus packs it up in vesicles and then it enters the follicle lumen by exocytosis. 2) Iodine uptake and transport: the phosphorylation of the kinase A protein increases the activity of the sodium/iodide basolateral sym-porter protein (Na+/I- symporter), driven by the Na + -K + -ATPase to get iodine out of the bloodstream to the thyrocites. Iodine diffuses from the basolateral surface to the apical surface of the cell, where it transports to the colloid through the pendrin transporter; 3) thyroglobulin iodination: the protein kinase A also phosphorylates and activates the thyroid peroxidase enzyme (TPO). The TPO has three main func-tions: oxidation, organification and coupling reaction. 4) Oxidation: the TPO uses hydrogen peroxide in order to oxidate iodide (I-) to iodine (I2). NADPH oxidase, an apical enzyme generates hydrogen peroxide for the TPO; 5) Organization: the TPO attaches the remainders of tyrosine from the thyroglobulin with the I2. It generates

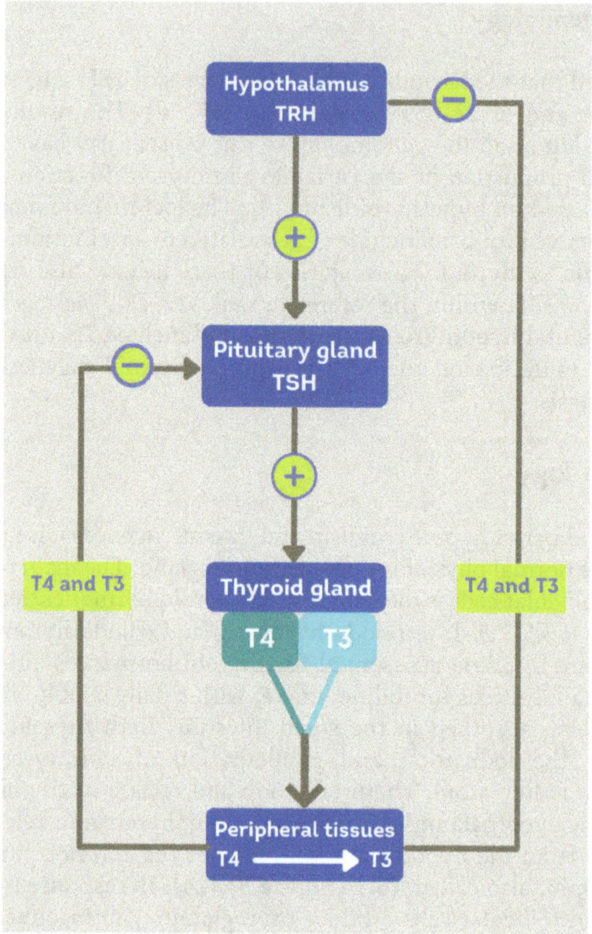

Figure 3.
Normal thyroid physiology.

monoiodityrosine (MIT) and diiodotyrosine (DIT) (**Figure 4**). MIT has only one remaining tyrosine with iodine and DIT has two remaining tyrosine with iodine; 6) mono and diiodotyrosine attachment; the TPO combines the remainers of iodated tyrosine to produce T3 and T4. [4] MIT and DIT combine to form T3 and two DIT molecules form T4; 5) Storing: thyroid hormones are attached to TG and are storage in the follicular lumen; and 6) secretion: the iodized thyroglobulin returns to the follicular cell, where the degradation of lysosomic proteases releases T3 and T4 in the fenestrated capillaries. Thyroid hormones travel through the bloodstream united to a binding protein called thyroxin [24]. The thyroxine-binding globuline (TBG), transthyretin (TTR) and albumin are proteins capable to bind to the thyroid hormone, thus becoming able to transport it through the bloodstream to their target sites [25].

Thyroid hormones are important for a variety of functions in the body, includ-ing development, growth and increase the basal metabolic rate (BMR) affecting circulation, corporal temperature, gluconeogenesis, lipolysis, proteolysis and glucose absorption [26]. It also increases systolic volume and heart rate, which increases cardiac output. In young populations it boosts growth and leads to bony maturation and the fusion of bone growth plates. It is essential for the maturity of the central nervous system (CNS) during fetal development [24]. All of these biochemical events that make the thyroid gland produce hormones is regulated by a negative feedback

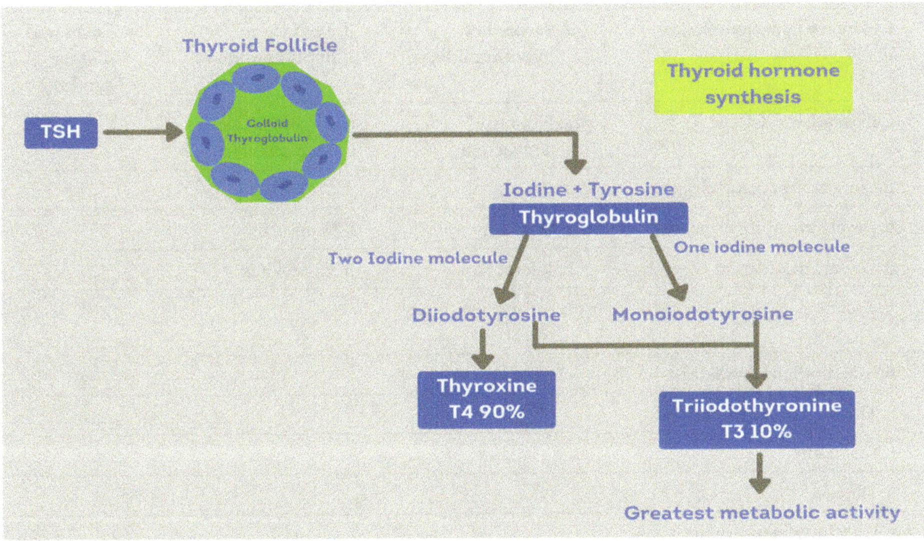

Figure 4.
Thyroid hormone formation.

system in which high levels of thyroid hormones, especially T3, inhibit the release of TSH from the anterior pituitary gland [24]. The counterforces of TRH and T3 allow our body to keep thyroid hormones levels relatively stable in healthy individuals [24]. Although, when alterations occur within this delicate system, severe and even fatal conditions can happen [27]. The most common cause of hypothyroidism is the incapacity of the thyroid gland to produce enough thyroid hormones, nevertheless, with less frequency the hypothalamus and the pituitary can also cause thyroid dysfunction. The half-life of a T4 molecule ranges from 6 to 12 days, in regards to T3 its half-life is of 24 hours, therefore, T4 is significantly more abundant (approximately 100–125 nmol/day) and T3 is found in less quantity, nevertheless T3 is two to tenfold more bioactive [4, 24], to counteract this difference target tissues contain 5′-iodinase, which converts T4 into T3 peripherally through deiodination 5′ [4]. The levels of T3 and T4, mostly T3, establish a negative feedback on the production of TRH and TSH. Alterations on the structure and function of any of these organs or axis can result in hypothyroidism. A decline in the production of T4 results in an increased secretion of TSH by the pituitary, which in turn causes hypertrophy and hyperplasia of the thyroid parenchyma, thus leading to an increase in the production of T3 [4].

5. Etiology

Hypothyroidism has numerous etiologies, some of them are originated on the thyroid itself and some others are of extrathyroid origin, with variable manifestations. **Table 1** resumes the principal causes of hypothyroidism. Hypothyroidism can be classified on primary hypothyroidism, secondary (central), tertiary and periph-eral. Primary hypothyroidism occurs when the thyroid gland is unable to produce adequate amounts of thyroid hormones. Secondary hypothyroidism occurs when the function of the thyroid gland is normal and the pathology is on the pituitary gland due to a deficiency of TSH. In tertiary hypothyroidism, the pathology is found in the hypothalamus, due to a deficiency of TRH. Central hypothyroidism (second-ary and tertiary) and peripheral hypothyroidism are less frequent and represent less than 1% of all cases [28].

1. Primary hypothyroidism	2. Secondary hypothyroidism	3. Tertiary hypothyroidism	4. Euthyroid disease syndrome
a. Biosynthetic defects:	a. Pituitary destruction	a. Hypothalamic destruction	
Innate errors of metabolism			
b. Development abnormalities	i. Neoplasms	i. Neoplasms	
c. Thyroid destruction	ii. Infarct	ii. Granuloma	
ii. Radiation	b. Isolated deficiency of TSH		
iii. Autoimmune thyroiditis			
d. Thyroid tissue replacement			
i. Neoplasms			
ii. Fibrosis			
e. Drug interference in the production and/or secretion of hormones			

Table 1.
Hypothyroidism causes.

5.1 Primary hypothyroidism

Thyroid autoimmune disease is the principal cause of primary hypothyroidism in the United States and in the geographical regions with enough iodine intake [10]. Hashimoto's thyroiditis (HT) is the most frequent etiology in the United States and has a strong association with the development of malignant neoplasms like papil-lary thyroid carcinoma (PTC) and lymphoma [29].

There are other causes of hypothyroidism induced by drugs, like amiodarone, thalidomide, tyrosine kinase inhibitors (sunitinib, imatinib), staduvine, inter-pheron, rifampicin, ethionamide, phenobarbital, phenytoin, carbamazepine, interleukin-2 and lithium. Therapy with radioactive iodine, thyroid surgery and radiotherapy to the head and neck area may be causes of hypothyroidism. Contrary to the previous, smoking and moderate alcohol consumption are related to a reduced risk of hypothyroidism [10].

Post-partum thyroiditis affects nearly 10% of women and generally occurs between 8 and 20 weeks after birth. Only few women will require hormonal treat-ment. However, some women have a higher risk of permanent hypothyroidism or recurrent thyroiditis postpartum in future pregnancies. The use of radioactive iodine in the treatment of Graves-Basedow disease generally results in permanent hypothyroidism in approximately 80–90% of patients between 8 and 20 weeks after treatment [10]. Treatment with radiation on the head and neck area can also induce hypothyroidism. A relatively infrequent cause of primary hypothyroidism is sub-acute granulomatous thyroiditis (Quervain's disease), it often arises in middle aged women and it tends to be an auto limited disease. Finally, Down syndrome and Turner syndrome patients have a higher risk of hypothyroidism [10].

5.2 Secondary and tertiary hypothyroidism (central)

Secondary and tertiary hypothyroidism, also known as central hypothyroid-ism, is caused by a defect in the hypothalamus-pituitary axis. Causes include:

pituitary tumors, tumors that compress the hypothalamus, Sheehan syndrome, resistance to TRH, TSH deficiency, lymphocytic hypophysitis, cerebral radiotherapy, drugs like dopamine, prednisone or opioids [10]. A new class of drugs against cancer, like the anti-CTLA-4 (ipilimumab) and anti-PD-L1/PD1 therapies (pembrolizumab and nivolumab) have been associated with primary or secondary hypothyroidism [30, 31]. In previous years, the use of immune check point inhibi-tors (ICPi) has improved the treatment and prognosis of different types of cancer. The ICPi are monoclonal antibodies associated with adverse effects pertaining the immune system. Thyroid dysfunction (thyrotoxicosis or hypothyroidism) are among the most common adverse consequences. These monoclonal antibodies inhibit immune checkpoints that are present in the surface of the T cells to assure immune auto-tolerance, which results in an increased T cell capacity to attack cancerous cells. The pathogenesis of the thyroid disorders associated to the use of ICPi is not fully understood. Data from observational studies suggest that thyroid dysfunction induced by ICPi is due to a destructive thyroiditis that can evolve to hypothyroidism. On the other hand, it is proposed that thyroid manifestations in patients with immunotherapy may represent an autoimmune phenomenon. Nevertheless, little is known about thyroid antibodies status during the course of the disease [30, 31].

5.3 Peripheral hypothyroidism

Consumptive hypothyroidism is caused by an aberrant expression of the enzyme type 3 iodothyronine deiodinase (D3), which inactive thyroid hormones in tumoral tissues. Even when rare, this overexpression can induce severe hypothyroidism. High concentrations of D3 was first described in a newborn with infantile hepatic hemangioma [32], but it can also occur in patients with vascular tumors and tumors of the gastrointestinal stroma [33].

6. Clinical presentation

Classic signs and symptoms of hypothyroidism are bradycardia, weight gain even when reducing food intake, cold intolerance, dry skin, sweat decrease, consti-pation, alopecia, hyporeflexia, slow talking and lethargy. Nonetheless, is important to keep high suspicion of hypothyroidism because signs and symptoms can be mild and non-specific. Chronic hypothyroidism also increases total cholesterol and low density lipoprotein and decreases high density lipoproteins, which increases the risk of cardiovascular mortality. Depression is also a common symptom of hypothyroid-ism and, therefore, it can be found in the medical history of patients that commit-ted suicide [10].

6.1 Subclinical hypothyroidism

A decrease in thyroid function is observed in subclinical hypothyroidism, defined by high levels of TSH and normal levels of free thyroid hormones [34]. Primary hypothyroidism is the most prevalent thyroid dysfunction in elderly population and subclinical hypothyroidism is found nearly in 20% of elderly people [10, 11]. Subclinical thyroid disease during pregnancy can be related to adverse results, including a lower than normal intellectual quotient in the offspring of the pregnant women. It is unknown if the treatment with levothyroxine in women identified with subclinical hypothyroidism or hypothyroxinemia during pregnancy improves cogni-tive function in their offspring [35].

6.2 Congenital hypothyroidism

Thyroid hormones are extremely important for the correct development of the central nervous system in the fetus. To ensure an adequate availability of thyroid hormones, human chorionic gonadotropin-β (hCG-β) coming from the placenta directly stimulates the maternal thyroid, which increases the production of T3 and T4 during the first trimester. After the first trimester the fetal thyroid converts in the principal source of hormonal thyroid. The placenta also expresses type 3 iodothyronine deiodinase, an enzyme that breaks down T4 into an inverse inactive T3 (rT3), as a protection against excessively high levels of thyroid hormones, however, regardless of this protection, abundant thyroid hormones cross to the fetus [36, 37].

Decreased levels of fetal thyroid hormones during the crucial period of neurologic development drives to severe mental retard, also called cretinism. Therefore, congenital hypothyroidism (CHT) is a pediatric condition that has to be treated with urgency. Other clinical features include musculoskeletal abnormalities, macroglossia and coarse facial features. This condition is irreversible if is not diagnosed early. Even when is not associated to mortality, CHT can be found on fetal and pediatric autopsy for other reasons and must be considered as a differential diagnosis with mental retardation. However, the natural history of CHT has drastically changed in previous years due to newborn screening (NS) programs that consist in detecting this disease in all apparently healthy newborns [36, 37]. In Mexico, the program of NS formally began in 1988 with the emission of the "technical norm 321.4" and in the present its realization is a mandatory action for all health centers that provide child and maternal care, according to the Norma Oficial Mexicana 007-SSA2–1993.5. (Mexican Official Norm 007-SSA2–1993.5) [11, 38].

The main causes that produce CHT are: a) aberrant or incomplete migration of the thyroid bud, which causes the formation of an ectopic gland without lateral lobes, this is also known as a thyroid nodule; b) deficient growth or differentiation that brings about thyroid agenesis or atyriosis, and c) defects on the biosynthesis of thyroid hormones or dyshormonogenesis with or without goiter. The first two entities are grouped under the name of thyroid dysgenesis, which are sporadic and have predominance for the female gender [36–38]. Female predominance is a characteristic particularly interesting in the epidemiology of CHT; although, it is not known if women are more susceptible to develop CHT or if female fetuses with CHT have higher uterine survival compared to masculine fetuses [36]. The molecular mechanisms implicated on thyroid cellular differentiation are not exactly known, yet, some mutations have been described in genes involved in thyroid growth and development, like TTF1, TTF2, PAX8 and TSHR among others [39, 40].

7. Pathology

Endocrine disorders can be difficult to diagnose based only on morphological features because endocrine manifestations are caused primarily by a hormonal imbalance. Nonetheless, anatomical findings, like organomegaly or nodules may suggest anomalies that should encourage further investigation through laboratory tests or microscopic evaluation. Thyroid gland disorders have a wide variety of clinical presentations and can affect many organs and systems.

7.1 Hashimoto's thyroiditis and hypothyroidism

Autoimmune chronic thyroiditis affects from 3 to 5 more times women than men, generally at a median age or older, as well as in children. HT is the most

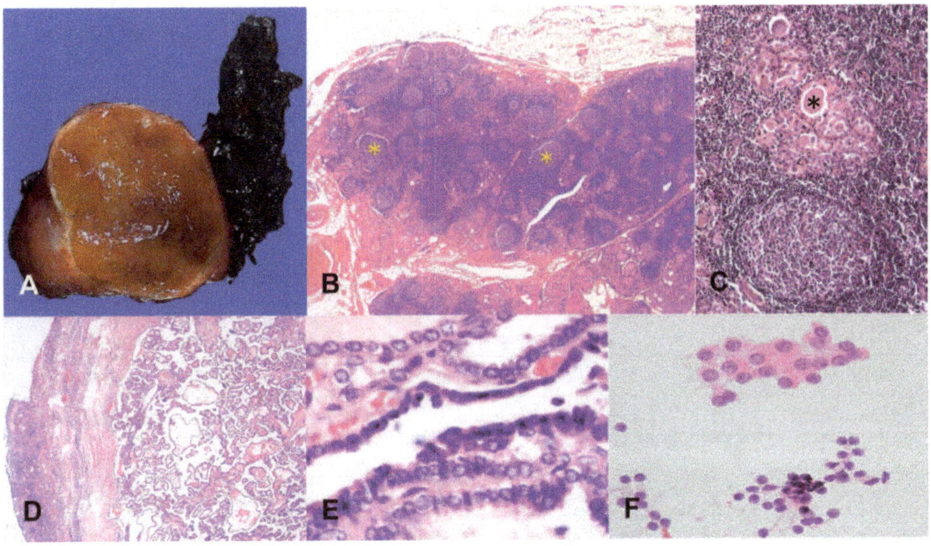

Figure 5.
Thyroid gland: A) product of a hemythyroidectomy with Hashimoto's thyroiditis associated to the presence of a solitary nodule; B) reactive lymphoid infiltrate with lymphoid folliculae with a germinal centers in the thyroid parenchyma (asterisk); C) Hürthle cells (asterisk); D) papillary thyroid carcinoma; E) neoplastic cells with nucleomegaly, open cromatine and nuclear clefts; F) biopsy with fine needle aspiration shows groups of cells with oxyphilic changes (Hürthle cells) and reactive lymphocytes compatible with Hashimoto's thyroiditis.

common autoimmune disease of the thyroid and is the main cause of autoimmune hypothyroidism. This disease was first described by Hakaru Hashimoto in 1912 as a "lymphomatous stroma" [40]. The global occurrence of HT is estimated to be between 0.3 and 1.5 cases for each 1000 individuals per year; predominantly in the female gender; it has a male-female ratio of 5:20 between the 30 and 50 year old population [41, 42]. There are two different clinical variants: the diffuse form and the nodular form. The nodular form is composed by a heterogeneous thyroid parenchyma that presents fibrosis, sclerosis and calcifications and is mainly associated to neoplasms, particularly PTC (**Figure 5A**). It is characterized by a lymphoid infiltrate capable to destroy the gland (**Figure 5B** and **C**), inducing fibrosis and hypothyroidism as a consequence [29]. Chronic inflammation of the thyroid parenchyma is regulated by an infiltrate of predominantly T lymphocytes [13]. The role of autoimmunity is backed by histological findings of diffuse lymphocytic thyroid infiltration and by specific circulating antibodies in almost all patients. Increased levels of anti-TPO antibodies are found in 95% of cases and anti-thyroglobulin antibodies are found in 60% of cases, these being higher in the atrophic form than in the goiter form of the disease [13]. Treatment is almost always non-surgical, surgery is indicated in cases of glandular enlargement with compressive symptoms, non-satisfactory pharmacological treatment and suspicion of neoplasm degeneration in one or more nodules. The association between HT and PTC, first described by Dailey et al. in 1955 [42, 43], is a controversial matter (**Figure 5D** and **E**).

8. Hypothyroidism detection

Even when there are no established guidelines for thyroid disease detection, the American Thyroid Association recommends to start detection at 35 years old and to continue screening each 5 years. Population with high risk of hypothyroidism include: women older than 60 years old, pregnant women, people with a history of head and neck radiation, patients with autoimmune disease, diabetes type 1,

Patients with 4 or more of the following symptoms	Patients with a history of	Pregnant patients with a history of
Fatigue	Sudden and unexplained thyroid growth	Family history of thyroid disease
Sleepiness	Dementia	Thyroid disease during pregnancy
Cold intolerance	Psychiatric diseases	Type 1 diabetes or autoinmune disease
Goiter	Genetic syndromes	
Dry skin	Neck radiation	Complicated pregnancy
Weight gain	Use of drugs that affect	Recurrent miscarriage
Constipation	thyroid function	Menstrual irregularities
	Dermatologic disease	Presence of anti-thyroid
	Hyperlipidemia	peroxidase antibodies
	Cardiovascular diseases	
	Autoimmune disease	
	Hypercholesterolemia	
	Infertility or menstrual	
	irregularities	

Table 2.
Guidelines for plasma TSH measurement and referral to a specialist in cases of hypothyroidism.

positive antibodies against thyroid peroxidase, and people with family history of hypothyroidism [44]. **Table 2** resumes the guidelines for hypothyroidism screening.

9. Treatment

The drug of choice in the treatment of hypothyroidism is the replacement of the thyroid hormones [6, 35, 45].

10. Prognosis

Hypothyroidism may have a higher risk of morbidity and mortality. It can eventually lead to coma or even death. In children, the lack of treatment can provoke severe mental retardation. One of the main causes of death in adults is cardiac failure. With treatment, the majority of patients have a good prognosis and symptoms normally revert within a few weeks or months [46].

10.1 Hypothyroidism complications

Even when is not common, in terminal stages, hypothyroidism, also known as myxedematous coma, is a medical emergency. First described by Sir William Gull in 1873, myxedematous coma has an estimated incidence of 0.22 per million per year [46]. It affects more frequently women older than 60 years old with a large history of hypothyroidism and it tends to occur during cold weather settings. Other trigger-ing factors include infection, cerebrovascular attack, myocardial infarction, trau-matism, pregnancy and the use of drugs containing lithium and amiodarone [46].

This affection is associated with a progressive deceleration of physical and mental skills as the disease advances. Initially, symptoms can simulate depression or early dementia, with fatigue, apathy, forgetfulness as the predominant complaints. If not treated, patients can develop severe hypothermia, urinary retention, respiratory depression, bradycardia, hypotension and arrhythmias that include cardiac blocks and torsade de pointes. Torsade de pointes is a non-frequent ventricular tachycardia that is found in a large QT syndrome, caused by an enlargement of the repolarization phase of the action potential. It exists a diffuse deposit of mucopolysaccharides that eventually leads to airway obstruction by affecting tongue and larynx, cardiac tamponade as a result of pericardial effusion and edema without skin and subcutaneous foveae. Electrolyte abnormalities are also produced, specifically hyponatremia and coagulopathies, including the acquired von Willebrand disease, which is associated with an increased mortality [11]. Altered mental state worsens from lethargy to stupor to coma, which increases the risk of aspiration pneumonia, urinary tract infection and sepsis. Mortality in patients with myxedematous coma is estimated to be around 20–25%, which represents a significant improvement respect previous reports of 60–70% due to a better acknowledge and treatment of this disease. Survival rates are worse in elderly patients and in those with severe and or persistent hypothermia, bradycardia, hypotension, lower coma Glasgow score and multiorganic disease. The most common immediate causes of death are sepsis, gastrointestinal hemorrhage secondary to coagulopathy and respiratory insuf-ficiency [10].

11. Directions for future investigation

Even when advances have been made regarding cause detection, knowledge of clinical implications, diagnosis and treatment of hypothyroidism, there are still many questions left without answers regarding diagnosis and treatment. Several risk factors have been identified for abnormal TSH concentrations, concentrations of free thyroxin and thyroid disease, but only a small proportion of these variability is explained. At the moment hypothyroidism diagnosis is based on reference ranks for TSH and free thyroxin. Due to the arbitrary nature of cut points that define mild and overt hypothyroidism, an alternative classification system has been proposed based on thyroid function tests.

Author details

Fernando Candanedo-Gonzalez[1*], Javier Rios-Valencia[1], Dafne Noemi Pacheco-Garcilazo[2], Wilfredo Valenzuela-Gonzalez[3] and Armando Gamboa-Dominguez[1,4]

1 Department of Pathology, INCMNSZ, Mexico

2 Student of the Faculty of Medicine, University City, UNAM, Mexico

3 Student of the Faculty of Medicine, Universidad Autonoma de Sinaloa, Mexico

4 Chief of the Department of Pathology, INCMNSZ, Mexico City, Mexico

*Address all correspondence to: fa_candanedo@yahoo.com.mx

References

[1] Beynon ME, Pinneri K. An overview of the thyroid gland and thyroid-related deaths for the forensic pathologist. Academic Forensic Pathology. 2016; **6**(2):217-236

[2] Ozguner G, Sulak O. Size and location of thyroid gland in the fetal period. Surgical and Radiologic Anatomy. 2014;**36**(4):359-367

[3] Takanashi Y, Honkura Y, Rodriguez-Vazquez JF, Murakami G, Kawase T, Katori Y. Pyramidal lobe of the thyroid gland and the thyroglossal duct remnant: A study using human fetal sections. Annals of Anatomy. 2015;**197**:29-37

[4] Mescher AL. Junqueira's basic histology text & atlas. In: Chapter 20, Endocrine Glands. 16th ed. New York: McGraw-Hill Lange; c2021. pp. 348-370

[5] Curling TB. Two cases of absence of the thyroid body, and symmetrical swelling of fat tissue at the sides of the neck, connected with defective cerebral development. Medico-Chirurgical Transactions. 1850;**33**:303-306

[6] Chaker L, Bianco AC, Jonklaas J, Peeters R. Hypothyroidism. Lancet. 2017;**390**:1550-1562

[7] Juárez-Cedillo T, Basurto-Acevedo L, Vega-García S, Sánchez-Rodríguez MA, Retana-Ugalde R, Juárez-Cedillo E, et al. Prevalence of thyroid dysfunction and its impact on cognition in older mexican adults: (SADEM study). Journal of Endocrinological Investigation. 2017; **40**(9):945-952

[8] Garber JR, Cobin RH, Gharib H, Hennessey JV, Klein I, Mechanick JI, et al. Clinical practice guidelines for hypothyroidism in adults: Cosponsored by the American Association of Clinical Endocrinologists and the American Thyroid Association. Thyroid. 2012;**22**(12):1200-1235

[9] Kostoglou-Athanassious I, Ntalles K. Hypothyroidims –new aspects of an old disease. Hippokratia. 2010;**14** (2):82-87

[10] Taylor PN, Albrecht D, Scholz A, Gutierrez-Buey G, Lazarus JH, Dayan CM, et al. Global epidemiology of hyperthyroidism and hypothyroidism. Nature Reviews. Endocrinology. 2018; **14**(5):301-316

[11] Asvold BO, Vatten LJ, BjØro T. Changes in the prevalence of hypothyroidism: The HUNT Study in Norway. European Journal of Endocrinology. 2013;**169**:613-620

[12] Aoki Y, Belin RM, Clickner R, Jeffries R, Philips L, Mahaffrey KR. Serum TSH and total T4 in the United States population and their association with participant characteristics: National Health and Nutrition Examination Survey (NHANES 1999-2002). Thyroid. 2007;**17**:1211-1223

[13] Canaris GJ, Manowitz NR, Mayor G, Ridgway EC. The Colorado thyroid disease prevalence study. Archives of Internal Medicine. 2000;**160**:526-534

[14] Garmendia Madariaga A, Santos Palacios S, Guillén-Grima F, Galofré JC. The incidence and prevalence of thyroid dysfunction in Europe: A meta-analysis. The Journal of Clinical Endocrinology and Metabolism. 2014;**99**:923-931

[15] Hollowell JG, Staehling NW, Flanders WD, Hannon WH, Gunter EW, Spencer CA, et al. Serum TSH, T(4), and thyroid antibodies in the United States population (1988 to 1994): National Health and Nutrition Examination Survey (NHANES III). The Journal of Clinical Endocrinology and Metabolism. 2002;**87**:489-499

[16] Lazarus JH. The importance of iodine in public health. Environmental Geochemistry and Health. 2015; **37**:605-618

[17] Hansen PS, Brix TH, Sorensen TI, Kyvik KO, Hegendüs L. Major genetic influence on the regulation of the pituitary-thyroid axis: A study of healthy Danish twins. The Journal of Clinical Endocrinology and Metabolism. 2004;**89**:1181-1187

[18] Panicker V, Wilson SG, Spector TD, Brown SJ, Falchi M, Richards JB, et al. Heritability of serum TSH, free T4 and free T3 concentrations: A study of a large UK twin cohort. Clinical Endocrinology. 2008;**68**:652-659

[19] Porcu E, Medici M, Pistis G, et al. A meta-analysis of thyroid-related traits reveals novel loci and gender-specific differences in the regulation of thyroid function. PLoS Genetics. 2013; **9**:e1003266

[20] Eriksson N, Tung JY, Kiefer AK, Hinds DA, Francke U, Mountain JL, et al. Novel associations for hypothyroidism include known autoinmune risk loci. PLoS One. 2012;**7**:e34442

[21] Denny JC, Crawford DC, Ritchie MD, et al. Variants near FOXE1 are associated with hypothyroidism and other thyroid conditions: Using electronic medical records for genome- and phenome-wide studies. American Journal of Human Genetics. 2011;**89**:529-542

[22] Pickrell JK, Berisa T, Liu JZ, Ségurel L, Tung JY, Hinds DA. Detection and interpretation of shared genetic influences on 42 human traits. Nature Genetics. 2016;**48**:709-717

[23] Medici M, Visser WE, Visser TJ, Peeters RP. Genetic determination of the hypothalamic - pituitary - thyroid axis: Where do we stand? Endocrine Reviews. 2015;**36**:214-244

[24] Costanzo LS. Thyroid hormones. In: Cicalese B, editor. Physiology. 4th ed. Philadelphia: Saunders Elsevier; 2010. pp. 401-409

[25] Dai G, Levy O, Carrasco N. Cloning and characterization of the thyroid iodide transporter. Nature. 1996;**379**:458-460

[26] Polikar R, Burger AG, Scherrer U, Nicod P. The thyroid and the heart. Circulation. 1993;**87**(5):1435-1441

[27] Zhang Y, Chang Y, Ryu S, Cho J, Lee WY, Rhee EJ, et al. Thyroid hormones and mortality risk in euthyroid individuals: The Kangbuk Samsung health study. The Journal of Clinical Endocrinology and Metabolism. 2014;**99**(7):2467-2476

[28] Persani L. Clinical review: Central hypothyroidism: Pathogenic, diagnostic, and therapeutic challenges. The Journal of Clinical Endocrinology and Metabolism. 2012;**97**:3068-3078

[29] Graceffa G, Patrone R, Vieni S, Campanella S, Calamia S, Laise I, et al. Association between Hashimoto's thyroiditis and papillary thyroid carcinoma: A retrospective analysis of 305 patients. BMC Endocrine Disorders. 2019;**19**(Suppl. 1):26

[30] Chang LS, Barroso-Sousa R, Tolaney SM, Hodi FS, Kaiser UB, Min L. Endocrine toxicity of cancer immunotherapy targeting immune checkpoints. Endocrine Reviews. 2019;**40**(1):17-65

[31] Sabbagh RE, Azar NS, Eid AA, Azar ST. Thyroid dysfunctions due to immune checkpoint inhibitors: A review. International Journal of General Medicine. 2020;**13**:1003-1009

[32] Huang SA, Tu HM, Harney JW, Venihaki M, Butte AJ, Kozahewich HP,

et al. Severe hypothyroidism caused by type 3 iodothyronine deiodinase in infantile hemangiomas. The New England Journal of Medicine. 2000;**343**:185-189

[33] Dumiltrescu AM, Refetoff S. The syndromes of reduced sensitivity of thyroid hormone. Biochimica et Biophysica Acta. 1830;**2013**:3987-4003

[34] Taylor P, Bianco AC. Urgent need for further research in subclinical hypothyroidism. Nature Reviews Endocrinology. 2019;**15**:503-504

[35] Casey BM, Thom EA, Peaceman AM, Varner MW, Sorokin Y, Hirtz DG, et al. Treatment of subclinical hypothyroidism or hypothyroxinemia in pregnancy. New England Journal of Medicine. 2017;**376**(9):815-825

[36] Vela-Amieva M, Gamboa-Cardiel S, Pérez-Andrade ME, Ortiz-Cortés J, González-Contreras CR, Ortega-Veláz - quez V. Epidemiología del hipotiroid - ismo congénito en México. Salud Púb Méx. 2004;**46**(2):141-148

[37] Castilla Peón MF. Congenital hypothyroidism. Boletín Médico del Hospital Infantil de México. 2015;**72**(2):140-148

[38] Flores-Robles CA, Coronado- Zarco IA, Ortega-González C, Arreola Ramírez G, Reyes-Muñoz E. Tamizaje neonatal de hipotiroidismo congénito, análisis de la evidencia actual y propuesta de tamizaje para la población mexicana. Perinatología y Reproducción Humana. 2018;**32**(1):43-52

[39] Di Magliano MP, Di Lauro R, Zannini M. Pax8 has a key role in thyroid cell differentiation. PNAS. 2000;**97**(24):13144-13149

[40] Postiglione MP, Parlato R, Rodriguez-Mallon A, Rosica A, Mithbaokar P, Maresca M, et al. Role of the thyroid-stimulating hormone receptor signaling in development and differentiation of the thyroid gland. PNAS. 2002;**99**(24):15462-15467

[41] Hashimoto H. Zur Kenntniss der lymphomatosen verandererung der schilddrusse strumal lymphomatosa. Arch Klin Chirugie. 1912;**97**:218-248

[42] Dailey ME, Lindsay S, Skahen R. Relation of thyroid neoplasms to Hashimoto's disease of the thyroid gland. Archives of Surgery. 1955; **70**:291-7. 6

[43] Davies L, Welch HG. Increasing incidence of thyroid cancer in United States, 1973-2002. Journal of the American Medical Association. 2006;**295**:2164-2167

[44] Alexander EK, Pearce EN, Brent GA, Brown RS, Chen H, Dosiou C, et al. 2017 Guidelines of the American Thyroid Association for the diagnosis and management of thyroid disease during pregnancy and the postpartum. Thyroid. 2017;**27**(3): 315-389

[45] Saudi Arabia and Gulf Thyroid Advisory Board. Protocol for Management of Hypothyroidism: A Consensus of GCC Countries. A Roadmap. London: Springer; 2018

[46] Rodríguez I, Fluiters E, Pérez-Méndez LF, Luna R, Páramoand C, García-Mayor RV. Factors associated with mortality of patients with myxoedemacoma: Prospective study in 11 cases treated in a single institution. The Journal of Endocrinology. 2004;**180**:347-350

Permissions

All chapters in this book were first published by InTech Open; hereby published with permission under the Creative Commons Attribution License or equivalent. Every chapter published in this book has been scrutinized by our experts. Their significance has been extensively debated. The topics covered herein carry significant findings which will fuel the growth of the discipline. They may even be implemented as practical applications or may be referred to as a beginning point for another development.

The contributors of this book come from diverse backgrounds, making this book a truly international effort. This book will bring forth new frontiers with its revolutionizing research information and detailed analysis of the nascent developments around the world.

We would like to thank all the contributing authors for lending their expertise to make the book truly unique. They have played a crucial role in the development of this book. Without their invaluable contributions this book wouldn't have been possible. They have made vital efforts to compile up to date information on the varied aspects of this subject to make this book a valuable addition to the collection of many professionals and students.

This book was conceptualized with the vision of imparting up-to-date information and advanced data in this field. To ensure the same, a matchless editorial board was set up. Every individual on the board went through rigorous rounds of assessment to prove their worth. After which they invested a large part of their time researching and compiling the most relevant data for our readers.

The editorial board has been involved in producing this book since its inception. They have spent rigorous hours researching and exploring the diverse topics which have resulted in the successful publishing of this book. They have passed on their knowledge of decades through this book. To expedite this challenging task, the publisher supported the team at every step. A small team of assistant editors was also appointed to further simplify the editing procedure and attain best results for the readers.

Apart from the editorial board, the designing team has also invested a significant amount of their time in understanding the subject and creating the most relevant covers. They scrutinized every image to scout for the most suitable representation of the subject and create an appropriate cover for the book.

The publishing team has been an ardent support to the editorial, designing and production team. Their endless efforts to recruit the best for this project, has resulted in the accomplishment of this book. They are a veteran in the field of academics and their pool of knowledge is as vast as their experience in printing. Their expertise and guidance has proved useful at every step. Their uncompromising quality standards have made this book an exceptional effort. Their encouragement from time to time has been an inspiration for everyone.

The publisher and the editorial board hope that this book will prove to be a valuable piece of knowledge for researchers, students, practitioners and scholars across the globe.

List of Contributors

Mauricio Alvarez Andrade
Endocrinólogo, Organización Sanitas Internacional, Keralty, Bogotá, Colombia

Oscar Rosero Olarte
Endocrinólogo, Instituto de Osteoporosis de los Llanos, Osteollanos, Regional Hospital, Villavicencio, Colombia

Valdete Topçiu-Shufta and Valdete Haxhibeqiri
Faculty of Medicine, Clinic of Medical Biochemistry, University Clinical Center of Kosova, University of Prishtina, Prishtina, Republic of Kosova

Armindo Miguel de Jesus Sousa de Araújo Ribeiro
Unidade Local de Saúde do Litoral Alentejano, Hospital Litoral Alentejano, Santiago do Cacém, Portugal
Universidad de Extremadura, Badajoz, Spain

Suzanne Fouad Soliman
Nutrition and Food Science Department, National Research Centre, Dokki, Giza, Egypt
Critical Care Medicine, Faculty of Medicine, Cairo University, Egypt

Yuan Lu
Case Western Reserve University School of Medicine, Cleveland, Ohio, USA
Harrington Heart and Vascular Institute, University Hospitals Cleveland Medical Center, Cleveland, Ohio, USA

Avital Schurr
Department of Anesthesiology and Perioperative Medicine, University of Louisville, School of Medicine, Louisville, KY, USA

Lilian Sanhueza
Diabetes Unit, Internal Medicine Service, San Juan de Dios Hospital, Santiago, Chile
Department of Medicine, Faculty of Medical Sciences, University of Santiago of Chile, Santiago, Chile

Karina Elgueta
Diabetes Unit, Internal Medicine Service, San Juan de Dios Hospital, Santiago, Chile
Adult Diabetes Unit, Department of Internal Medicine Las Condes Clinic, Santiago, Chile

Pilar Durruty
Diabetes Unit, Department of Medicine, Faculty of Medicine, San Juan de Dios Hospital, University of Chile, Santiago, Chile
Endocrinonology and Diabetes Section, Department of Medicine, Faculty of Medicine, Clinical Hospital University of Chile, Santiago, Chile

Cecilia Vargas
Endocrinonology and Diabetes Section, Department of Medicine, Faculty of Medicine, Clinical Hospital University of Chile, Santiago, Chile

Paulina Vignolo
Diabetes Unit, Department of Internal Medicine, Padre Hurtado Hospital, Santiago, Chile
Faculty of Medicine, Development University, Santiago, Chile

Gaffar S. Zaman
Department of Clinical Laboratory Sciences, College of Applied Medical Science, King Khalid University, Abha, Kingdom of Saudi Arabia

Georg Bollig
Palliative Care Team, Medical Department Sønderborg/Tønder, South Jutland Hospital, Sønderborg, Denmark
Medical Research Unit, Institute of Regional Health Research, University of Southern Denmark, Odense, Denmark

Andrea L. Gropman
Division of Neurogenetics and Developmental Pediatrics, Neurology and Pediatrics, Children's National Medical Center, N.W. Washington, United States of America

Ágnes Kinyó
Department of Dermatology, Venereology and Oncodermatology, University of Pécs, Pécs, Hungary

Gabriela Roman
Clinical Center of Diabetes, Nutrition and Metabolic Diseases, "Iuliu Hatieganu" University of Medicine and Pharmacy, Cluj-Napoca, Romania

Anca Pantea Stoian
Department of Diabetes, Nutrition and Metabolic Diseases, "Carol Davila" University of Medicine and Pharmacy, Bucharest, Romania

Dafne Noemi Pacheco-Garcilazo
Student of the Faculty of Medicine, University City, UNAM, Mexico

Wissal Abassi
Research Unit, Sportive Performance and Physical Rehabilitation, High Institute of Sports and Physical Education of Kef, University of Jendouba, El Kef, Tunisia
Higher Institute of Sport and Physical Education of Ksar Said, University of "La Manouba", Manouba, Tunisia

Nejmeddine Ouerghi
Research Unit, Sportive Performance and Physical Rehabilitation, High Institute of Sports and Physical Education of Kef, University of Jendouba, El Kef, Tunisia
University of Tunis El Manar, Faculty of Medicine of Tunis, Rabta Hospital, Tunis, Tunisia

Anissa Bouassida
Research Unit, Sportive Performance and Physical Rehabilitation, High Institute of Sports and Physical Education of Kef, University of Jendouba, El Kef, Tunisia

Maria Grazia Dalfrà, Silvia Burlina and Annunziara Lapolla
DPT Medicine-Padova University, Padova, Italy

Fernando Candanedo-Gonzalez and Javier Rios-Valencia
Department of Pathology, INCMNSZ, Mexico

Wilfredo Valenzuela-Gonzalez
Student of the Faculty of Medicine, Universidad Autonoma de Sinaloa, Mexico

Armando Gamboa-Dominguez
Department of Pathology, INCMNSZ, Mexico
Chief of the Department of Pathology, INCMNSZ, Mexico City, Mexico

Index

A

Acquired Hypothyroidism, 203, 210

Adenosine Triphosphate, 13

Adipose Tissue, 43-44, 69-70, 88-108, 114-115, 129-138, 143-144, 150-153, 160-163, 217, 221, 224

Aerobic Exercise, 31, 36-39

Ammonia, 14-15, 164-169, 171, 173-177, 179-183

Anaerobic Exercise, 31, 36-39

Anemia, 2-4, 49-50, 202, 208-209, 214

Anti-inflammatory Effects, 49-50, 77-78

Autoimmune Diabetes, 109, 120-123, 126-127

B

Blood Glucose, 23, 25, 45-50, 116-119, 129-134, 145-149, 152-153, 189-190, 200, 225

Brown Adipocyte, 88-96, 99-100

Brown Adipose Tissue, 88-91, 93, 95, 97, 99-108, 135-136, 221

Bullous Pemphigoid, 57-67

C

Cardiac Output, 137-138, 206, 230

Cardiovascular Disease, 43-54, 68-70, 79-82, 133-134, 139-142, 184-185, 187-188, 196-198, 204-205, 207, 215

Cardiovascular Risk, 68-70, 128-132, 139-140, 184-186, 195-198, 204-205

Cerebral Energy Metabolism, 13-14, 16-22, 25-26, 28

Cerebral Metabolic Rate, 13, 18-21, 25

Chronic Autoimmune Thyroiditis, 4

Chronic Uremia, 68, 73-82, 87

Chylomicron Remnants, 69-70, 75-76

Chylomicrons, 69-76, 103-104

Circadian Rhythm, 42-44, 48, 53-54

Compensatory Hyperinsulinemia, 145

Congenital Hypothyroidism, 2-3, 202-203, 210, 229, 234

Coronary Artery Disease, 125, 185, 187, 189, 202, 206

D

Diabetes, 4-5, 8-9, 30, 36-39, 42-54, 58-67, 84-88, 96, 99-100, 103-132, 139-153, 155-163, 184-193, 195-201, 212-213, 215, 217-221, 223-225, 235-236

Differential Diagnosis, 4, 10-11, 234

Dyslipidemia, 42-50, 53-56, 68-70, 83-85, 128-132, 134-140, 156-157, 184-186, 189, 194, 196, 207

E

Endothelial Dysfunction, 55-56, 84-85, 124-125, 135-136, 143, 184, 188

Endurance Exercise, 31, 36-37

Enzyme Lipoprotein Lipase, 69-72, 75-76

Eosinophil, 57, 60-63, 228

Epidemiology, 8-9, 81-82, 120-121, 128, 131-132, 139-140, 156-159, 185, 196, 225-226, 228-229, 234, 238

Erythema, 57-63

Etiology, 2-6, 10, 110-111, 131-132, 173-174, 177-178, 226, 231-232

F

Fetal Complications, 118-119, 215

Free Fatty Acids, 89-90, 93-94, 131-132, 137-138, 152-153

Functional Foods, 42, 49-50, 55-56

Functional Thermoactivity, 88, 97-98

G

Genetic Defects, 112-115, 137-138

Gestational Diabetes, 120-121, 146-149, 212, 215, 218-219, 221, 223-225

Gestational Diabetes Mellitus, 212, 215, 218, 221, 223-225

Gliptins, 57-61

Glucose Tolerance Testing, 145, 152-153

Glycolysis, 13-23, 25-30, 93-94, 99-100, 105-108

Granulomatous Thyroiditis, 4-5, 232

H

Hashimoto's Disease, 1, 6, 240

Heart Failure, 7, 45-46, 141-142, 184-189, 192-193, 196-198, 200, 206

Hypertension, 42-44, 47-54, 125, 128-132, 137-144, 148-149, 184, 186, 188-189, 191, 193-194, 196

Hypocaloric Diet, 42, 47-48, 55-56

Hypothyroidism, 1-12, 202-214, 226, 228-229, 231-240

Hypoxia, 25, 28-29, 93-94, 133-138, 143-144, 221

I

Impaired Glucose Tolerance, 150-151, 153, 158-161

Insulin Resistance, 36-37, 42-49, 51-56, 75-76, 88, 96, 107-108, 112-118, 120-125, 128-153, 155-163, 184-185, 188, 215, 217, 219, 223-224

K

Kidney Disease, 68, 81-87, 116-117, 137-138, 186, 188, 192-194, 198

L

Lactate, 13-23, 25-30, 34-35, 93-96, 106, 165-166

Latent Autoimmune Diabetes In Adults, 109, 126

Levothyroxine, 1, 6-7, 10-11, 202-203, 210-212, 214, 233

Lifestyle Modification, 42

Lipid Metabolism, 43-48, 69-70, 89-92, 97-98, 100-102, 113, 135-136, 141-142, 215

Lipoprotein Lipase, 69-72, 75-76, 84-85, 93-94, 103-104, 150, 160-161

Lipoprotein Metabolism, 69-72, 76, 81-82, 84-85, 91-92

Lipoproteins, 53-54, 68-78, 81-85, 91-92, 103-104, 131-132, 158-159, 185, 233

M

Metabolic Disorders, 55-56, 68, 88, 94-98, 139-142, 165-166, 181-182

Metabolic Syndrome, 42-56, 104, 112-113, 128-133, 135-144, 152-153, 185, 188, 196, 198, 215

Mitochondrial Diabetes, 109, 116-117, 120-121

Modified Insulin Suppression Test, 145, 152-153

Monogenic Diabetes, 109, 116-117, 124-125, 127

Muscle Strength Exercise, 31

Myxoedema Coma, 10, 202, 205

N

Neurocognitive Deficits, 173-174, 178, 203

Neuroimaging, 164-166, 169-172

O

Obesity, 42-56, 88, 96-104, 107-108, 112-117, 122, 126-147, 150 151, 156-163, 184-186, 188-190, 196, 201-202, 207-208, 213, 215, 219-221, 223-224

Oxidative Stress, 28-29, 43-46, 53-56, 68-70, 73-74, 77-78, 80-82, 86-87, 101-102, 137-138, 145, 173-176, 184-185, 189, 215

P

Paradigm Shift, 13, 16-17, 25-29

Pathophysiology, 42, 84-85, 110-111, 120-121, 128, 131-134, 137-138, 155-157, 162-163, 165-166, 179-180, 213, 226, 229

Peripheral Tissue, 43-44, 75-76, 91-92

Pharmacotherapy, 192-194

Plasma Glucose, 45-46, 146-149, 152-153, 158-159, 219-221

Polarography, 13, 18-19, 22-23, 28-29

Prognosis, 34-35, 188, 198, 226, 233, 236

R

Renal Failure, 36-37, 40, 68-70, 73-78, 81-82, 84-86

Rhabdomyolysis, 31, 34-37, 40

S

Sports Medicine, 31-35, 38-39, 41

Sympathetic Nervous System, 43-44, 137-138

T

Thyroid Dysfunction, 8-11, 209, 211-212, 228, 231, 233, 238

Thyroid Hormones, 1-4, 6-7, 203, 209, 229-231, 233-234, 236, 239

U

Urea Cycle Disorders, 164-168, 171-172, 179-183

V

Vasoconstriction, 137-138, 184

www.ingramcontent.com/pod-product-compliance
Lightning Source LLC
LaVergne TN
LVHW080918210926
848485LV00004B/679